THE INTERNATIONAL ENERGY EXPERIENCE

Markets, **R**egulation and the **E**nvironment

THE INTERNATIONAL ENERGY EXPERIENCE

Markets, Regulation and the Environment

Editors

G MacKerron
University of Sussex

P Pearson
Imperial College

ICP

Imperial College Press

Published by

Imperial College Press
57 Shelton Street
Covent Garden
London WC2H 9HE

Distributed by

World Scientific Publishing Co. Pte. Ltd.

P O Box 128, Farrer Road, Singapore 912805

USA office: Suite 1B, 1060 Main Street, River Edge, NJ 07661

UK office: 57 Shelton Street, Covent Garden, London WC2H 9HE

British Library Cataloguing-in-Publication Data
A catalogue record for this book is available from the British Library.

THE INTERNATIONAL ENERGY EXPERIENCE: MARKETS, REGULATION AND THE ENVIRONMENT

ISBN 1-86094-197-4

Printed in Singapore by Uto-Print

NA

FOREWORD

ANDREW BARTON
BP Amoco
Chairman, British Institute of Energy Economics

A primary role of The British Institute of Energy Economics (BIEE) is to encourage high level debate on contemporary energy issues between business, Government and the academic community. One of the ways in which we fulfil this role is to convene a major, mainly academic, conference on contemporary issues every 18 to 24 months. The first such conference at Warwick University was held in December 1995 and produced a remarkable array of chapters, all of which were subsequently published by Imperial College Press[1].

Such was the success of this first effort that a second conference was held, also at Warwick, in December 1997, this time focussing on the wider international experience of liberalisation, regulation and environmental control in the energy sector. This second conference was particularly interested in the interplay between market forces and the continuing role of Governments in energy matters, often now re-packaged as 'regulation' - either economic or environmental.

Some 140 participants attended from more than 15 countries, with excellent representation from business and the academic community internationally. The debate, as at the first conference, was always stimulating and occasionally lively. Many participants welcomed current changes in policy directions towards energy market liberalisation, although there were others who were less convinced and favoured existing public utility structures for their countries. I am particularly grateful to Marilyn Hall of BP Amoco, Mary Scanlan of BIEE, Professor Catherine Waddams and Dr Monica Giulietti of the University of Warwick, for making the conference such a success at all levels.

The BIEE was again committed to helping to sponsor publication, but this time the editors, Gordon MacKerron and Peter Pearson, took the different approach of selecting about half of all the papers presented, offering their authors time to make corrections, and then presenting them under six main themes of particular contemporary relevance. The present volume is the result. It represents a fascinating view of a wide range of critical issues, and shows the continuing relevance of applied economics in understanding complex problems in energy, environment and regulation.

By the time this preface is published, the BIEE will have held a third conference at St John's College, Oxford in September 1999. In the meantime I commend this book as an excellent survey of current thinking on the interplay between markets, regulation and environmental protection in the energy business.

---

[1] G. MacKerron and P.Pearson *The UK Energy Experience: a Model or a Warning?* Imperial College Press, 1996.

CONTENTS

FIGURES

INTRODUCTION

GORDON MACKERRON

Head of the Energy Programme, SPRU
University of Sussex, Falmer, Brighton BN1 9RF, UK
E-mail: gmackerron@mistral.co.uk

PETER PEARSON

Director, Environmental Policy & Management Group
T H Huxley School, Imperial College, 48 Prince's Gardens, London SW7 2PE
E-mail: p.j.pearson@ic.ac.uk

The British Institute of Energy Economics (BIEE) has a broadly educational remit and in recent years this has included the sponsoring of inexpensive and mainly academic conferences on current issues in energy economics and policy. This has proved an excellent way of bringing a quite disparate academic community together, and adding to it a large number of interested industry and Government people to discuss contemporary issues (see Andrew Barton's foreword to this volume). Imperial College Press published the whole proceedings of the first such conference,[1] amounting to 36 papers. This was the first such academic volume to emerge from the UK for some years.

The second BIEE conference was held at Warwick in December 1997, and attracted even more papers - more than 50 in all. This time the focus was on a wider international experience, but still based to a degree on an assessment of the UK model of change in the energy sector: in what ways, and how well had this model (or set of models) travelled elsewhere in Europe and to the rest of the world?

In approaching the issue of publication, it was impractical to try and publish more than 50 papers from the conference. Instead we decided to select around half of the best papers, partly influenced by what we believed were the main intellectual and pragmatic themes that emerged at the conference. What we present here is therefore not a conference proceedings but rather a selection of readings on a range of distinct topics.

Because this book was not to be simply a proceedings volume, we also decided to allow authors a reasonable time in which to consider making changes or improvements to the versions submitted to the conference itself. All authors were therefore given until at least autumn 1998 to make revisions. In the case of a few chapters, which referred to empirical realities (such as the international oil price)

[1] G. MacKerron and P. Pearson (1996), *The UK Energy Experience: a Model or a Warning?* Imperial College Press, London.

which were changing very rapidly, we decided to allow authors to May 1999 to make their final corrections.

The theme of the conference had been the international experience of change in energy systems in recent years, with a particular focus on the interplay between markets, regulation (the new guise of much of what used to be known as 'policy') and environmental protection. Papers came from all over the world, and the selection published here covers eight nationalities and three Continents, with a particular emphasis on Europe.

In thinking about this volume and its organisation, we needed to establish some themes which had both run through the conference and which sounded a wider intellectual and practical resonance. We also had to ensure that there were enough strong chapters to illustrate relevant issues within those themes. In the end we chose six main thematic areas, forming the first six sections of this volume. For all but the first theme there are at least four chapters.

Theme 1. World Energy Markets

The essential scene-setting is undertaken in the first section, which contains a single overview chapter on world energy markets from Peter Davies of BP Amoco. This is a comprehensive and lucid piece, covering a wide territory for the period since 1986, the point at which oil and other primary fuel prices started an unmistakable and relatively long-term decline. Essentially the chapter asks what this means, dividing the world into the Former Soviet Union (FSU), where experience has been mostly affected by the break-up of the old central planning system and the uneven process of moving towards a market economy, and the rest of the world, where low fossil fuel prices, a strong trend to liberalised markets and an increasing concern with environmental targets have been the predominant forces.

Davies suggests that the link between primary energy consumption and world GDP has (away from FSU) been firmly re-established since 1986. This immediately poses problems for Governments, which wish to ensure that GDP continues to grow, but which also (in the industrialised countries) have since Kyoto accepted commitments to reducing greenhouse gas, especially carbon dioxide, emissions, by 2008-2012. Contrary to the expectations of the 1970s, the supply of all fossil fuels has held up well at falling cost and price, and reserves seem more extensive now that they did fifteen years ago. Technology has played a large role, both in helping less apparently economic deposits of fossil fuels to be re-classified as reserves, and in cheapening the extraction process. The large growth area has been gas, and FSU energy exports seem set to rise. The analysis in Davies' chapter is carefully marshalled and is a well-supported example of a widespread industry view that energy supply will remain plentiful.

Theme 2. National Studies of Energy Structure and Reform

In the 1990s there has been reform activity, some of it radical, in the energy industries of almost all countries. The main activity has been in electricity systems, possibly because in a general climate of worldwide liberalisation, electricity often seemed to have the greatest distance to travel - electricity industries have most often been state-owned and heavily monopolistic. While there are important natural monopoly elements in electricity (the network or transport functions of transmission and distribution) many countries organised their electricity supply as vertically integrated monopolies from production or generation right through to retail supply. A variety of pressures has induced virtually all countries, whether OECD, transitional or developing, to move some way towards liberalisation.

For the industrialised countries, the motivations have been a mixture of a desire to raise cash for Government from privatisation and a conviction that liberalisation would bring benefits from improved economic efficiency. For transitional economies, reform to the energy industries has been part of a wider movement towards the market economy. In the case of developing countries, shortage of state resources for high rates of required investment have often been primary motives, together with pressure from lending institutions like the World Bank to adopt a more market-oriented economy.

While electricity has undoubtedly been in the lead, other energy sectors, including oil and gas, have also been subject to reform, though market pressures were formerly more evident in these energy markets, where trade and competition issues have always been important. In particular, the national oil monopolies of a number of non-OECD producing countries have come under increasing pressure to allow new entrants, and tax regimes have been simplified in a range of countries.

The four chapters in this section of the book illustrate some of these themes. Robinson offers a retrospective on a decade of UK experience, especially in relation to regulation; Arocena explains and analyses the series of reforms to Spain's electricity system; Jasinski provides detailed insights into Poland's electricity reforms; while Andrews-Speed and Dow look at the very wide range of problems affecting the regulation of China's energy system as a whole.

Theme 3. Electricity and Gas: Markets and Regulation

In North America, much of the debate about liberalising the energy utility industries has been conducted around the idea of deregulation. The hope has been that market forces will largely replace not only former structures of Governmental control, but also render the newer (for Europe) 'regulatory' forms of intervention as marginal as possible. Most experiences of liberalisation in electricity and gas have not found it possible to follow such a free market model. Even in countries such as the UK with a strong ideological commitment to deregulation and the flourishing of the free market, it has not proved possible to allow the State to withdraw to the passive role that had been hoped. While there has been progress towards making markets substantially more competitive both in the production or generation of energy and in its retail sale, most electricity and gas markets remain quite heavily regulated.

There seem to be a number of reasons for the difficulty experienced in moving towards 'free' as opposed to competitive markets. The first is the obvious one that the transport functions for both electricity and gas remain natural monopolies and are therefore inevitably subject to quite strict regulation (of profits, prices or some hybrid). A second is that in most countries the companies involved in the potentially competitive parts of the energy utility businesses retain large amounts of market power, so that for orthodox competition policy reasons, public intervention remains important here as well. However there are social as well as narrowly economic reasons for continued public control of gas and electricity. One is the sensitivity of many electorates to increases in the retail price of gas and electricity, increases that often seem required by the need to make the industries financially viable, especially in transition and developing countries. The other, much more strongly evident in the OECD countries, is that both gas and electricity - but especially electricity - are deeply implicated in major environmental issues, notably acid deposition and currently climate change. This implies close environmental regulation.

The four chapters in this section of the book tackle different aspects of these questions. Morrison expounds the various possible ways in which wholesale electricity trading can be organised (and controlled); Dixon and Easaw explore some theoretical questions in the interaction of regulation and competition in the UK contract gas market; Acutt and Elliott explore some problems for regulation of low levels of competition in electricity generation markets; and Haas offers the provocative forecast that liberalisation is unlikely in the longer run to make electricity cheaper for consumers.

Theme 4. Oil Markets and Regulation

Despite the large international market power of the oil majors, oil markets are often thought of as highly competitive, and more 'free' than those for the supply of gas and electricity to final consumers. While such views have weight, it is also true that oil markets are subject to a substantial body of regulatory activity, much of it concerned with unique sectoral tax policies but also covering a range of other issues including the licensing of exploration (administrative versus auctioning rights) and the ability of the majors to compete in protected national oil markets especially in the developing world.

In addition, as Peter Davies has noted in the opening chapter, there has been a very rapid rate of technological change in the oil industry in recent years, both in exploration and in production. The result of this intensive activity, relatively scantily reported in the public domain, has been that the production costs of oil and gas, especially in offshore waters, have fallen dramatically. This has shielded the industry at least in part from the impacts of much lower wholesale oil and gas prices. It has also rendered small, previously uneconomic deposits worth exploiting even in this regime of lower prices. Technology change and falling prices have also made many governments re-consider their fiscal policy towards the industry. This has been true both in producing countries, where the pressure has been to relax the stringency of earlier tax regimes as the level of economic rents falls sharply, and in consuming countries, where falling wholesale prices have given headroom for higher tax takes, often justified in environmental policy terms.

The four chapters under this theme look at different elements of all the issues just described. Watkins provides a statistical analysis of trends in supply functions as between OPEC and non-OPEC countries; Frewer explores efficiency and other issues around administrative discretion versus auctions as a basis for allocating exploration rights; Martin tries to explain developments in UK oil production in terms of the roles of technological change and reform in fiscal incentives; while Kemp analyses interactions between investment activity and taxation under different oil and gas prices.

Theme 5. Renewable Energy

Renewable energy has traditionally meant either hydro-electricity (usually in large schemes) or the use of biomass (often quite inefficiently) in many rural parts of the world, especially the developing countries. However, a combination of new interest in a much wider range of renewable energy sources as one response to a need for a less carbon-intensive energy economy, and some technological advances in renewables such as wind and photovoltaics, has led to a much more prominent role for renewable energy in the 1990s, and the prospects for further expansion in the 21st century.

While renewable forms of energy have been getting cheaper, sometimes quite rapidly, most are not yet competitive with fossil fuel-based energy systems. This means that the growth in investment in renewables has been heavily dependent on various forms of subsidy. A wide range of policy instruments has been developed in the last decade in an attempt to bring new forms of renewable energy into the market place at low cost and in ways which provide incentives for costs to fall as rapidly as possible. Bearing in mind that most renewable energy has developed in forms where it is converted to electricity there have also been important techno-economic and regulatory questions about the possible problems of integrating intermittent renewables into grid systems where reliability is vital. More recently, such debates have moved in a new direction in which the possible advantages to grid systems of small plants located close to load centres ('distributed generation', not necessarily renewables) have been explored.

The much expanded interest in renewable energy is reflected in the presence of four chapters in this section. Mitchell asks how well the innovative UK policy to support renewables (based on a competitive model) has worked; Skytte looks at some grid integration issues for renewable energy; Fouquet explores the issue of the extent to which there is a definable demand for specifically renewable energy and how large this demand might be; and Madlener engages in modelling to explore the potential contribution of renewable energy in the UK to 2020 and what first-order economic and environmental impacts might flow from such a development.

Theme 6. Environment and Energy Efficiency

In the course of the last 15 years or so, energy policy has been increasingly colonised by environmental policy. For the OECD countries especially, environmental policy has come to occupy a more and more prominent role in the general domain of public policy, and because so many environmental problems derive largely from the energy system - especially those involving air pollution (including acid deposition and climate change) - environmental needs have rivalled liberalisation as a major driver of energy policies.

Environmental (and ecological) economics has been a major growth area as a consequence of the increasing prominence of environmental policy, and because of the focus on energy systems, environmental economics and energy economics have moved closer together. The issues to be analysed are immensely complex, and only a few are described here. The general umbrella idea of sustainability is a starting point for many of the more specific questions. Climate change has now moved to centre stage, and this has created large interest in mechanisms (especially emissions trading) that might efficiently allow large reductions in greenhouse gas emissions. At the level of political economy, it also raises questions of the utility of reducing emissions in OECD countries while developing countries such as China and India remain outside the control regime. Complementarities or conflicts between the liberalisation agenda and the environmental control agendas are also a major issue. The extent to which nuclear power is a sustainable option also excites much controversy, and the issue of environmental valuation remains a large and difficult area.

The six chapters in this section illustrate some of these themes and raise other issues as well. Bailey and Clarke examine the large question of the relationship between China's energy development and its environmental impact; Bristow tackles the difficult issue of appropriate ways to evaluate transport investment in the context of environmental objectives; Pearson looks at the interactions between liberalisation and environmental policy (especially air pollution) in the UK; Stirling raises basic issues of appropriate ways of evaluating environmental damage; MacKerron and Sadnicki ask whether or not current nuclear clean-up policy in the UK satisfies practical sustainability criteria; while Eyre tackles the question of whether or not liberalisation can be expected to help or harm movements towards greater energy efficiency (also one of the more obvious ways to reduce pollutant emissions).

The final section of this book is a masterly summing up of some major themes from the chapters in this book from David Newbery. It would be redundant to provide here a summary of a summary; but for a personal and highly readable analysis of some of the principal issues emerging across all our themes, the reader is recommended to turn to Chapter 25.

SECTION 1

WORLD MARKETS

CHAPTER 1

WORLD ENERGY MARKETS TRENDS & CHANGES[1]

PETER DAVIES

BP Amoco plc, Britannic House, 1 Finsbury Circus, London EC2M 7BA
and the University of Dundee[2]
Email: daviespa@bp.com

Keywords: gas; liberalisation; oil; prices; regulation; technology.

1 Introduction

This chapter reviews the broad trends in world markets for primary commercial energy. It assesses, in particular, developments and changes in all primary energies in the decade from the oil price decline in 1986 to 1996. It also presents data on the growth of CO_2 emissions from the use of fossil fuels over the same period. The chapter then assesses the hiatus in energy markets in 1998 before considering the future by querying whether the recent trends can continue into the long term. A range of economic, technological and political forces are assessed.

2 World energy markets 1986-96

2.1 Introduction

World primary energy developments since 1986 need to be considered in two parts: firstly, the world excluding the Former Soviet Union (FSU); and secondly, developments in the FSU itself. There have been a series of fundamental changes in the energy economy of the countries of the Former Soviet Union (FSU) where energy production and consumption have both declined from peaks in 1989/90 by some 30%. This has changed the fundamental structure of the FSU energy economy.

However, in the intervening period the net impact of the FSU upon the rest of the world's energy markets has been very limited. On an oil equivalent basis, net FSU energy exports were almost the same in 1996 as in 1986, although there was

---

[1] Paul Appleby, Andrew Barton and Gavin Attridge of the BP Amoco Economics Unit all contributed to this chapter.

[2] The author is Chief Economist of BP Amoco plc, London and Honorary Visiting Professor at the Centre for Energy, Petroleum & Mineral Law and Policy at the University of Dundee.

some shift between fuels (more gas and less oil exports). Thus, much of the historical analysis in this chapter distinguishes between the world excluding the FSU and the FSU energy markets.

2.2 *Total energy consumption*

2.2.1 World outside the FSU

The link between world primary energy consumption and world GDP (both excluding the FSU) has been restored since the energy price declines of 1986. The correlation co-efficient for 1986-96 has been 99%, with an average income elasticity of 0.9. The average annual growth rate has been 2.5%. To a large degree the error term can be explained by weather volatility. For the previous decade the correlation was only 88% and elasticity 0.55. By implication this indicates that primary energy consumption is cyclical - in line with the world economy.

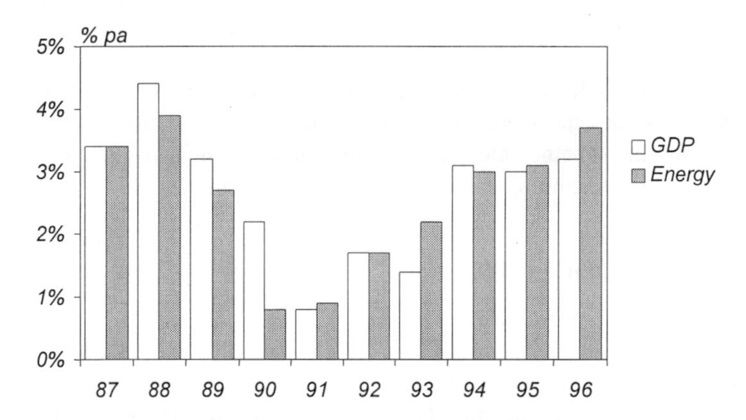

Figure 1.1: World energy demand exc. FSU

2.2.2 FSU energy

The FSU energy economy shows a contrasting picture. FSU GDP peaked in 1989 and primary energy consumption in 1990. Both have fallen continuously from their peaks, but both are showing signs of stabilising. Primary energy consumption has fallen by 34% from its peak. GDP has declined by 40%. Nevertheless, the relationship between GDP growth and primary energy consumption growth remains strong. Since 1986 the correlation has been 97.5% and the income elasticity 0.56.

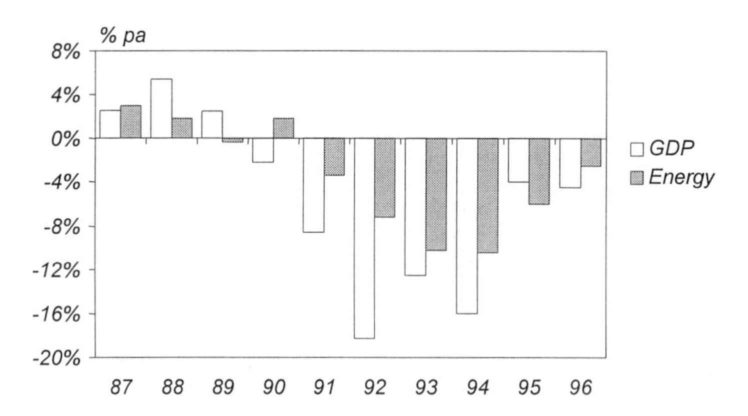

Figure 1.2: FSU energy demand

This tends to imply that FSU primary energy consumption will begin to rise again in the future once positive economic growth returns. FSU gas production and consumption began to increase in 1996. Nevertheless, given that FSU primary energy consumption only comprises 11% of total world energy consumption, it will be the trends in energy consumption in the rest of the world that are likely to dominate future global trends.

2.2.3 Regional energy trends

Within these global aggregates there have been important regional differences. N. America remains the largest continental energy market with 29.4% of world primary energy consumption, but the rapid growth of Asian energy consumption in the last decade (4.8% average annual growth as against 2% in N America) has resulted in Asia almost matching N America in terms of absolute size today. Asia generated over 50% of world energy consumption growth outside the FSU from 1986-96. Europe, previously the second largest world energy market, contributed only 6% to world energy consumption growth over the decade. The 1986-96 decade was thus driven strongly by Asian growth. Asian volume growth has been diverse: 36% has come from China alone, but the rest has been fairly evenly distributed between Japan (16%), ASEAN (15%), S Korea (13%) and India (12%).

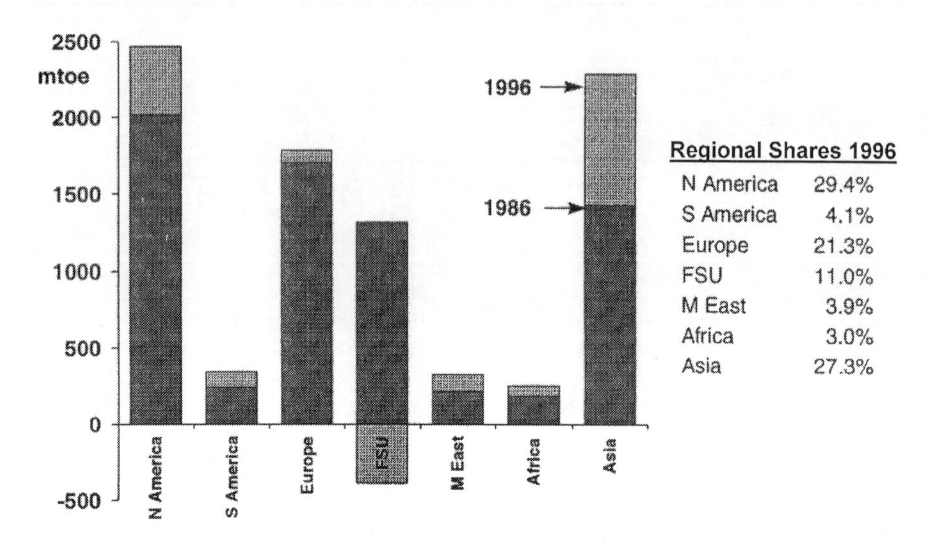

Figure 1.3: World primary energy consumption: regional shares

2.2.4 Fuel shares

Fuel shares showed a remarkable steadiness over the decade in contrast to the fundamental shifts in the 1973-86 period when oil lost share to both coal and nuclear. Oil consumption growth has averaged about 2.3% - almost in line with total primary energy - and has thus retained global market share at around 40%. Oil continues to lose the under-the-boiler market, especially to gas, but this is offset by the relatively fast growth of transportation demand, where oil still has an effective monopoly. Gas is gaining share as it is preferred on both economic and environmental grounds. The efficiency of combined cycle gas turbines (CCGTs) has resulted in gas rapidly gaining share in electricity generation. Coal has begun to lose share with declines in coal use in Europe offsetting rapid growth in much of Asia. Nuclear power has continued to gain share but the trend is flattening as nuclear programmes in many OECD nations remain in abeyance.

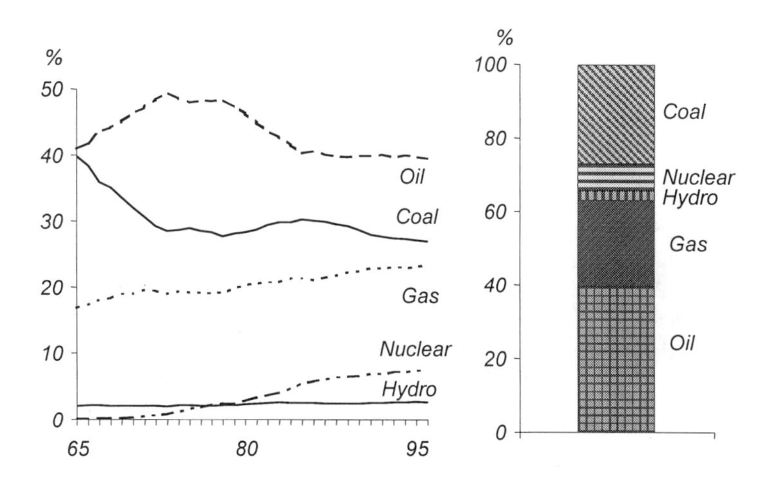

Figure 1.4: Fuel shares of total energy

2.3 Energy consumption by fuel

These forces can equally be observed by considering energy consumption growth by fuel. Oil consumption growth averaged 2.3% p.a. Year-by-year volatility in the rate of growth is common, largely as a result of weather variability. Oil products are still used for heating in Japan and parts of continental Europe. Oil is also the swing fuel in power generation in Japan, Korea and parts of the United States. Gas consumption growth is, in contrast, both higher and accelerating. Gas has become the fuel of preference and expansion of pipeline distribution networks is increasing penetration. Gas is gaining share rapidly in Europe. UK growth has been triggered by deregulation. German growth has been assisted by the reunification of the country and availability of increased volumes of imported gas from Russia, Norway etc. Gas is slowly regaining share in N America, assisted by deregulation. LNG growth has generated rapid gas consumption growth in Japan, S Korea and Taiwan. Domestic availability has created rapid gas growth in markets such as Malaysia, Argentina and Venezuela. Coal, in contrast, struggles in aggregate at a global level.

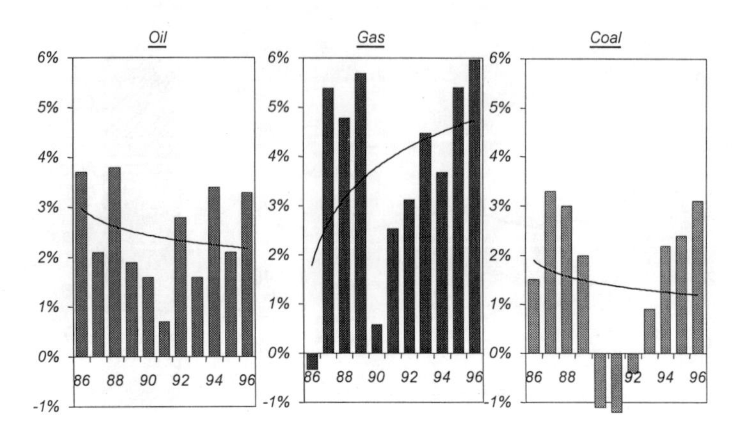

Figure 1.5: Energy consumption growth by fuel 1986-96 (Exc. FSU)

2.4 *Energy supply*

However, the decade of 1986-96 is better understood by reviewing supply developments. Even though the world in the mid-80s was still preoccupied with the availability of energy resources and adjustment to changing energy prices, the subsequent decade has in fact been characterised instead by rising energy supplies that have been delivered at declining cost and price. Over the decade it has become increasingly clear that energy supply is not a problem: commercial responses to market-based opportunities have been prevalent. The decade has been a period of transition from an era concerned about scarcity, to an era increasingly concerned about the environment.

2.4.1 Oil

The oil market has been a case in point. The late 80s witnessed rising oil consumption on the back of the strength of the global economy and lower oil prices. This was initially supplied predominantly by OPEC producers who had low cost spare production capacity and were accordingly both able and willing to increase supply. Non-OPEC producers, especially in the US and UK were struggling to adjust to lower oil prices. Over the period 1986-91 OPEC countries supplied 80% of the growth of oil consumption outside the FSU. Russian oil production had meanwhile, independently, begun its dramatic decline.

But the period since 1991 has been very different and reveals an era of oil supply push. Since 1991 the balance has changed. In the last five years (1991-6) non-OPEC has supplied 55% of the growth in oil production and OPEC 45%. The non-OPEC performance was substantial and critical. In total, non-OPEC production grew by 202.1 mtoe (4.25 million b/d), despite US production declining by 40.6 mtoe (775,000 b/d). The growth in non-OPEC, excluding the US, was equivalent to 26.6% of the increase in the world's total energy supply outside the FSU. It was a supply driven increase, resulting from the ability of the international industry to invest and produce at costs below the prevailing price. It represented a step change downward in costs in adjustment to the lower oil prices of the post 1986 period - with about a 5 year adjustment period. It involved a wide range of technological advances (including deepwater technology, 3D and 4D seismic, computerised data processing, directional drilling etc.) and fundamentally new operational and contractual methods of working. Improved fiscal terms also played an important role and had the effect of reducing the sensitivity of oil production to the price of oil.

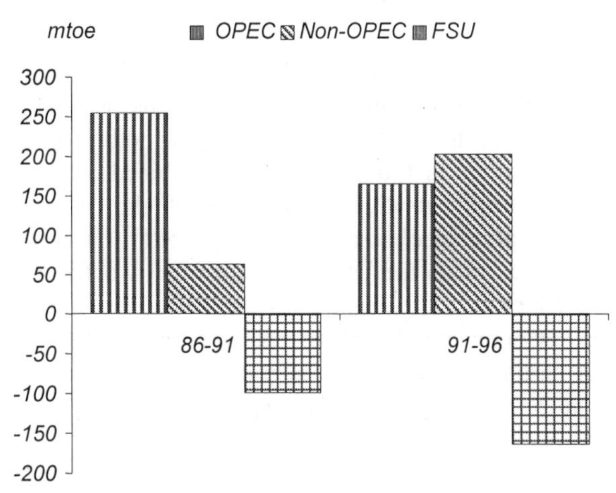

Figure 1.6: Oil supply growth 1986-96

These forces were widely dispersed across many producing countries. 13 non-OPEC countries each increased production by over 5 mtoe (100,000 b/d) over the 5 years. Norway was the largest, contributing 30.8% of the non-OPEC growth. In three instances - China[3], Mexico and Brazil - the growth was achieved by state oil companies operating without participation from international investment.

[3] International companies were active in the Chinese offshore. Activity is now taking place onshore also but no significant production has been achieved.

Over the same period (91-6) OPEC oil production also increased, by 165.1 mtoe - 3.4 million b/d, but the composition differed critically from that during the 1986-91 period. 58.9% of the OPEC increase was supplied by Kuwait as it restored its production to pre-Gulf War levels. There was virtually no growth in production amongst the rest of the core-Gulf producers. Meanwhile, the rest of OPEC (or non-core OPEC as they have started to be known) began to make an important contribution to growth, especially in 1996. The biggest contributor was Venezuela where production grew by 33.2 mtoe but output has also begun to increase in Algeria, Indonesia, Nigeria and Qatar, partly in response to the impact of international investment in those countries. Iraqi output began to rise in late 1996 as the UN embargo was partially eased.

2.4.2 Gas

Gas markets have also begun to experience new supply forces. World gas output outside the FSU accelerated with growth increasing from 3.3% in the 1986-91 period to 4.3% in 1991-96. The acceleration of world gas supply (outside the FSU) has been an important phenomenon that has occurred across all continents. It has been broadly in response to the attraction of gas both commercially and environmentally in light of the fact that newly built combined cycle plants produce electricity cheaply and cleanly, relative to other fossil fuels. It has progressively become the fossil fuel of choice in the electricity sector. Deregulation in N. American gas markets has made the US and Canada the two fastest growing gas producers (in absolute terms). Deregulation has also been the dominant force in the UK and Argentina. Continental European demand for gas has grown rapidly. Deregulation is only partial and varied but gas grids have been extended and gas has begun to enter the power sector. As a result, exports of gas from Norway, Algeria and the Netherlands - as well as Russia - have all increased rapidly.

But deregulation has not been the only factor during this trend. Growth of demand for LNG in Asia - Japan, S. Korea and Taiwan - has generated large volume growth in Malaysia, Indonesia, UAE and Australia. Several major oil producers, mainly in OPEC, have rapidly increased gas output for domestic consumption in order to substitute for exportable oil and to feed domestic petrochemical industries. This group includes Iran, Venezuela and Saudi Arabia as well as non-OPEC Malaysia.

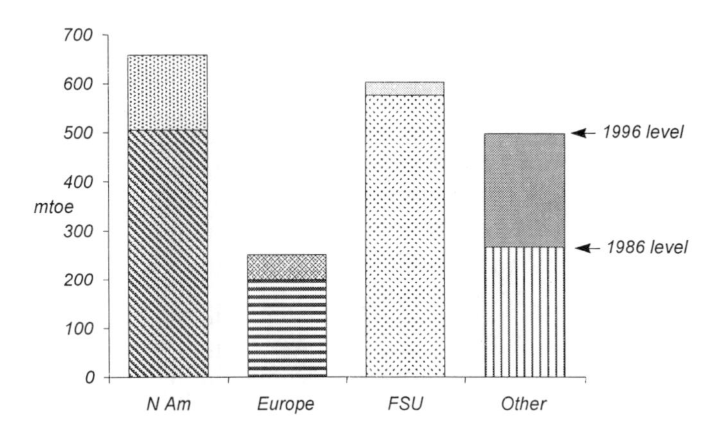

Figure 1.7: Gas supply growth 1986-96

2.4.3 Coal

The coal experience has been different but still confirms the broad point that energy supply is increasingly available at declining cost and price. European coal production has declined as subsidies have been reduced, especially in the UK and Germany. US coal output continues to increase in a highly and increasingly competitive market. US coal production costs have declined over a 20 year period driven by technological advance (e.g., longwall mining) and railroad deregulation (see Ellerman, 1997). Nevertheless it also has to compete with lower cost deregulated natural gas, which has contained its growth. The US remains the world's swing coal exporter. Meanwhile Chinese coal production increased by 50% over the decade and alone contributed to 18% of the world's increase in energy supply (outside the FSU). This was 'demand pull' as the Chinese economy boomed and the surging electricity demand was met largely by domestic coal. Elsewhere, the world's major coal exporters - the US, Australia, South Africa, Indonesia etc. - were able to expand capacity and sales at competitive prices. Asia was the main market for international coal sales with new coal fired generation being constructed in Japan, S. Korea, Taiwan, etc.

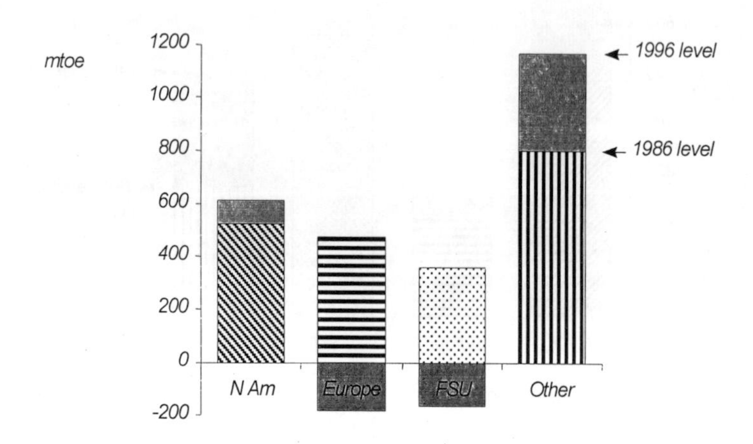

Figure 1.8: Coal supply growth 1986-96

2.4.4 Nuclear and hydroelectricity

Finally, there are the non fossil fuels, nuclear power and hydroelectricity. Nuclear power is the largest source of non-fossil energy. Its contribution to total energy has grown to today's level of 7.5% outside the FSU. It has been the fastest growing source of energy over the last 30 years. However, its rate of growth is slowing steadily. Nuclear power continues to grow, but on a declining trend. Trend growth is now in the region of 3.5%. Nuclear growth is being progressively confined to a fairly small group of countries where Governments continue to favour and promote nuclear power. Nuclear power is not prospering, in terms of new investment and growth, in competitive power markets. There is no sign that this trend will be reversed.

Hydroelectricity remains a minor source of total primary energy in the world, with a share of 2.6% in 1996. The dominant force for change on a year by year basis remains the weather. However, it is growing: it has held its share in total primary energy. Nevertheless, major new schemes are increasingly rare in light of concerns about their socio-environmental impact.

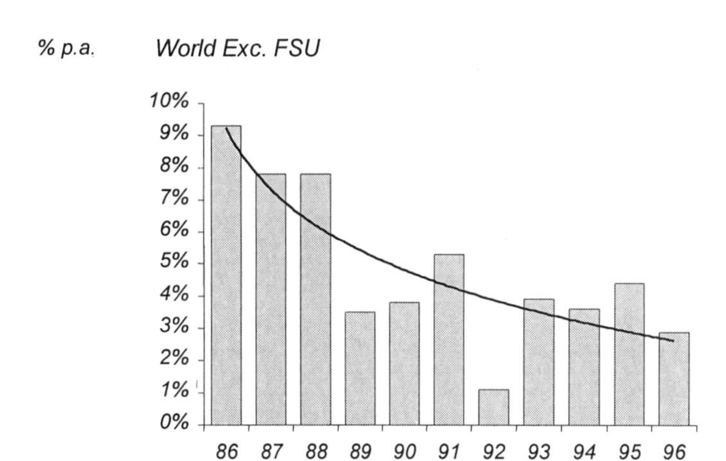

Figure 1.9: Nuclear growth

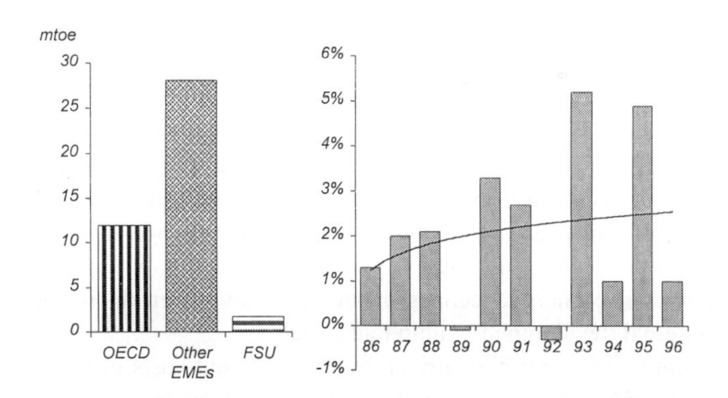

Figure 1.10: Hydro supply growth 1986-96

2.5 *Supply or demand driven?*

This leaves the final question: have energy trends over the decade been predominantly driven by supply or by demand growth? The analysis so far has cited elements of both. The fact that real energy prices have declined over the period indicates the importance of the supply effect. The market prices of oil, gas in the US, Europe and Asian LNG markets and for internationally traded coal have all fallen on a trend basis in real terms over the decade. If energy demand growth had been the sole or dominant force, energy prices would have tended to increase. These declining prices relate to producer prices of energy. Consumer prices of many energies have increased as consumer governments have increased taxation.

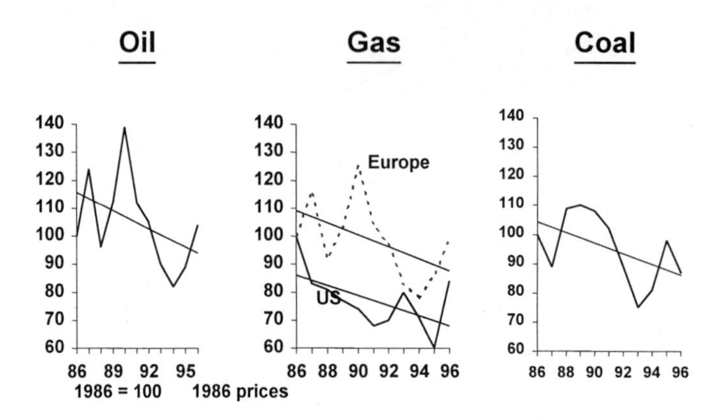

Figure 1.11: Real energy price trends 1986-96

This analysis of the trends in primary commercial energy over the last decade therefore lead us to some key conclusions:

- Energy supply growth has been the driving force in world energy markets. Supply has consistently exceeded expectations and has been delivered at declining cost and price. This has developed in a decade that followed hard upon a major decline in energy prices and fears about supply cost and availability. The energy declines in the FSU have been an independent and largely internal development within that region.
- Energy supply has itself been driven by efforts to cut costs in face of the price declines of 1986, deregulation and technological advances.
- Gas has become the fastest growing primary energy, largely at the expense of coal. Oil is holding share due to the strength of transportation demand. Nuclear growth is slowing.
- There has been a progressive transition away from concerns over resource scarcity and supply security to concerns over the natural environment.

2.6 CO_2 emissions

The analysis of the 1986-96 period concluded that there has been a transition over the last decade and that one of the key elements of this transition is that concerns in energy markets and industries have shifted from fears of resource depletion and supply availability towards increasing attention to the environment. In light of this what has happened to atmospheric emissions from the use of primary commercial energy? The only meaningful global data that can be estimated is for CO_2 . This can be estimated from global and regional energy data using standard carbon emission factors for the fossil fuels[4], as used by the IPCC.

Some caveats are required. First, CO_2 is not the only Greenhouse Gas, and energy use is not the only source of CO_2. Also, the data excludes biomass, and the burning of waste products, so some energy-related carbon emissions are not captured here. Likewise not all hydrocarbons are burned - some become plastics or bitumen, for example - so not all the carbon is released into the atmosphere. For these reasons it is not appropriate to draw too detailed conclusions from these estimates, nor to disaggregate them below the regional level.

Carbon emissions from fossil fuel use were virtually flat from 1990 to 1994. Since then emissions grew by 1.6% in 1995 and then accelerated to 2.8% in 1996. The principal cause of the period of flatness from 1990-94 was the collapse of energy consumption in the FSU and Central Europe, which offset the growth in energy use in other EMEs. Another contributor in the early part of the period was the very low growth rate of emissions from the OECD, primarily owing to weak economic growth.

The strong growth in emissions in 1996 in part reflects the diminishing effect of the FSU and Central Europe, where fossil fuel use continues to decline but at a much slower rate than in previous years. It also reflects the above trend increase in energy consumption in the OECD, where fossil fuel use grew by 3% in 1996, almost double the average rate of growth over the past ten years. This was largely driven by the weather and should be regarded as exceptional.

---

[4] **Carbon Emission Factors:**

	tic/toe	Kg C/GJ
Oil	0.84	20.0
Gas	0.64	15.3
Coal	1.08	25.8

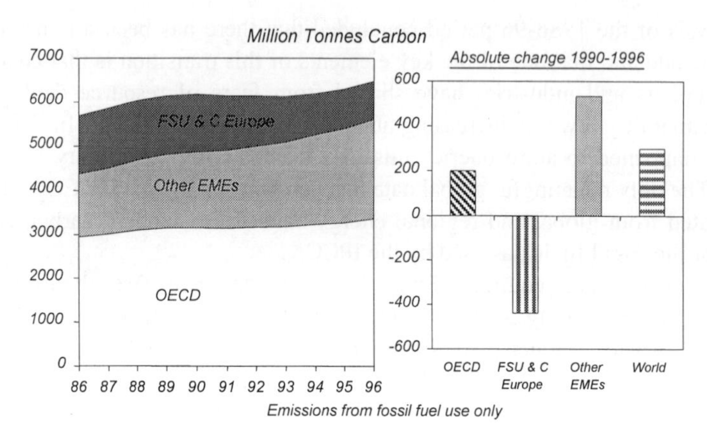

Figure 1.12: CO_2 emissions

Compared to the benchmark year of 1990, global carbon emissions are up by 4.8%, or in volume terms by 300 million tonnes of carbon. The contraction of FSU and Central European energy use caused a fall in emissions of 440 million tonnes of carbon, almost but not quite matching the increase in other EMEs (540 million tonnes of carbon). For the OECD countries in aggregate, emissions rose 6.4%, or by 200 million tonnes of carbon since 1990.

It is also possible to disaggregate the data one level further in order to understand regional trends. The picture that this generates is:

1. Europe is the only region, apart from the FSU, where emissions have not grown since 1990.
2. All other regions have exhibited some growth with Asia-Pacific, which contains 66% of the world's population[5], increasing estimated emissions by 450 mtC - or by 65% of the growth in the world outside the FSU.
3. Emissions in N. America have grown by 8.8% - or 150 mtC - since 1990. It is estimated that N America is the second largest regional source of global emissions[6] contributing 28.4% in 1996.

The growth of carbon emissions is closely related to the growth of total primary energy consumption, but tends to grow at a slightly slower rate. This is owing to:

---

5 Although only about 26% of world GDP at current exchange rates.

6 N. America is estimated to have emitted 1830 mtC in 1996. Asia-Pacific, in total, emitted an estimated 1998 mtC.

(i) switching away from fossil fuels, to nuclear or hydro, and to
(ii) switching within fossil fuels from coal to gas.

The effect can be seen by tracing the carbon intensity of energy consumption, that is the tonnes of carbon emitted per tonne of oil equivalent energy consumed. This has been on a steady downward trend for at least three decades. However, the rate of decline is falling. This largely reflects the slowing growth of nuclear power in the global energy mix, despite the acceleration of gas.

The effect of fuel-switching can be illustrated by assuming that the carbon intensity had stabilised at its 1990 level. In this case carbon emissions would have been about 2% higher in 1996 than they actually were, or about 120 million tonnes higher. That is roughly equivalent to saving the emissions of an energy market the size of Italy or South Korea. To put this into context, the overall growth in energy consumption over the same period has added more than three South Koreas to world energy consumption. If anything the rate of decline of carbon intensity is likely to slow further as the rate of growth of nuclear power continues its trend decline. In the longer term, non conventional sources of energy (such as solar) will increasingly enter the energy mix and help to reduce the carbon intensity.

To estimate a trend rate of carbon emissions growth then, we need to combine a trend rate of energy consumption growth (around 2-2.5% pa) with an estimate of the trend decline in carbon intensity (0.2-0.3% pa). That yields a trend estimate of about 2% pa.

This trend growth looks like continuing into the next century.

2.7 Conclusions

The analysis so far has described a world of fairly resilient trends in world energy markets. They can be summarised as follows:

1. Steadily rising energy consumption in line with GDP.
2. A rising energy share for gas, largely at the expense of coal. Nuclear growth is steadily slowing.
3. Trend growth of about 2% in CO_2 emissions from fossil fuels.
4. Energy supply growth has at least matched the growth in demand and has been sufficiently strong to lead to a trend decline in real energy prices.
5. Geopolitical developments have caused volatility but, outside the FSU, have not altered trends.

3 1998: A hiatus in energy trends

The historical analysis above reviews the decade following the oil price decline of 1986. 1998, however, witnessed a hiatus, a significant deviation from these trends. While global energy data has yet to be finalised for 1998, the shape and form of the hiatus has been apparent in terms of energy supply, demand and prices.

The most obvious manifestations of this hiatus have been the *contraction* of Asian energy consumption during 1998 and the dramatic weakness of oil and gas prices between December 1997 and March 1999. The two developments are partly interrelated.

Asian energy consumption is preliminarily believed to have declined in 1998 as a result of the Asian financial crisis that began in Thailand in mid-1997 and later spread to Indonesia, Malaysia, South Korea and, to a lesser degree, the Philippines. Japan, at the same time, was in prolonged recession. Absolute declines in economic activity reduced primary energy consumption with oil the most adversely affected due to its swing-fuel role. Oil and energy consumption continued to grow in China and India, but, in total, Asian consumption fell back. This clearly breaks the trend of rapid growth of the last decade, although the correlation between economic growth and energy consumption appears to have remained relatively intact.

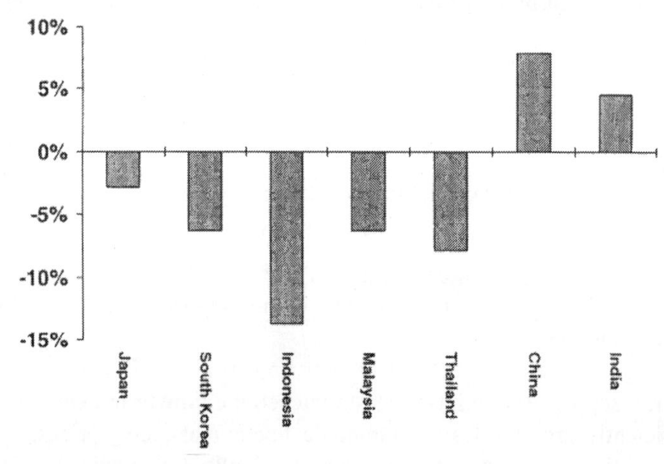

Figure 1.13: Asian economic growth 1998

The oil price decline was also inconsistent with previous trends and other behavioural characteristics. From 1986-97 oil prices declined in real terms but were behaviourally regular in nominal terms. Excluding the Gulf War, the price series was normally distributed with a mean of $18 Brent. Prices remained between $15 and $21 for 80% of the time. The price decline of 1998 pushed prices outside this range for some 15 months.

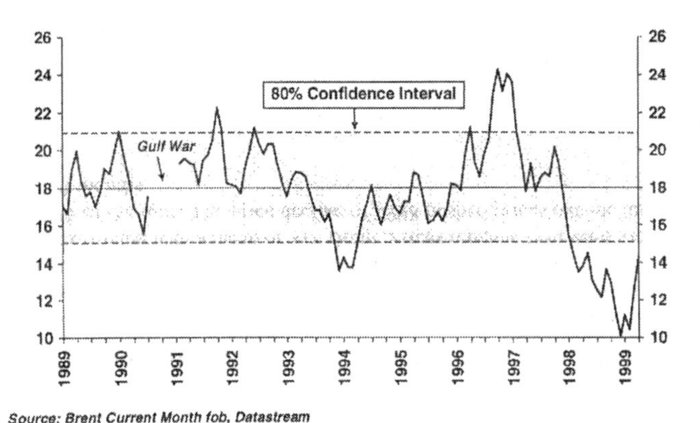

Figure 1.14: Oil price range 1988-1999 (excluding Gulf War)

The price decline was caused by a combination of factors, of which the Asian crisis was but one. Oil supply had increased. As analysed above, non-OPEC production growth was strong, partly as a result of a reduction in costs and increases in investment. Iraqi exports grew rapidly under UN auspices and other OPEC production also accelerated. At the same time El Niño produced a warm winter and resulting weak consumption. The Asian Crisis was the final straw. Oil prices only recovered in March 1999 following a series of three production cuts by OPEC producers together with a small group of non-OPEC producers.

By May 1999 the Asian economies and energy consumption had begun a slow and partial recovery. Oil prices had crept into the bottom of the previous price range ($16).

4 Long term prospects

Longer term prospects accordingly pose the question as to whether the developments of 1998 were merely a deviation from trend as well as a series of additional issues:

4.1 Will future world GDP growth trends differ from those of the last decade?

The most obvious uncertainty relates to the prospects for economic growth in East Asia. Current evidence indicates that a recovery has commenced, especially in

South Korea. Some economic and financial reforms have been enacted. While uncertainties are rife, on balance it would seem improbable that Asia could return and sustain the high growth rates of 1986-96.

Meanwhile, the other major structural macroeconomic change is occurring in the FSU. The financial crisis of August 1998 set Russia back again after a period of progress in terms of reform and partial recovery. It seems unlikely that rapid economic growth can be achieved in the next several years. In due course, if and when growth re-emerges, energy consumption growth should also re-emerge after a (short) lag, even if only slowly due to the expected steady scrapping of some of the least energy efficient industrial infrastructure. (FSU export prospects are discussed below).

World financial markets have also begun to question whether trend rates of growth can be expected to increase in the United States in particular. While financial markets have been driven up and US economic growth has been strong, there is insufficient evidence at this stage to conclude that productivity trends have shifted upwards.

4.2 Are there adequate energy resources to meet future needs?

The current evidence is overwhelming in confirming that there is no shortage of reserves of fossil fuels and that there is unlikely to be any shortage in the foreseeable future. Data on proved reserves for oil, gas and coal all show high reserve to production ratios (42 years for oil, 62 years for gas and 224 years for coal). Such ratios have increased over the last 20 years. However, there is much acrimonious debate (especially between economists and geoscientists and engineers[7]) about the precise validity of such data. The evidence still points clearly to the fact that reserves are being replaced consistently. The recent period of oversupply of oil, in particular, casts doubt upon the worst depletion fears.

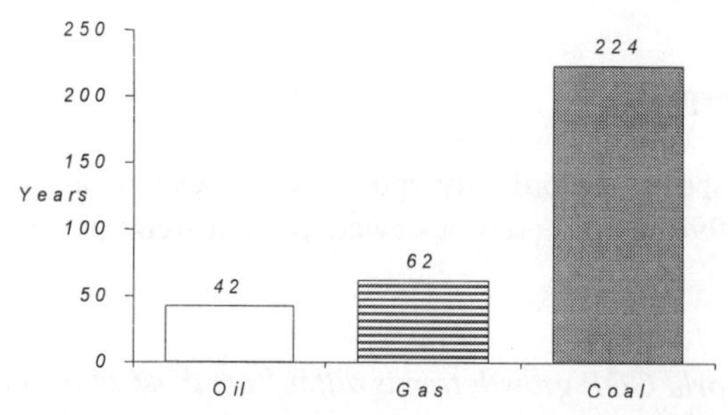

Figure 1.15: Energy resources - 1996 reserve to production ratios based upon proved reserves

[7] See Campbell & Lahererre (1995), Campbell (1997) and Lynch (1996).

4.3 Will energy supply/demand dynamics cause a future imbalance?

Energy products remain homogeneous commodities produced in highly capital intensive industries. As such, it is probable that energy prices will be both volatile and cyclical. Oil markets and US natural gas markets both experienced price strength in 1996 and, to a lesser degree, in 1997 before the price crash of 1998. There were no signs that the initial price strength was structural. For it to have been structural we would tend to have to experience rising supply costs. Oil production costs outside OPEC continue to fall due to cost efficiencies and technological advance. It is more probable that the cost reductions are an indicator of a structural oil price decline. The case of US natural gas prices may be partially different due to the geographical constraints of the market. Structural US gas price weakness is less likely than in the case of oil. In coal markets, the US has a continual overhang of production capacity, especially in the Western states, that can be produced and delivered at costs close to prevailing world traded coal prices.

In all these instances, there is some evidence that the relevant supply curves are highly elastic at prices close to those currently prevailing. It tends to imply that a long term upside structural change in producer prices of energy is unlikely. Rising consumer taxation of energy may, however, increase consumer energy prices, possibly substantially.

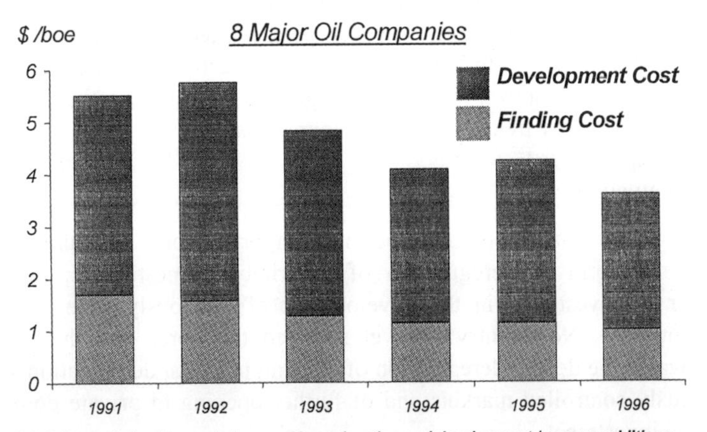

Figure 1.16: Oil production costs

4.4 What exogenous forces could hit energy markets?

There appear to be four potential exogenous forces that could drive energy markets in the future:

4.4.1 Environmental policies

It is highly probable that clean air legislation will be continually tightened in both OECD and emerging market economies. To date the impact of such has been limited upon demand. The incidence has fallen upon processors (refiners) and final consumers. The demand effect has been impossible to detect. Refining profitability has been poor.

Policy responses to the challenges of climate change pose a more serious risk to energy market trends. As discussed above, the global trend growth in CO_2 emissions is 2%. This appears to be inconsistent with objectives to either stabilise or reduce emissions and with the Kyoto Agreement.

There appears to be a growing acceptance and belief among scientists and others who are both concerned and well informed that human activity is influencing the earth's climate through the emission of CO_2 as a result of the burning of fossil fuels. This does not mean that there are no disagreements, nor does it mean that there are no uncertainties. It does, however, say that there is an important issue and that many now consider that the time has come to begin to take action.

Policy initiatives to date have, however, been limited in global terms and, as a result, can only be expected to have a significant impact on global energy trends in the longer term. Nevertheless, if they are implemented rigorously, the implication is that recent global energy trends will in due course change.

4.4.2 Deregulation

Energy deregulation has been an increasingly important force over the last decade whether in the form of deregulation of electricity, domestic gas, international pipelines or of investment in the development of previously state monopolised natural resources. No country has yet reversed this force and the momentum remains towards the deeper deregulation of existing markets, deregulation spreading into previously controlled markets and of further opening to private domestic and international investment.

As seen above the result of deregulation to date has tended to be to improve allocative efficiency, reduce costs and, especially where financial or technological constraints prevailed, raise investment. This has tended to stimulate supply - shift the supply curve downwards and to the right. Only in previously subsidised coal markets (in Europe) has it reduced supply. There has also tended to be a shift in the demand curve in cases where demand was previously constrained - as in some gas markets - together with a move along the demand curve. The net effect has been a tendency to increase the amount demanded with prices, if anything falling.

Some of the large energy markets have already been deregulated, e.g., the US and UK. As such many of the largest gains have already been achieved. Nevertheless, the process of deregulation in the US is continuing, especially in the electricity sector. Meanwhile we have yet to achieve the gains from the European Union's deregulating Electricity and Gas Directives. Other major energy markets such as Japan have only just begun energy deregulation.

Accordingly, deregulation is likely to continue and to continue to stimulate both supply availability and, in some cases, consumption growth. Deregulation looks set to reinforce rather than reverse the trends of the last decade.

4.4.3 Supply security

Government and public concerns about the availability of energy supply have been among the most powerful driving forces of energy policy, especially in the 1973-86 period. It remains a key factor in many energy importing nations. Potentially responses to such concerns could drive new energy policies in the future and partially reverse some of the trends that have been generated in the last decade.

The last decade has witnessed no serious energy supply interruptions despite the Gulf War, the energy supply contractions in the FSU and embargoes upon Iraq, Libya and Iran. A renewed confidence has developed, especially after the Gulf War, which reassured many energy consumers about future energy supply availability. Energy markets 'worked' in 1990/91. Meanwhile energy producers world-wide have shown that they do not intend to rely upon future price rises. Producers have shown a preference for higher volumes, even at the expense of weaker prices. Geopolitical disruption remains a possibility at any time. However, consumers and market participants are tending to believe that any future disruptive effect will be only temporary.

Nevertheless, the International Energy Agency (IEA) still exists and has an increasing membership. It continues to seek to promote energy policies that will secure energy availability in the future, even if they impose a cost on consumers. Most member countries of the IEA still have concerns about future energy supply and retain policies which at least partially address these concerns. However, none are reinforcing these policies and it would seem to be very unlikely that energy security concerns will influence energy markets significantly in the absence of some new geopolitical change in supplying nations. The exceptions could be in Asia where consuming and importing nations such as China are showing concern over their increasing import dependency. Some restrictive policies are in place, but the balance of new policies is more towards investing in and 'directly securing' new energy supplies.

To some degree the concept of energy security is changing. In some of the newly industrialising countries the concerns are more over the future availability of electric power. The concern is that industry will be constrained, consumers cut off, computers crash and air conditioning switched off. In this situation the issue is to create the investment in new capacity and to ensure grid reliability. Energy security then becomes a regulatory, financial and technological issue rather than one driven by geopolitics. It does not constrain consumption. Rather it ensures efficient supplies are available. It reinforces rather than distorts or reverses recent energy trends.

4.4.4 OPEC policies

The rise in oil prices in 1973/4 and 1979/80 and resulting government and consumer perceptions were important determinants of the energy trends of the 1973-86 period which differed so greatly from those from 1986-97. If the OPEC member countries were to reimplement oil production and investment policies that were to restrain future oil supplies and raise prices, the world could enter a new energy era.

The oil price experience of 1998/9 gives some indication as to the prospects for the future. Over this period it became clear that OPEC can still coalesce when prices fall far enough and for long enough. Despite evidence that all members wish to expand their capacity and production in the long term, the key producers eschewed long term market share in return for shorter term price recovery. Nevertheless, it seems unlikely that OPEC members would either be willing or able to enact a policy that drives up prices above those that have prevailed since 1986. Instead, it seems more likely that the challenge will be to prevent long term price erosion in face of declining non-OPEC production costs.

4.5 FSU energy markets

Energy developments in the FSU are critical in determining future global energy trends. As shown above, the massive disruptions of the 1990s have, somewhat surprisingly, had little effect upon underlying global trends. But, given that both energy consumption and production in the FSU have bottomed out and look set to increase again, could the FSU energy sector be a determinant of a new energy world beyond 2000?

There is little doubt that the region is well endowed with energy resources, especially of oil and gas. It contains 40.4% of the world's proved gas reserves and with 65 billion barrels of proved oil reserves, 6.4% of world oil. The potential for increases in reserves of both oil and gas is widely considered to be substantial.

The prospects for domestic energy consumption are particularly uncertain given the depth of economic change. Nevertheless there is virtually no prospect that the region will become a net energy importer of any fuel, even though there may be some trans-regional oil trading. The key global questions are:

1. What impact will the FSU have on global emissions of CO_2?
2. How large and in what direction will new energy exports flow?

As noted above, the energy contractions in the FSU served to flatten CO_2 growth in the early 1990s. In 1996 FSU emissions of CO_2 were an estimated 663 mtC, which represented 8.8% of global emissions. Any growth in consumption of fossil fuels will increase global emissions of CO_2. However, given its relatively modest global share, developments in the CO_2 emissions in the FSU are unlikely to have a major global impact.

Figure 1.17: Future of FSU energy exports

It seems very likely that the FSU will be an increasing exporter of both oil and gas from at least four areas:

- Gas sales from Russia to Western and Central Europe are set to increase. The new EuroPol gas pipeline from Russia through Poland to Germany is one key step. Domestic gas pipelines up to the prolific Yamal peninsula are also set to be expanded. Russian gas will be an important force throughout Europe.
- Caspian (Azerbaijani and Kazakh) exports of oil have begun and are set to expand rapidly in light of ongoing international investment in production capacity and pipelines. Final export routes have yet to be determined.

- East Siberian gas exports to China and possibly other parts of NE Asia are planned and have unanimous support from the relevant parties. Lead times, however, may be long.
- Plans are advancing for the development of the oil and gas resources of the Sakhalin area for export to Japan, S Korea and other regional markets.

A range of other plans are also being considered including Turkmen gas sales to Turkey through Iran and the construction of an export route from the C Asian states to China through the Tarim Basin.

There are substantial resources, even if they cannot be quantified without further exploration. The greatest challenge will be to ensure that long distance transportation options are cost competitive. Net, it seems certain that the FSU will be a contributor to the global availability of oil and gas well into the next century. Again this seems set to reinforce rather than change recent global energy trends of growth and availability.

5 Conclusions

The analysis in this chapter has concluded that the global energy markets over the 1986-96 decade exhibited clear trends, namely:

- Global energy consumption (outside the FSU) was driven by global economic growth with deviations around the trend mainly caused by weather volatility.
- Energy developments in the Former Soviet Union were substantive but did not have a structural impact upon energy markets in the rest of the world.
- Energy supply push was the dominant driver of global energy markets with the result that real energy prices declined on a trend basis.
- Global CO_2 emissions have been growing at a trend rate of about 2% and this rate will continue into the next century.

The trends of the 1986-96 decade have been tested by the Asian financial crisis and the oil price weakness of 1998/9. Nevertheless, apart from the likelihood of slower Asian growth, most of the underlying forces of the last decade seem set to continue. There are no apparent resource constraints and supply/demand dynamics point to a sufficiency of supply over at least the medium term.

The major risks of a discontinuity or change in trend in energy markets stem from potential exogenous changes with uncertain and indeterminate probabilities. As ever, there is always a risk of oil markets being driven by a new supply push from OPEC, although there is no clear evidence today that there has been a fundamental change in strategy. Alternatively, if governments decide that they wish to enact policies to reduce global greenhouse gas emissions, this will require policies that lead the world energy economy into a new era, although present indications are that any such impacts will only emerge in the longer term.

References

BP (1997), *Statistical Review of World Energy*, London 1997.
Campbell C.J. and J.H. Laherrere (1995), *Supply of World Oil 1930-2050*, Petroconsultants: Geneva.
Campbell C.J. (1997), *The Coming Oil Crisis*, Brentwood Multiscience Publishing.
Ellerman D.A. (1995), *The World Price of Coal*, MIT CEEPR Reprint series No. 122.
Ellerman D.A. (1997), *Longwall Mining in the United States*, MIT CEEPR Working Paper.
Lynch M.C. (1996), Analysis of Forecasting of Petroleum Supply: Sources of Error and Bias: *Energy Watches VII*, CEED: Boulder Co.

SECTION 2

NATIONAL STUDIES OF ENERGY STRUCTURE AND REFORM

CHAPTER 2

BRITAIN'S REGULATORY REGIME IN PERSPECTIVE

COLIN ROBINSON

Professor of Economics, University of Surrey
Editorial Director, Institute of Economic Affairs, 2 Lord North Street, London SW1P 3LB
Email: crobinson@iea.org.uk

Keywords: Britain; competition; market failure; privatisation; regulation; utilities.

In this chapter I want to take a rather broad look at the regulatory system for Britain's utilities[1]. I begin from some issues of principle about the nature of competition and the nature of regulation and then discuss to what extent Britain has devised a regulatory regime which is an advance on other such regimes. *En passant*, I shall say a few words about British utility privatisation schemes because they were important determinants of the regulatory system. I shall conclude with some observations on the review of regulation which the government is conducting[2].

1 The nature of competition

Many discussions of economic policy issues are unproductive because participants are talking at cross-purposes about the nature of competition. In particular, there is confusion between perfectly competitive markets and what I shall label 'real-world competitive markets'. So I shall set out what seems to me to be the essence of a real world competitive market. Then I can explain more easily why such markets are generally to be preferred to those in which regulation prevails.

Economists' views about the nature of competition have changed radically since about the time of the 'marginalist revolution' of the 1870s. As Mark Blaug has explained (Blaug 1987), the:

> '...tendency throughout the history of economic thought to place the accent on the end-state of competitive equilibrium rather than the process of disequilibrium adjustments leading up to it...became remorseless after 1870 or thereabouts'.

Blaug goes on to explain that perfect competition is:

> '...foreign to the classical conception of competition as a process of rivalry in the search for unrealised profit opportunities.'

The distinction between competition as an end-state and competition as a process is extremely significant when it comes to judging the extent to which market regulation is likely to be successful in real-world markets. Mainstream economists

---

[1] An earlier version of some of the ideas in this chapter is in Robinson (1997).

[2] This chapter was written in November 1997 when the government's review was under way. By July 1998, at the time of this revision, two papers on the results of the review had been published: *A Fair Deal for Consumers: Modernising the Framework for Utility Regulation*, Green Paper, March 1998, and *Response to Consultation*, DTI, July 1998.

seem, explicitly or implicitly, to cherish the state model whereas writers in the older classical tradition or modern-day 'Austrians' see competition essentially as the disequilibrium adjustment process to which Blaug refers.

Superficially, perfect competition appears an ideal market form. The force of competition is so strong that in the long run all 'excess' profits are eliminated: the only firms which remain in the industry earn just enough to stop their productive factors leaving. Price is equal to marginal cost in long run equilibrium. In this idealised state, Pareto optimality is attained so that it is impossible to make anyone better off without making someone else worse off.

At least since the time of Pigou (Pigou 1929, 1932), mainstream economists have tended to assume that there is something perfect about perfect competition. Consequently, faced with any real-world market they examine it to determine whether or not it lives up to the perfectly competitive ideal. If it does not - and invariably it will not - the standard approach is to recommend government action to remedy perceived 'imperfections' and 'failures'. Perhaps the market is imperfect because one or more companies appears to have market power. Or perhaps it fails because there are externalities such as environmental spillovers.

This approach to economic policy-making - neatly labelled 'Nirvana economics' (Demsetz 1989) - leads inevitably to widespread intervention by governments or regulators. The market failure approach is very persistent in economics even though it has been subject to devastating criticisms. For instance, the Lipsey-Lancaster 'second-best' critique of forty years ago (Lipsey and Lancaster 1956) undermines this kind of piecemeal welfare economics. Second, there is a serious problem in using the Pareto criterion in practice since it is rare indeed to find moves which make some better off without making others worse off (if only because people judge their satisfactions relative to others): most policy moves result in both winners and losers. Third, public choice theory demonstrates that government failure is pervasive (Tullock 1976). Thus intervention to correct supposed market failures cannot be assumed to be beneficial.

Despite such criticisms, many economists refuse to abandon the perfectly competitive paradigm. It is well known that such paradigms, once established in scientific communities, are resistant to attack because established interests support and maintain them (Kuhn 1970). So it is in the economics community with the market failure approach, based on the perfectly competitive paradigm. Mainstream economists have too much intellectual capital invested in it for it to be abandoned readily. Moreover, it is an approach dear to many policy-makers because it provides an intellectual backing for regulation in all its varieties: it is convenient to have economic arguments which appear to justify intervention in markets to move them closer to the perfectly competitive or to mimic the perfectly competitive outcome.

An even more fundamental problem with the perfectly competitive paradigm - when used for policy-making rather than merely for pedagogic purposes - is one to which I want to draw special attention because it is quite crucial when considering the role of regulation. Reverting to my earlier discussion of different concepts of competition, can competition usefully be regarded as a state or should it be seen as a process taking place over time?

Even if the criticisms I have already mentioned did not apply, the long run equilibrium of perfect competition would be a totally inappropriate policy target. A fundamental difficulty is that it does not represent competition at all. It is an end-state in which competitive forces have been exhausted: the process by which equilibrium is achieved is simply disregarded. The market is at rest whereas the essence of competition is disequilibrium characterised by continuous change (Kirzner 1997).

A real-world competitive market, by contrast, is the kind of market envisaged by modern Austrian economists. A process of competitive discovery amongst entrepreneurs is under way. Companies in the market feel the threat of actual and potential competition (because entry to such a competitive market is free) and so bend their efforts not just to holding down costs but to finding ways of retaining existing customers and capturing new ones. Consumers have choice of supplier and therefore the power of exit. Discovery by entrepreneurs of new ways of doing things and consequent knowledge-creation are the essence of the competitive process (Hayek 1948). It is quite different from a perfectly competitive market where knowledge is costlessly available or, in more sophisticated versions, where knowledge will be obtained if the benefits exceed the costs of searching for it. The 'search' model assumes there is a pool of knowledge which is readily available and that market participants know what they need: more succinctly, it is assumed those participants know what they do not know[3].

2 Implications for regulation

These distinctions between notions of competition have significant implications for regulation. Starting from the assumption that competition is a discovery process, driven by entrepreneurs, with markets always in disequilibrium - rather than an equilibrium state in which all relevant knowledge is available to be tapped - leads to quite different conclusions about the role of regulation.

Traditional views about the case for regulation arose from a preoccupation with the outcome of perfect competition. In an 'imperfect' market, there is an evident case for regulation, either of the market-improving or market-displacing variety, to produce an outcome closer to that of the long run equilibrium of perfect competition. Regulation is a means of achieving the results of perfect competition without the messy and apparently wasteful process of competition itself.

But if real-world markets are a key part of a process which produces knowledge it is not possible to achieve the results of competition in the absence of the process. In other words, it is impossible to determine, *ex ante*, what the outcome of a

[3] Kirzner, op cit.

competitive market will be. Only after the event, when new knowledge has appeared as a result of the actions of rivals, can the outcome be seen. Even then, 'equilibrium' is not a helpful description of what can be observed since real-world markets are characterised by disequilibrium and continuous change as entrepreneurs seek unexploited profit opportunities[4]. Equilibrium never exists and so can never be observed. Models of regulatory systems which are founded, explicitly or implicitly, on the perfectly competitive paradigm, are unhelpful as guides to practical action.

Of course, regulators will act. Once in office, they will make decisions which change market outcomes. But there is no reason to believe that the results of these actions will be an improvement on what would otherwise have happened. Indeed, regulation which aims to remove 'imperfections' from markets may undermine the entrepreneurial discovery process which is the prime source of innovations of all kinds. As Schumpeter saw, the ability of companies to retain 'excess' profits for a period is probably one of the principal driving forces of innovation. If regulators prevent such retention, adverse effects on innovation seem likely (Schumpeter 1976).

The principal point I want to make is that once the traditional anchor of the perfectly competitive market is removed, the basis for regulation becomes elusive. If market outcomes can never be known in advance, regulators can by definition never have the knowledge to simulate market outcomes. What then is their purpose? Is it just to make piecemeal adjustments to deal with apparent market failures? The awful possibility is that, over a period of time, well-intentioned regulation will seriously hamper market processes, making the outcomes worse rather than better. Another anchor is needed.

3 British-style regulation

Fortunately, in Britain there has been some success in finding this other anchor, though it has been partly accidental. In most of Britain's utility markets, the danger that regulation will stifle entrepreneurship and innovation has been circumvented because of the emphasis placed on pro-competition regulation. That is not true of all the utilities. Water is the main exception, as I have explained elsewhere: it operates under a regime where there is virtually no competition and evidently little prospect of there being any[5]. It is the closest of any of the utility markets to an old-style nationalised regime - but instead of state ownership has been substituted control by government and the regulator. Elsewhere regulators have been harnessed to the cause of stimulating competitive processes rather than hindering them.

When the early regulatory regimes were being devised, those involved were aware of the dangers of two types of systems (Beesley and Littlechild 1983). Of one

---

[4] Ibid.

[5] Robinson (1997), *Introducing Competition into Water*, op cit.

of them Britain had plenty of experience - nationalisation embodied a form of regulation carried on behind closed doors by politicians and civil servants on ill-defined rules which removed responsibility from managements of the corporations and stopped the industries evolving in market-led directions. The other was US-style rate-of-return regulation, which failed to provide incentives to innovate and reduce costs and was subject to capture by producer and other interests.

These perceptive ideas about appropriate kinds of regulation were unfortunately for a time overwhelmed by the deficiencies of the privatisation schemes. The utility privatisations, in particular, were seriously flawed (Robinson 1992). To be brief, the principal underlying problem was that the government, heavily influenced by the main producer pressure groups (managements of the nationalised corporations, their unions and the City), gave priority to revenue-raising and widening share ownership at the expense of market liberalisation. There was, at the time, no constituency for market liberalisation apart from unorganised consumers. All the powerful organised groups were against.

Now that we have a short historical perspective on utility privatisation we can see that in the early years regulators have been forced to spend time fighting to overcome these deficiencies. It would have been much better if the privatisation schemes had made a cleaner separation between 'natural monopoly' sectors and others and provided structures which would have permitted competition to flourish in the potentially competitive sectors. It would have been better also if capital markets had been allowed to operate from the beginning, placing the companies in the market for corporate control, instead of sheltering them with 'golden shares'.

Nevertheless, regulatory regimes developed since privatisation of the first tranche of British Telecom in 1984 are a considerable improvement both on regulation by nationalisation and on US regulation. Their distinctive feature (the water industry apart) is the emphasis on stimulating entry and promoting competition. Rather than relying on the stylised notions of perfect competition as a model, the underlying objective - whether conscious or not - appears to be to start competitive processes working wherever it is feasible to do so. That seems an entirely sensible idea which gives a strength to the British utility regulation regime as compared with regimes elsewhere. Indeed, the British system is beginning to be copied and adapted as countries seek appropriate regulatory systems in the wave of privatisations now sweeping the world.

4 Advantages of the British system

I would identify the specific strengths of the British system of regulating utilities as follows.

First, is the independence of regulatory offices from direct political control. That is a very significant advance on nationalisation when there was constant interference in state-owned corporations, generally for short-term political reasons, and managerial incentives were sadly lacking. There are complaints that regulation now is not as open as it should be, but it is a far cry from the old regime under which some industries were more instruments of state industrial policy than commercial enterprises. The state-owned electricity industry, for example, was used as a support mechanism for the state-owned coal industry, for the British nuclear power industry and for the manufacturers of heavy electrical equipment (Robinson 1996). Of course, Ministers still try to intervene but it is more difficult than it used to be.

Second, price cap regulation is more closely consistent with competitive markets than rate of return regulation and is a vast improvement on price control under nationalisation when price changes were, on occasions under both Conservative and Labour governments, determined in Cabinet. There were occasional attempts, such as in the 1967 White Paper[6], to introduce some economic principles into pricing. But governments were never willing to give up interfering with the prices of goods and services prominent in the retail price index. Compared with rate of return regulation, price caps have advantages though there is a tendency to revert to rate of return because regulators tend to look at profits and returns when setting X and because of pressures on regulators to review controls frequently. Price caps are of course to some extent arbitrary, as is all regulatory price control, but if reviews are relatively infrequent (say, every five years) they have a very desirable property. Because, between reviews, regulated companies can retain unanticipated cost savings, there are good efficiency incentives. In such circumstances, a price cap is probably the nearest regulatory equivalent to Schumpeterian competition. Instead of Schumpeter's 'gale' extinguishing 'excess' profits, the task is performed by the next regulatory review (Beesley 1996).

The *third* innovative feature of British utility regulation is the pro-competition duty which, in one form or another, the utility regulators have. There was an accidental element in the imposition of this duty (Robinson 1994). It can also be argued that it has proved so significant mainly because of the deficiencies of the privatisation schemes: had markets been liberalised more at the beginning, there would have been correspondingly less need for liberalising action by regulators. However, whether or not the government realised what it was doing, the pro-competition duty has become a key feature of the British regulatory system. The emphasis it has placed on stimulating entry and starting competitive processes outside natural monopoly sectors has allowed regulators to take the initiative in promoting entry to markets in very unpromising circumstances (as Sir James McKinnon did in the early days of gas privatisation). Moreover, it has changed the incentives of regulators so that capture by producers is not the issue here it has been in the United States. When one of a regulator's principal duties is to promote or facilitate competition, the interests of consumers come naturally to the fore and there is relatively little chance he or she will bend to the lobbying of producers.

---

[6] *Nationalised Industries: A Review of Financial and Economic Objectives*, Cmnd.3437, 1967.

These key elements of British regulation have helped to avoid the worst features of regulation. Taken together, they provide an emphasis on market entry and liberalisation and increase the chances that the scope and scale of regulation will diminish over time as it is displaced by competition. The institution of competition in production and supply in the major British utilities (water apart), leaving an open access transportation network as the only 'natural monopoly' sector, has been a genuine innovation, now being copied in many countries. After all, until very recent years, the conventional wisdom was that utilities such as electricity, gas, telecommunications and the railways were 'natural monopoly industries', lacking any potentially competitive areas. But competition is now spreading to all potentially competitive sectors and regulatory activity should decline.

If you accept my view about the importance of setting competitive processes to work, the British system's emphasis on entry and liberalisation is extremely important. It is a great improvement on traditional regulatory systems which do not explicitly promote competition and where there is no tendency for regulation to wither away: such systems usually breed more regulation because of empire-building tendencies and the influence of pressure groups.

5 Pro-competition regulation and the influence of large consumers

Pro-competition regulation has had another effect which has not been widely noticed - large consumers of goods and services supplied by the utilities have reinforced the regulator's attempts to introduce competition.

It is generally argued that government policies are dominated by producer interests. There is a well-known conclusion of public choice theory that, because the benefits of successful lobbying are concentrated whereas the costs are dispersed, producer groups have powerful incentives to invest resources in persuading governments to accept their views: the prospective returns from lobbying to a company or group appear very high compared with other uses of its resources. Vote-seeking politicians have an incentive to yield to such lobbying because organised groups, which often have near-monopolies of information on a subject, can provide convincing cases in support of their views and often appear able to deliver substantial numbers of votes. Consumers, by contrast, are unorganised and their views carry little weight with government.

Such claims are true enough as far as small consumers are concerned. But companies acting in their consumer (rather than producer) role are a different matter. What utility privatisation and subsequent regulation have revealed is that large companies, acting individually or jointly with others, can be powerful forces for liberalisation. For example, both in gas (where large consumers stimulated an MMC enquiry within a year after privatisation) and in electricity, large consumers which felt short-changed by privatisation schemes which put little emphasis on competition pressed regulators to stimulate competition so as to bring down prices and improve service standards. Regulators have generally been happy to accede to such demands, where feasible, because of their pro-competition duties. Liberalisation has occurred first in markets for larger consumers not just because it is easier to introduce competition there, but because that is where the powerful pressure groups are.

Small consumers remain unorganised and unable to exert such significant pressure. But the demonstration of what competition can do helps regulators in pressing ahead with schemes (such as '1998' in gas and electricity) which give households freedom to choose supplier.

6 Reviewing regulation

To summarise, I have argued that, despite an unpromising beginning resulting from seriously flawed privatisation schemes, Britain has a regulatory system which is in many respects better than the alternatives. Contained within it is an implicit recognition of the dynamic nature of competition. That, I think, is a significant advance on traditional regulatory systems which are ultimately founded on a static view of competition from which flows the belief that regulation is mainly about correcting supposed market 'failures' which can readily be detected and remedied.

However, I do not wish to sound too sanguine about Britain's system of regulation. Its biggest deficiency, in my view, is one which stems from flaws in the privatisation schemes, which often left incumbents with too much market power. As a consequence, there has been less competition than there should have been - and too much regulation of potentially competitive sectors of the utilities (such as electricity generation). I would like to see competition extended and the scope of regulation correspondingly reduced.

For that reason, I would suggest the best outcome of the review of regulation which the government is now conducting would be for it to do nothing which significantly disturbs the present system. Competition should be extended to all domestic consumers of gas and electricity in 1998, as planned, and any other opportunities to enhance competition should be seized. Within a few years, it should be possible to confine regulation to strictly-defined 'natural monopoly' networks, otherwise relying on competition (with the new competition laws in the background as a safeguard). The original aim of 'light touch' regulation would then be achieved, at least in the sense that the scale and scope of regulation would be much reduced compared with the present.

From leaks and statements made by some of the interested parties it appears that the review is dealing with regulatory processes - should there be colleges of regulators rather than a single regulator for each utility? Should each regulator be required to have a formal advisory panel? Should regulators give more information about the reasons for their decisions? Should OFFER and OFGAS, and perhaps other regulatory offices, be merged? There is also some interest in replacing the price cap system by some form of profit sharing mechanism. Some government statements have also hinted that the regulators might be given additional duties (for example, to protect the environment, to encourage energy efficiency, to help the 'disadvantaged' or, more generally, to protect consumers).

I am very doubtful about imposing on regulators duties to protect particular sections of the community, or to safeguard the environment or to promote energy efficiency. That route leads back towards the kind of muddle which used to exist under nationalisation when managers were unable to manage because of confusion over whether they were supposed to pursue 'commercial' or 'public interest' objectives. As for duties to protect consumers in general, competition is by far the most effective way of achieving that and the regulators already have duties to encourage competition.

More generally, the underlying danger in any such review is that the government will feel that it will be criticised if it is not seen to be doing something which distinguishes it from its predecessor. It may therefore decide to increase the scale and scope of regulation, perhaps pleading the need to protect consumers and/or the environment. That seems to me precisely the opposite of the direction in which we should be heading - which is towards more competition and less regulation. I have dwelt on the strengths of the existing regime. But it could, given a few modifications, be used for interventionist purposes. The apparatus of utility regulation could, with some changes in the duties of regulators (for instance, reducing the emphasis on competition-promotion) and perhaps a few changes of regulators, be used to seize control without ownership, thus re-politicising the industries.

The outcome of the regulatory review is one possible danger[7]. Another stems from events likely to coincide with the final stages of the review - the introduction of choice of supplier for domestic electricity consumers and the extension of choice to domestic gas consumers in more areas. The government evidently wishes domestic market competition to go ahead - and indeed there should be benefits for the party which is in office when consumers find prices falling and service standards rising, as they should do. But past experience suggests temporary problems (such as errors in billing) are almost certain to arise. A test of the government will be whether it is willing to let domestic market competition go forward despite the complaints there will be. It will be very tempted to intervene either directly, or indirectly via the regulators. Its response to incipient problems in the water industry in the summer of 1997 is not encouraging: Mr Prescott called a 'water summit' at which he exhorted the companies as though the industry were still nationalised. Subsequently the regulator has set quantitative leakage targets for the companies in a rather obvious demonstration of the lack of economic incentives in the industry.

Perhaps I am too pessimistic. My view is quite simply that, partly by accident and partly by design, Britain's regulatory system - with its emphasis on pro-

[7] Some of the results of the review were published in March and July 1998. See footnote 2.

competition regulation - has evolved into one of the better regulatory regimes in the world. It would be a pity if the good work done so far were to be upset just when we are on the verge of realising the benefits likely to emerge from a bold and far-sighted extension of competition to the small consumer.

References

Beesley M.E. and S.C. Littlechild (1983), 'Privatisation: Principles, Problems and Priorities', *Lloyds Bank Review*, July.
Beesley M.E. (1996), 'RPI-X: Principles and their Application to Gas', in Beesley M.E. (ed), *Regulating Utilities: A Time for Change?*, Readings 44, Institute of Economic Affairs.
Blaug M. (1987), 'Classical Economics', in Eatwell, Milgate and Newman (eds), *The New Palgrave - A Dictionary of Economics*, Vol. 1, Macmillan.
Demsetz H. (1989), *Efficiency, Competition and Policy*, Blackwell.
Hayek F.A. (1948), 'The Meaning of Competition', *in Individualism and Economic Order, George Routledge and Sons.*
Kirzner I.M. (1997), *How Markets Work: Disequilibrium, Entrepreneurship and Discovery*, Hobart Paper 133, Institute of Economic Affairs.
Kuhn T. (1970), *The Structure of Scientific Revolutions*, University of Chicago Press, 2nd edition.
Lipsey R.G. and K. Lancaster (1956), 'The General Theory of Second Best', *The Review of Economic Studies*, 24(1), October, 11-32.
Pigou A.C. (1932), *The Economics of Welfare*, Macmillan, 1929, Fourth Edition.
Robinson C. (1992), 'Privatising the British Energy Industries: the lessons to be learned', *Metroeconomica*, Vol.43, Nos. 1-2.
Robinson C. (1994), 'Gas: What to do after the MMC Verdict', in M.E. Beesley (ed), *Regulating Utilities: the way forward*, Readings 41, Institute of Economic Affairs.
Robinson C. (1996), 'Profit, Discovery and the Role of Entry, in Beesley M.E. (ed.), *Regulating Utilities: a Time for Change?* Readings 44, Institute of Economic Affairs.
Robinson C. (1997), 'Introducing Competition into Water', in Beesley M.E. (ed), *Utility Regulation: Broadening the Debate*, Readings 46, Institute of Economic Affairs.
Schumpeter J.A. (1976), *Capitalism, Socialism and Democracy,* London: Allen & Unwin, 5th Edition, especially Chapter VII.
Tullock G. (1976), *The Vote Motive,* Hobart Paperback 9, Institute of Economic Affairs.

CHAPTER 3

LIBERALISING THE SPANISH ELECTRICITY MARKET: CAN COMPETITION WORK?

PABLO AROCENA[1]

Universidad Pública de Navarra, Dpto. Gestion de Empresas,
Campus de Arrosadia s/n, 31006 Pamplona (Navarra), Spain
Email: pablo@upna.es

Keywords: competition; electricity; independent power projects (IPPs); regulation; Spain; stranded assets.

1 Introduction: the industry before the liberalisation

The recent approval of the new Electricity Law in November 1997 has meant the opening up of the Spanish electricity sector to competition. The model of liberalisation established in the Law is consistent with the goals of the European Directive and places Spain among the countries with the most liberalised electricity sectors. The purpose of this chapter is to discuss the basic features of this process and to offer a preliminary assessment of the reform. It particularly underlines the most critical aspects that could impede and inevitably delay the development of the competition within the Spanish market. But first, the present section briefly summarises the basic characteristics of the Spanish electricity industry and the regulatory regime existing before the liberalisation in order to illustrate the background to the reform.

The Spanish electrical power industry is the fifth largest in the European Union with 42,859 MW of installed capacity and 156,245 GWh demanded in 1996 (REE, 1996). Table 3.1 below shows the total production and the installed capacity at the end of 1996 as well as the annual average energy shares over the 1987-1996 time period. One can observe the well diversified composition of the generation mix, resulting from the national resource endowment and the energy policy decisions made at the end of the 1970s.

---

[1] Centre for Management under Regulation, Warwick Business School, University of Warwick, Coventry CV4 7AL. Tel +44 (0)1203 522987; fax +44 (0)1203 524965; email: wbsrbpar@razor.wbs.warwick.ac.uk
I would like to thank Catherine Waddams for her helpful comments and the supportive facilities at the Centre for Management under Regulation, Warwick Business School. All opinions expressed in this chapter are those of the author.

Table 3.1: Capacity (1996) and generation by fuel type (1987-96)

	Capacity (MW)		Production (GWh)	
Hydro	16,549	(39 %)	37,692	(19 %)
Coal	10,674	(25%)	52,395	(41 %)
Nuclear	7,422	(17 %)	56,329	(38 %)
Oil + Gas	8,214	(19 %)	2,149	(2 %)
Total	42,859		148,565	

Source: REE (1996) and CSEN (1996a)

Spain has significant hydroelectric resources, which represent 39% of total installed capacity. However, because of variation in precipitation, there is substantial variation in hydro production between dry and wet years. Hence during the 1987-1996 dry period the average annual hydro production only accounted for around 19% of total output whereas it was around 30% of total production from 1977 to 1987. On the other hand, like many other European countries, Spain has practically no domestic reserves of petroleum and natural gas, coal being the only indigenous fossil fuel available. However, oil-based generation represented around 40% of the total production in 1974/76. The oil crisis of 1973 and 1979 made the traditionally high Spanish oil bill more expensive, so that the efforts of successive governments were intended to reduce the dependence on external energy sources. As a result, investment focused on the construction of coal-fired and nuclear plants. Between 1974 and 1986 five nuclear reactors came into operation with a joint capacity of 4,600 MW and twenty-one coal generating units (8,300 MW). This investment program doubled the existing capacity in 1974 with the aim of meeting the demand growth and eliminating oil in the generation of electricity. Coal-fired plants generated 41% of total electricity production during the 1987-1996 period, whereas the role of oil declined to around 1.5%. In 1997 the contribution of oil has been virtually negligible, representing 0.1% of total production.

Nevertheless, as in the UK and Germany, domestic coal is of low quality and very costly to extract, elevating the price substantially above the price of imported coal. The Government has traditionally maintained a stable policy of supporting the national coal industry by requiring electricity generators to use it extensively. The resulting economic distortions are substantial: Sánchez (1992) estimates that electricity tariffs would be about 8% lower if all coal were acquired in international markets.

Finally, natural gas is expected to be increasingly used in electricity production in the medium term. The last National Energy Plan (PEN) estimated that natural gas should account for 11% of the electricity production by the year 2000[2]. The flexibility and high-efficiency of the new combined-cycle gas turbine generator sets (CCGT), the necessity of reducing the sulphur dioxide emissions and the construction of the new pipeline permitting the supply of gas from Algeria are factors that might encourage the Spanish "dash for gas".

In 1984, the industry comprised eleven vertically integrated companies operating in generation, transmission and distribution, and one state-owned company (Endesa) exclusively involved in generation. As a result of the process of mergers and acquisitions carried out over the last ten years, in 1997 these firms are clustered into four generation and distribution groups (see Table 3.2), with two firms, Endesa and Iberdrola, having a dominant position both in the generation and the distribution market. However the sources of power vary greatly across the companies. Thus, Endesa owns 60% of fossil fuel fired plants whereas Iberdrola controls 50% of the total hydro capacity. The industry has been a mixed system of public and private ownership until the recent privatisation of Endesa in October 1997.

Table 3.2: Sector shares of Spanish electricity companies (1996)

	Capacity Share	Generation Share	Distribution Share
ENDESA GROUP	45	51	41
Endesa	18	30	--
Fecsa	9.5	8	11
Sevillana	9	6	15
Enher	4	3	7
Viesgo	2	2	2
Hidruña	1.5	1	3
Erz	1	1	3
IBERDROLA	38	32	39
UNIÓN FENOSA	12.5	12	16
H.CANTÁBRICO	4.5	5	4

Source: Adapted from CSEN (1996a,b)

[2] In 1997 gas based generation has reached the historical record of 4% of total production.

In the early 1980s, a number of changes were introduced in the Spanish electricity system, partially motivated by the financial crisis in the industry. The exaggerated demand increase forecast by the Government had led to an excessive investment program and electrical companies got into substantial foreign debt. The increase in interest rates and the fall of the Spanish exchange rate caused severe financial deterioration. In 1984, the review of the PEN imposed a nuclear moratorium on five of the nuclear plants under construction. Likewise, in 1985 the major utilities agreed a swapping of assets, so that the financially strongest companies (including ENDESA) acquired assets previously owned by financially weaker companies. Also in 1985, a national grid company was created, *Red Eléctrica Española* (REE) with the transmission assets of all firms in the industry. REE is responsible for the management, design and maintenance of the high voltage transmission network and for national central dispatch. This process of reorganisation of the industry concluded with the introduction of a new regulatory regime, the so-called *Marco Legal Estable* (hereafter MLE) or Stable Legal Framework. The MLE established a regulatory model, led by the Royal Decree 1538/1987 of 11 December 1987, that has been in operation until the end of 1997.

Under the MLE the industry operated as an integrated system with respect to key decision-making both long term and short term. On the one hand, the Ministry of Industry and Energy was responsible for the long-term planning of the generation investments. These decisions were published in the National Energy Plans, approved by Parliament, the last of which covered the 1991-2000 time period. These documents indicated the demand forecast, the generation capacity needs required to meet the demand and the type of plants and fuels to be used. An important consequence of this type of national planning was that, since the Government decided what amount and mix of plant had to be built, it should also bear the long-term associated risks. On the other hand, the state-owned grid company (REE) was responsible of the unified dispatch system. REE dispatched all plants in a merit order of variable costs. Firms managed the availability of plants, but production was determined by REE, irrespective of the demand of the clients of each firm and regardless of plant ownership.

The MLE replaced a previous regulatory regime, based on a cost of service regulation where the Ministry of Industry approved the tariffs with the aim of covering the costs declared by the firms. Given this discretionary and non-transparent method, it is not surprising that there existed a deep concern about the firms' level of productive efficiency. For this reason, a basic aspect of the economic regulation of the firms was the introduction of incentive mechanisms to promote the efficient behaviour of the firms. The MLE established the "standard costs", which constituted the key feature of the regulatory regime. Each cost item necessary to supply power had a cost allowance or standard cost, established separately for generation and for distribution. Firms were remunerated with the standard costs, irrespective of the actual costs incurred by the company. If the company's actual costs were less (greater) than standard costs, the company kept (absorbed) the difference. Although there were differences by customer category (load, voltage, interruptibility) and time, there was no regional differentiation in tariffs. The national average tariff was calculated by dividing the sum of the total standard costs by the forecast demand. This procedure ensured that total income of the system covered the standard costs of the entire sector.

The allowed revenues were based on standard cost of the inputs, regardless of the costs of any individual firm, and were indeed fixed without reference to the average costs of the firms. Additionally, they were annually increased by the retail price index, without setting any adjustment clause for potential productivity improvement[3]. The scheme thus had all the incentive properties of price cap regulation. The design of the standard cost formulae gave incentives to increase plant availability and the life of the assets as well as to reduce plant minimal load, plant's internal consumption and distribution losses. The evidence available confirms significant efficiency improvements in the electricity sector under the MLE (Sánchez, 1994; Arocena and Rodríguez, 1996). At the same time, several studies show that electricity prices in Spain are among the highest in the European Union (CSEN 1997a).

In summary, the MLE guaranteed a stable remuneration to the electricity companies, without the uncertainty of the previous system. At the same time, the incentive mechanisms promoted efficient behaviour of the firms. That is, investors were provided with an assurance that they would be able to recover their investment over the life of the plant, provided that the companies operate efficiently. The vertical integration between generation distribution and retailing generated a lack of transparency that prevented the identification and the true cost allocation among activities, allowing the existence of cross-subsidies. On the other hand, the government increased its control of the electricity sector, directly via the PEN, or indirectly, through Endesa or REE. In consequence, under MLE an opaque industry was created with high levels of administrative intervention.

2 The process of liberalisation

The first step towards a more flexible and liberalised regime was taken in December 1994 with the approval of the *Ley de Ordenación del Sistema Eléctrico Nacional* (hereafter, LOSEN). In order to promote some degree of competition the LOSEN introduced a system similar to the single buyer procedure proposed by the European Directive (EC, 1997). The most salient measure of this legislative reform was the

[3] Only in 1996 an RPI-X formula was introduced for the actualisation of some generation and distribution standard costs.

creation of the National Electricity Regulatory Commission (CNSE, originally CSEN). The CNSE was created as an independent regulatory agency with the aim of guaranteeing the transparency and objectivity of the operation of the entire system. The creation of the CNSE should be considered as the first attempt to reduce the direct control that the government has traditionally imposed on the industry. However, although the Law assigned multiple tasks to CNSE, and this has played a very dynamic role in the industry since its creation, it is mainly a consultative agency and not properly a regulator. The CNSE had some regulatory functions but mainly advises and consults the Ministry of Industry about new rules related to investment planning, authorisation of investments, the calculation of the remuneration, or the necessary rules to develop the regulatory regime. If the ultimate purpose was to turn the CNSE into the key element of the system it would have been necessary to invest it with more regulatory powers, which were instead assigned to the Ministry, as the Spanish Competition Authority suggested (TDC, 1995).

The LOSEN also ordered the accounting and legal unbundling of generation and distribution businesses and opened the possibility for future development of competitive bidding procedures for the procurement of new generating capacity, although it did not state any criteria related to the design and implementation of these mechanisms. The aim of the Law was to establish a dual system: on the one hand, the "integrated system", characterised by the principles of centralised planning, unified dispatch, nation-wide tariff and public service character with the duty of guaranteeing the supply of power to all users; on the other hand, the "independent system" based on the freedom of contracting. The final purpose was to promote the growth of the Independent system by reducing progressively the participation constraints of different types of producers and consumers. In the long term, after a period of coexistence of both systems the integrated system should be substituted for the independent system. Nevertheless, the LOSEN did not specify the details for the initial configuration and the future development of the independent system, only establishing very general rules.

In summary the LOSEN introduced a set of measures intended to gradually enhance the transparency and the role of the market forces in the industry while the key features of the MLE were maintained. However, as Kahn (1996) points out, it was too cautious and vague in many aspects. Since the applicable rules of the LOSEN never were developed, the sector continued operating under the MLE until the recent approval of the Electricity Law 54/1997 of 27 November 1997. The current Law is based on an agreement that the Ministry had previously signed with the electricity companies in December 1996, the so-called "Protocol".

The Law establishes a more liberalised model of organisation of the market by extending the introduction of competition to generation and supply. The central feature is the creation of a competitive wholesale electricity spot market based on generators' and consumers' bids. Therefore, unlike other electricity pools (for example in England and Wales) demand side bidding is allowed. The wholesale market is organised in several markets: daily, intra-daily, constraint management and ancillary services. At the same time, both financial and bilateral physical contracts will be developed. An Independent System Operator runs the physical national electricity grid and a Market Operator determines the power exchange and the hourly market-clearing prices. Both operators will be private regulated firms. Red Eléctrica Española will run the operation of the transmission system, although it has the responsibility for serving as the market operator until the creation of the new company that performs this function. The regulatory scheme of the charges for the use of the grids has still to be determined.

The new Law reiterates the legal separation between regulated and non-regulated activities imposed by the LOSEN, although it does not require ownership separation. Particularly, the generation and distribution activities carried out currently by vertically integrated companies must be performed by different legal entities by the end of 2000.

The National Electricity Regulatory Commission is confirmed as the regulator of the system, with a number of functions related to regulatory supervision while the Ministry of Industry and Energy retains its administrative authority to intervene. The CNSE is managed by a Board of Directors, comprising a chairman and eight members who are appointed for a six-year term. However, four or five members will be alternatively renewed every three years. The Ministry or one representative can attend the meetings of the Board of Directors although without having a vote. The CNSE is assisted by an Advisory Board consisting of 34 members representing the State, the Autonomous Communities, the Nuclear Security Council, the electric companies, the market and system operators, consumers and social and environmental interest associations.

The Law establishes the complete liberalisation of the supply activities by the year 2007 and specifies a transitory period during which the degree of deregulation will be much more limited. Until 2000 only consumers with more than 15 GWh annual electricity consumption will be free to choose from competing suppliers. This threshold will be reduced to a level of 9 GWh by the year 2000, to 5 GWh by 2002 and to 1 GWh in 2004.

Likewise, the Law establishes that all firms operating under the MLE before the end of December 1997 have the right to receive a payment, in terms of PTAs/kWh, in order to compensate the "costs of transition to the competitive regime". The compensation will reflect the difference between the average revenue obtained through the tariff regime and the remuneration under the spot market regime. The government can establish the maximum annual value during at most 10 years period. The present value of the total amount of such payments will not exceed to 1,988,561 millions pesetas, including the incentives related to the use of domestic coal. This amount was fixed by the Protocol and also specified how this payment must be shared among the companies. The support for national coal remains in place over the transitory period in order to ensure supply up to 15% of the market.

Finally, the government decided to privatise the state-owned ENDESA. In the run-up to privatisation the government had strengthened the company by acquiring the majority stake of the companies FECSA and Sevillana. The regulator argued strongly against this operation because of the substantial increase in the degree of market concentration, but the government did not take into consideration the arguments of the regulator.

3 An assessment of the regulatory reform

As shown in the previous section, the Spanish reform extends the liberalisation of the electrical sector beyond that required by the European Directive. It introduces competition through an advanced design of the wholesale market defining an ambitious range of power trading arrangements. However despite the potentially competitive virtues of this market, a preliminary analysis of the industry structure and of some aspects of the process of reform reveals many difficulties for the workability of competition. I discuss these aspects below.

3.1 The market structure

The high level of concentration in the generation market leaves the industry very vulnerable to the exercise of market power. The problems caused by the lack of a sufficient number of competitors are well known. Thus, it is widely recognised that the main failure attributed to the British process is that the former state monopoly Central Electricity Generating Board was insufficiently fragmented before its privatisation (Green and Newbery 1992). The resulting concentrated structure caused problems in the British Pool because of the ability of the two major generators to influence prices. Moreover, despite the massive entrance of new independent producers (mainly CCGTs), five years after privatisation the market was not fully competitive (Littlechild, 1995), with Pool prices remaining unjustifiably high. Hence, the Office of Electricity Regulation (OFFER, 1994) agreed with the two companies the divestiture of 6 GW of existing mid-merit plants, which was completed in June 1996.

In order to compare the Spanish and British market conditions Table 3.3 reports for both countries the concentration index C2, defined as the sum of the two biggest firms' market shares, as well as the Herfindahl index, which is constructed by summing the squares of the market shares of all firms in the market. It is evident that the concentration in the Spanish electricity industry is massive, both in generation and distribution/supply. The degree of concentration in Spain is higher than in England and Wales not only in 1996, before the mentioned divestment, but also at the time of British restructuring. Moreover, the problem of market power is even more serious than in Britain, since Endesa and Iberdrola, control almost all mid-merit and peaking plants that determine the system marginal price.

Table 3.3: Concentration in the Spanish and British electricity industry

Capacity	*Spain, 1996*	*E&W 90/91, (95/96)*
Herfindahl	0.36	0.34 (0.25)
C2	83 %	78 % (59%)
Generation market		
Herfindahl	0.38	0.32 (0.22)
C2	83 %	74 % (55%)
Distribution market		
Herfindahl	0.35	0.09 (0.09)
C2	80 %	23 % (23 %)

Source: Adapted from CSEN (1996a,b) and CRI (1997)

With regard to the unbundling of activities, the Law does not go further than the previous regime and it does not require ownership separation, permitting the creation of holdings-owning firms operating in both activities. The small number of competitors existing in the market turn the legal unbundling into a very weak type of vertical separation, since vertical integration implies that the same firms that are bidding in the supply side as generators are bidding in the demand side as distributors/retailers. In such circumstances the spot market becomes a simple transfer between both sides. Another point of concern is that the lack of comparators in distribution makes extremely difficult some type of regulation by yardstick competition.

On the other hand, the actual capacity of the interconnector with France, and in consequence, with the rest of the European market, is extremely small. Particularly, Table 3.4 shows that in 1996 the Spanish electricity market is even more isolated than was the British one at vesting. Therefore, any eventual rapid increase of competition through the Pyrenees is at present seriously restricted.

Table 3.4: Interconnectors with France in Spain and Britain

	Capacity (MW)	% Maximum Load
Spain (1996)	700	2.7 %
England and Wales (1990)	2000	4.2 %

Source: CSEN (1996b) and DTI (1997)

Although theory and practice show that the partial divestiture of the dominant firms is the most effective measure to increase competition (Green, 1996 and DTI, 1997), this type of intervention has not been contemplated in Spain. The big size of the state-owned ENDESA (see Table 3.2) offered the opportunity of far-reaching structural reform without having to face private ownership rights. Splitting up the company and its later sale to different owners would have eased the creation of a more competitive structure and mitigated the problems related to the market power. Despite the international experience and the reiterated recommendation made by the regulatory commission, the government decided to sell off the entire group without any horizontal or vertical separation. Moreover, as mentioned above, the public group was strengthened the year before its privatisation. In summary, the priority given to revenue raising at the expense of market restructuring imposes a heavy burden that unfortunately, constitutes a long-term obstacle for the competitive operation of the electricity market in Spain. As Newbery (1997) points out "opportunities for restructuring are rare and hard to reverse" (page 358).

New entry is a longer-term means for increasing the number of independent competitors. As stated before it is expected that new plants will be based on gas-fired CCGT technology so it is essential to ensure competitive access to this input. Since Spain has no gas fields, all fuel requirements must be imported by sea in liquefied state or through the new pipeline from Algeria. Gas Natural is the private monopoly that owns and operates the Maghribian pipeline as well as the internal transport and distribution network. Iberdrola have already signed one strategic alliance with the oil company Repsol and Gas Natural with the objective of developing joint projects of gas-based electricity generation. Endesa signed a similar agreement with Cepsa, the main competitor of Repsol in the oil industry and smaller competitor of Gas Natural in gas supply. Thus, the incumbents have substantial advantage over the new entrants in gaining access to gas supplies. Therefore, it is unlikely that new entry will erode significantly the dominant market share of the duopoly of Endesa and Iberdrola.

3.2 *The liberalisation of supply*

Given the extreme concentration in the generation market, competition in supply is indispensable to enable buyers to put pressure on the generators, in order to achieve further reductions in costs and prices. The Law introduces liberalisation slightly faster than the proposals of the European Directive, which defines *minimum* degrees of opening for national markets. Competition in the supply of electricity will be progressively introduced over a ten-year transitional period. Until 2000 the market share opened up to competition only accounts for 28 per cent and will increase to about 36 percent at the end of 2001. Therefore, the liberalisation of the competitive supply market is slower than in Britain, where 55 per cent of the market was opened up to competition after the first four years of transition and fully liberalised after eight years. This delay in the liberalisation of supply implies the minimisation of the competitive pressure on the spot market, so that competition in supply has less effect in counteracting the market power of the generators.

3.3 *The remuneration for the competitive transition costs*

As mentioned above, most past investment decisions were attributable to government energy policy, so some form of compensation to the companies is required for those sunk costs associated with stranded investments. The Law establishes a maximum recoverable amount of 1,988 billion pesetas under the heading of competitive transition costs, previously agreed in the Protocol. A comparison with equity for the whole sector - about 2,702 billion - (UNESA, 1996) reveals the outstanding significance of this amount. The Law (and the Protocol) has adopted the "lost revenue approach" to calculate recoverable stranded costs. That is, transition costs are estimated as the difference between the average revenue that would have been obtained under the previous regulated regime and the revenue under the competitive regime.

The determination of the amount of stranded costs entirely depends upon the definition of stranded assets, the methodology selected and the design of stranded cost recovery mechanisms; in consequence different assumptions can generate widely divergent results[4]. There does not exist a universally accepted method, though alternative methods lead to lower and more favourable valuations to consumers. Given its magnitude, the recovery of stranded costs is an issue of great importance to both company shareholders and ratepayers and every party involved obviously has an opinion on how these costs should be treated. But it is also

[4] For example, in the USA, the magnitude of stranded cost estimates have range from $10 billion (American Public Power Association) to $500 billion (NERA) nationwide, with a Moody's most likely scenario of $135 billion (DOE, 1996).

relevant for potential entrants, since excessive remuneration would give an unfair advantage to incumbent firms by handicapping entry over the transition period.

Neither the Law nor the Protocol specifies the assumptions employed to obtain this amount, and since the total value was estimated without the participation of consumers, it is reasonable to think that it might reflect a generous computation for the companies. The report conducted by the electricity commission (CSEN, 1997b) confirms this hypothesis, since its estimates under different assumptions are in the order of 35% to 97% of the value established in the Law, even although the same approach of lost revenue was used.

3.4 *The procedure for the regulatory reform*

Many of the problems outlined before have their origin in the procedure in which the reform has been conducted. The independence of regulatory institutions is crucial to guarantee transparency and minimise the risk of regulator capture in their pro-competitive task. As the British example shows, this becomes especially important in the event of failures in the privatisation process. According to the model adopted by the LOSEN, the CNSE should have been the independent regulatory body that initiates and manages the proposal of new regulation with the participation not only of the electric companies but also the consumers associations, self-producers, small distributors, and the rest of the agents in the sector. This would have encouraged the transparency, openness and credibility of the process. Instead, the reform was born following the "old style", with one agreement, the Protocol, between the government and the electric companies about the basis for the future regulation of the industry. In the same way, most of the recommendations made by the regulator with respect to the horizontal and vertical market structure were not taken into consideration. In summary, the process has been marked by some politicised and short-term decisions that help to explain the creation of a starting-point for competition that is heavily company-biased.

4 Conclusions

The process of reform carried out in the Spanish power sector has led to the liberalisation of the electricity market. The new legal framework introduces an advanced and ambitious design of market arrangements that offers a wide range of competitive possibilities. Nevertheless, in this chapter it is argued that the massive level of horizontal market concentration, the vertical integration of the companies and the limited scope of the liberalisation of supply make the development of effective competition very difficult within the Spanish market in the medium term. In this respect, the privatisation programme should be considered a lost opportunity to create a more competitive structure. At the same time, the imperfections in the gas market and the quotas of domestic coal consumption agreed over the transitory period may deter significant new entry and competitive impact in the medium term.

The lack of structural reforms has limited considerably the scope of the competitive measures introduced by the reform and relies on regulatory intervention to promote competition in the Spanish electricity sector. However, the possibility that this pro-competitive duty can be managed with the necessary independence and transparency largely depends on the minimisation of government interference that traditionally has characterised the regulation of the Spanish electrical sector.

References

Arocena, P. and L. Rodríguez (1996), 'La regulación por precios máximos y el crecimiento productivo en la generación termoeléctrica en España'. Documento de Trabajo, DT 30-98. Dpto. de Gestión de Empresas, Universidad Pública de Navarra, Pamplona.

CRI (1997), *The UK Electricity Industry. Financial and Operating Review 1995/1996*, Centre for Study of Regulated Industries, CIPFA, London.

CSEN (1996a), *Información básica del sector eléctrico*, Comisión del Sistema Eléctrico Nacional, Madrid.

CSEN (1996b), *Informe sobre las consecuencias que las diferentes formas de venta de las participaciones del estado en las empresas eléctricas pueden tener en el precio de la energía eléctrica en España en los próximos años,* P 005/96, Junio, Comisión del Sistema Electrico Nacional, Madrid.

CSEN (1997a), *Comparación de precios de la electricidad en el entorno europeo*, OI 002/97 Comisión del Sistema Eléctrico Nacional, Madrid.

CSEN (1997b), *Informe sobre el cálculo de costes hundidos en diversas hipótesis*, IE 002/97 Comisión del Sistema Eléctrico Nacional, Madrid.

DOE (1996), The Changing Structure of the Electric Power Industry: An Update. Energy Information Administration, DOE/EIA-0562(96), U.S. Department of Energy.

DTI (1997), *The Energy Report*, Vol. 1, Department of Trade and Industry, HMSO, London.

EC (1997), 'Directive 96/92EC of the European Parliament and of the Council Concerning the Common Rules for the Internal Electricity Market', Official Journal L27 of the 1/30/1997, Luxemburg: European Commission.

Green R. (1996), 'Increasing competition in the British Electricity Spot Market', *Journal of Industrial Economics*, 44 (2), 205-16.

Green R. and D. Newbery (1992), 'Competition in the British Electricity Spot Market', *Journal of Political Economy*, 100 (5), 929-53.

Kahn E. (1996), 'The Electricity Industry in Spain', *The Electricity Journal*, 9 (2), 46-55.

Littlechild S. (1995), 'Competition in Electricity: Retrospect and Prospect' in *Utility Regulation: Challenge and Response*, Institute of Economic Affairs, London.
MIE (1996), *Protocolo para el Establecimiento de una Nueva Regulación del Sistema Eléctrico Nacional*, Ministerio de Industria y Energía, Madrid.
Newbery D.M. (1997), 'Privatisation and liberalisation of network utilities', *European Economic Review*, 41, 357-383.
OFFER (1994), *Decision on a Monopolies and Mergers Commission Reference*. Office of Electricity Regulation, Birmingham.
REE (1996), *Explotación del Sistema Eléctrico. Informe 1996*, Red Eléctrica de España.
Sánchez P. (1992), *Informe económico del sector energético de la electricidad. Análisis de competencia*, December, Tribunal de Defensa de la Competencia, Madrid.
Sánchez P. (1994), *La eficiencia en el sector eléctrico en el periodo 1979-1991*, Centro de Estudios Monetarios y Financieros, Madrid.
TDC (1995), *La competencia en España: balance y nuevas propuestas*, Tribunal de Defensa de la Competencia, Madrid.
UNESA (1996), *Annual Statistical Report,* 1995, Unidad Eléctrica, Madrid.

63-76

P31
L33
L94
L98

CHAPTER 4

THE ELECTRICITY SUPPLY INDUSTRY IN POLAND: THE NEW LEGAL FRAMEWORK AND PRIVATISATION

PIOTR JASINKSKI

Campion Hall, Oxford University, Oxford OX1 1QS, UK and Oxecon Ltd
Email: Piotr.Jasinski@economics.oxford.ac.uk

Keywords: competition; efficiency; electricity; law; Poland; regulation.

1 The electricity supply industry in Poland

Although the electricity supply industry in Poland has undergone a remarkable change since the political and economic breakthrough of 1989, there is still a lot to be done to complete the pro-market reforms. Until now the efforts concentrated on creating an electricity market. Now there is no doubt that the new Energy Law, recently passed after many years of delay, is an important step forward. The law sets a legal framework for the operation of the sector and for its regulation. It is, however, quite general and in many places it refers to ordinances to be issued either by one of the ministers (usually the Minister of the Economy) or by the Prime Minister. Until the content of these ordinances is known, there is a risk that potentially radical market solutions, possible under the new law, will boil down to an only half-hearted liberalisation, and the sector become suffocated by bureaucratic interferences. On the other hand, if the pool starts to play a much more important role and supply becomes more clearly separated from distribution, Poland could easily have one of the most liberal electricity markets in Europe.

Therefore one must hope that the position of Leszek Juchniewicz, the newly appointed president of the Energy Regulatory Office, will not be weakened by the fact that he is an appointee of the previous government, and that the powers of his office will be used to promote competition, which is one of the duties of the regulator. However, since competition is difficult to imagine without privatisation, the overall success will depend on the speed with which the new Minister of the State Treasury is prepared to dispose of the electricity supply industry (ESI) assets. As usual, budgetary pressures in general and the reform of the social security system, recently started by passing a few acts of Parliament, offer some hope for the sector, in which in the seven years of ownership transformation only one enterprise was transferred from the public sector to the private one (or so it was claimed as the buyer of this CHP station in Cracow was Electricité de France). There is no doubt that since the general election of 21 September 1997 led to a major realignment in

Polish political life, a certain degree of uncertainty will persist for some time to come.

In this chapter I shall concentrate above all on the issue of the new legal framework of the electricity supply industry, but the chapter will end with a few remarks on the prospects of privatisation.

Until 1989, the organisation of the Polish electricity supply industry was based upon public ownership, central planning and high degrees of both horizontal and vertical integration. When the communist era came to an end, Poland inherited a very centralised structure responsible for production and transmission of electricity, called *Wspólnota Energetyki i W gla Brunatnego* (WEWB). Almost immediately after the Mazowiecki government was formed, i.e. in the Autumn of 1989, a team composed of parliamentary, government, and trade union representatives started to work on how to reform the energy sector, and in June 1991 the Council of Ministers approved the principles of the reforms, which later on were expressed in the so-called Letter of Intent addressed to the World Bank. However, by February 1990 the Polish Parliament decided that on 30 September of that year WEWB would cease to exist as an organisational structure. In effect the PSE (*Polskie Sieci Elektorenergetyczne S.A.*), the Polish national power grid company, was created, and after having taken over the whole of the high voltage grid and the dispatch system as well as the pumped storage power stations it became a monopsony buyer of almost all electricity produced and a monopoly seller to *Zak ady Energetyczne*, 33 regional distribution companies. At the same time, all power and larger CHP stations connected to the national grid gained full autonomy. In their relationships with PSE, long- and medium-term contracts, used as a collateral in external financing of necessary investments, are beginning to play a more important role, and a pool is operational on an experimental basis. A competitive electricity market is envisaged. In October 1995 Poland was successfully connected to the UCPTE system.

There were however limits to what could be done under the current legislation (the Energy Management Act of 6 April 1984), and therefore preparing and passing a new Energy Law had long become one of the most urgent tasks. What was needed was indeed a fundamental overhaul of the governance structure of the sector, and in practice the task consisted in separating the different roles that the government was until then playing in the energy industry: policy maker, owner and regulator. The most important part of the whole concept was to establish a separate, independent regulatory agency responsible for economic regulation of the energy sector.

2 Scope and objectives of the new Act

The government started to work on the new law governing this sector in 1989. Obligations undertaken by the government of the Republic of Poland and expressed in the letter of intent of 14 June 1991, presented to the World Bank, specifying directions of changes in the Polish electricity supply industry, their scope and speed of implementation, constituted one of the most important points of reference for that job. International points of reference, relevant for the drafting of the new energy law, resulted also from the Treaty establishing an association with the European Communities (mainly Articles 65, 68-69 and 78-79), from the White Paper of the European Commission on Preparation on the Associated Countries of CEE for Integration into the Internal Market of the Union (mainly Chapter 20, but also Chapters 3 and 8), and from the Energy Charter and the Energy Charter Treaty of 1994.

After a few years of the government's, the Sejm's and the Senate's work on this new act, the new Energy Law, covering the electricity supply and gas industries as well as district heating, was finally adopted on 10 April 1997, signed by the President on 14 May 1997, and promulgated on 4 June 1997. The Act was to enter into force six months after its promulgation (*vacatio legis*), i.e. at the beginning of December 1997, except for the provision (Article 21) establishing the Energy Regulatory Office (Urz d Regulacji Energetyki), which entered into force on the day of the Act's promulgation.

The scope of the Act has been delineated very broadly. Firstly, it relates to all the energy sectors, with the exception of extraction of fuels from their deposits and of storage thereof as well as of the use of nuclear energy (Article 2). Therefore it covers energy management and management of energy carriers in electricity, gas and district heating subsectors; detailed regulations regarding individual sectors will be contained in secondary legislation (ordinances issued by the Ministry of Economy) and in licences given on the basis of these ordinances. Secondly, the Act sets the framework for economic activity regarding the use of fuels (solid, liquid and gaseous fuels, which are carriers of chemical energy), energy (i.e. processed energy in any form) and heat (heat energy in hot water, steam or other energy carriers), with respect to all technological stages of energy processes (generation, processing, storage, distribution and use), as well as to retail and wholesale trade in fuels and energy (Article 1, in conjunction with Article 3). Thirdly, the Act comprises not only substantive provisions of economic law (rules of formulating energy policy, terms and conditions of fuels and energy supply and use, and activities of energy undertakings), but also provisions relating to the structure of the Polish government (bodies competent in the matters of fuels and energy management) (Article 1).

The Act, however, fails to regulate any questions relating to the privatisation of the energy sector and the principles and the mode of access to any real estate necessary for the construction and maintenance of energy equipment used for generation of electricity as well as for its transmission and distribution through networks (Article 60 of the Act).

The main objective of the Act (Article 1 (2)) is to create conditions for:

(a) the balanced development of the country;
(b) the provision of the country's energy security, understood as such a state of the economy which enables the current and future supplying of customers with fuels and energy in a technologically and economically justified way, in accordance with the requirements relating to environmental protection (Article 3 (16));
(c) the assurance of economical and rational use of fuels and energy;
(d) the development of competition and prevention of negative consequences of natural monopolies;
(e) the requirements of environmental protection to be taken into account;
(f) the protection of customers' interests and the minimisation of costs;
(g) the fulfilment of obligations resulting from international agreements.

In order to realise these objective there were created statutory grounds of the energy law, based on principles relating to (1) the functioning of the energy industry and (2) the role of the State in this sector. These principles will be presented below with respect to the electricity supply subsector.

3 Basic principles of organisation and functioning of the electricity supply industry

3.1 The principle of separation of generation, transmission, distribution and sale of electricity

This principle has not been expressed explicitly, but it may be inferred from the definitions contained in the Act (particularly from Article 3 (7) and (12)), as well as from the provisions of Article 4 of the Act, which article contains particular provisions regarding the status (the duties) of the energy enterprises operating in the sphere of transmission and distribution of electricity through networks (Article 3 (4) and (5)). Electricity generation (as well as processing and storage) is subjected to that part of the Act (Chapters 5 and 6), which regulates the principles and the mode of granting concessions and securing the required reliability of electrical equipment (technical equipment used in energy processes - Article 3 (9)) and installations (the pieces of equipment together with the systems of links between them - Article 3 (10)). More and more often, however, one distinguished a fourth and final stage of the production cycle in this industry, namely the sale or supply of electricity. This stage is not dealt with by the Act separately. It is contained in the provision on trade understood as an economic activity consisting in wholesale and retail trade of electricity (Article 3 (6)), which activity is subject to the duty of obtaining a concession according to principles specified in the Act (Article 32 (1) (4). Since the Act defines the category of a "consumer" very broadly, namely as anybody who takes electricity from an electricity enterprise on the basis of a contract (Article 3 (13)), this act indirectly allows for the existence of energy undertakings dealing exclusively with retail sale of electricity bought from distribution enterprises (Article 3 (12)). Therefore, a separate retail seller (being an intermediary between a distribution enterprise and the final consumer) may be a consumer.

3.2 The principle of operation of transmission and distribution enterprises as public utilities

The new Polish Energy Law has solved the basic conflict between competition and the principles of the market economy on the one hand, and the ideas of natural monopoly and public utilities, on the other hand, by reducing the scope of the latter to activities performed through networks, i.e. installations connected with each other and operating jointly, used for transmission or distribution of fuels or energy (objective restriction), and owned by energy enterprises (subjective restriction) (Article 3 (11)). Therefore, energy production (fuels included), which may and ideally should be performed in competitive conditions, has been explicitly excluded from the idea of public utilities; the licensing practice will determine whether or not there will be competition in the realm of sales (supply) of electricity.

Article 4 (1) of the Act establishes the principle that transmission and distribution enterprises operate as public utilities (network monopolies). The Act charges them with responsibility for the capability of premises, installations and equipment to supply electricity in an uninterrupted and reliable way, in accordance with the qualitative requirements. It means, therefore, that the solution put forward in the government draft has been accepted, and that the idea, proposed in the deputies' draft, statutorily to grant a "public utility character" to the activity of all the electricity enterprises and to have that category statutorily defined, was rejected.

Above all, the provisions regarding the obligation of energy enterprises which deal with transmission and distribution of electricity aim at the realisation of the idea of public utility, as contained in Article 4 (1). These obligations are to:

(a) conclude agreements relating to the sale of electricity or to the provision of transmission services with consumers or entities applying to be connected to the network. Such an obligation does not exist: when there are no technical or economic conditions for the supply; the party requesting the agreement to be concluded does not meet the requirements of becoming connected and receiving electricity; or fails to have any legal title to use the premises to which electricity is to be supplied (Article 7 (1) and (2));

(b) fulfil technical conditions required to supply electricity, specified under separate regulations or in the concession (Article 7 (3));

(c) ensure the realisation and financing of construction and extension of the network, including connections to consumers, provided that such networks are included in local plan of spatial development (Article 7 (4));
(d) maintain a reserve of energy fuels in the amount sufficient to secure continuous supply of electricity to customers (Article 10);
(e) work out (for the serviced area) plans for satisfying the current and prospective demand for electricity; drafts of such plans must be approved by the Energy Regulatory Office (Article 16 (3) and Article 23 (2) (3)).

3.3 The principle of limited third party access to the network

The introduction of competition in the generation and sale (supplies) of electricity is made possible by giving some economic entities the right of non-discriminatory access to electricity networks. The principle of universal and equal access of "third parties" (power stations on the one hand, and, for example, large consumers of electricity, on the other hand) to the networks used for transmission and distribution of electricity doubtless constitutes the most important instrument for forcing higher efficiency by competition and liberalisation of national and international trade in electricity. Although the Act establishes this principle, imposing the obligation to provide services consisting in transmission of electricity (Article 4 (2)) upon electricity enterprises operating in transmission and distribution subsectors of this industry, the scope of applying this principle is restricted in three ways.

Firstly, access of third parties to the network is allowed only according to the principles agreed by the parties under a contract, taking into account the technical and economic conditions of providing such services (Article 4 (2)); therefore in Poland one deals with the so called negotiated third party access.

Secondly, in accordance with Article 4 (2) of the new Energy Law, unlike under the European Union law, the obligation to provide transmission services relates only to electricity produced in Poland, which provision is in contravention of the provisions establishing the Single European Electricity Market in the EU and will have to be abolished at the latest on the day Poland enters the EU.

Thirdly, in accordance with the Act, provision of transmission and distribution services to third parties must reduce neither the reliability of supplies, nor the quality of electricity below the level specified under separate regulations; neither must it cause any change for the worse in prices or in the extent of electricity supplies to other entities connected to the network (Article 4 (3)).

Therefore the scope of third party access to the electricity network will depend on secondary legislation (provided for under Article 9 (1) and Article 65 of the Act). Of particular importance is the latter provision imposing upon the Minister of Economy the obligation to determine in an ordinance a timetable, covering a period no longer than 8 years from the day on which the Act enters into force, of gaining by particular groups of consumers the right to use transmission services, realising the principle of negotiated Third party access to electricity network; these groups of consumers will be determined according to the quantity of electricity that they purchase annually (Article 4 (2)).

Until this ordinance is issued, and then until the end of time periods established by this ordinance, the scope of the implementation of negotiated third party access to electricity networks will depend on the interpretation of the assumptions of this principle by transmission and distribution enterprises. The final decision in these matters (e.g. in cases of controversies relating to terms and conditions of provision of such services) will lie with the President of the Energy Regulatory Office (Article 8).

3.4 The principle of supplying electricity on the basis of a contract

The principle of supplying electricity on the basis of a contract has been introduced by Article 5 (1) of the Act. Taking into account the specific character of the subject matter of such contracts (supplies through a network) and the significant market dominance of the supplier over the customer, the legislator imposes on energy enterprises supplying electricity many restrictions relating to freedom to conclude contracts with consumers and to shape their contents. Therefore:

(a) specific contracts for the supply of electricity must take into account the principles specified under the contracts or concessions, and must include at least the provisions listed in the Act (Article 5 (2));
(b) as mentioned above, the energy enterprises dealing with transmission or distribution of electricity are obliged to sign contracts for connection to the network and sale of electricity (Article 7 (2));
(c) freedom of contracting in the field of electricity supply is also limited by the possibility of the Government introducing restrictions regarding sales of solid or liquid fuels or the supply and use of electric power (Article 11), motivated by the requirements of the State's energy security.

3.5 The principle of securing rational and economical use of fuels and electricity in the processes of designing, producing, importing, constructing and using equipment, installations and networks

From the point of view of the objectives of the Act, of particular importance is securing rational and economical use of fuels and electricity in designing, producing, importing, constructing and using equipment, installations and networks, specified in Article 51.

The following provisions aim at the implementation of that principle:

(a) specification and disclosure of "the electric efficiency" of equipment manufactured and traded (Article 52 (1) and (2)), and the establishment of a prohibition of trading, on the territory of Poland, of equipment that does not meet the requirements. A certificate obtained under the Testing and Granting Certificates Act (Article 52 (3)) is a document certifying that the said requirements have been met;
(b) the possession of qualifications to operate networks, equipment and installations confirmed by a certificate issued by a special commission (Article 54 (1)); it is prohibited to employ persons not having such qualifications (Article 54 (2)); certificates stating possession of the required qualifications, issued under provisions until now in force, remain valid in the period for which they were issued but no longer than until the end of 1999 (Article 70 (2)).

4 Basic principles of influencing the electricity supply industry by the State

4.1 The principle of separation of ownership, political and regulatory functions

There is no doubt that the Act is based on the principle of separation of the State's ownership, political and regulatory functions in relation to electricity enterprises, although it does not regulate the issue of ownership of energy infrastructure, nor does it provide for the way in which the State Treasury can exercise its property rights. However, it clearly divides the functions of energy policy and regulation and specifies in detail the bodies and the instruments of the exercise of these functions (Chapters 3 - 5).

The Act does not violate the rights of other public authorities as far as their exerting influence on electricity enterprises is concerned, including, among others, the rights of the Office for Competition and Consumer Protection pertaining to the matters not subjected to energy regulation.

4.2 The principle of uniformity of energy policy

The necessity to create and implement a single energy policy in Poland is not questioned by anybody. It is obvious that such a policy can be secured even where there are many bodies dealing with energy policy, although the achievement of this objective would be certainly more difficult than if one minister were charged with the function of the principal organ of the State administration in matters regarding energy policy. Following numerous debates it was decided that the Minister of the Economy would be the principal organ of the State administration in matters regarding energy policy (i.e. with respect to all sectors of the energy industry) (Article 13).

The new Energy Law determines the responsibilities of the principal organ of the State administration for electricity industry policy. These responsibilities include in particular the preparation of a document containing a framework foundation of energy policy (Article 13 (2)), containing, among others, a long term (for the period of 15 years) forecast of the prospect for the development of the energy economy in Poland, based on the assessment of energy security (Articles 14 and 15). The legislator attaches great significance to that document. That is proved by the fact that the assumptions of energy policy are determined, upon the request of the Minister of the Economy, by the Council of Ministers (Article 13 (1)). The Minister of the Economy, instead, is obliged to present the assessment of implementation of the framework foundations of energy policy, together with possible suggestions for correcting them, as well as a short term forecast, i.e. for a period not exceeding 5 years (Article 13 (2)).

Another task of the principal body of the State administration of energy policy is the determination of particular terms and conditions of operation and planning of electricity supply systems and supervision of the functioning of the domestic electricity supply industry system and co-ordination of international co-operation (Article 13 (2)).

4.3 The principle of electricity industry regulation independence

By energy regulation the Act understands the application of legal measures provided under the Act, including those relating to issuing concessions. These legal measures aim at securing correct fuel and energy economy as well as at protecting consumers' interests (Article 3, 15)). Nobody questions the fact that a foundation of true and effective energy regulation is independence of the organs entrusted with its exercise. Such independence can be more easily secured if there is only one strong organ, the scope of activities of which embraces all energy enterprises (and therefore also the energy enterprises which supply district heating).

The independence of the Energy Regulatory Office (URE), finds its expression above all in the placing of its President as the principal organ of the State administration, appointed by the President of the Council of Ministers for a 5 year long term of office and removed during his term of office only in case of illness making it impossible for him to perform his duties, violation of his duties, commitment of criminal act, substantiated by a valid court judgement, or his resignation (Article 21 (2) and (3)). This independence is strengthened by full authority in organisational and personal matters with respect to URE's local branches (Article 22), a separate (outside the budget) system of employees' remuneration, taking into consideration the level of remuneration in the fuel and energy sector (Article 29), as well as by the independent mode of financing URE in the period of its organisation (Article 63).

Such solutions make energy regulation independent, mainly of the sphere of politics. Organs of government and State administration may influence the President of URE solely by new acts of Parliament and by the energy policy framework foundations (Article 23 (1)). However, being a law-applying individual, operating under the Code of Administrative Procedure (Article 30 (1)), the President is subjected to judicial review; the legislators decided that the typical administrative decisions made by the URE President (mainly decisions regarding concessions) will be under the control of a special common court - the Antimonopoly Court (Article 30 (2)), instead of the Administrative Court, clearly applying the pattern established in cases related to the counteracting of monopolistic practices.

The practice in other States makes us aware that the pressure exerted by interest groups concerned, mainly by energy enterprises, the majority of which will remain public undertakings for a long time, very often constitutes a real threat to the independence of energy regulation. In order to prevent this kind of threat to the independence of energy regulation, the legislator imposed on the URE President an obligation to balance the interests of energy enterprises and of consumers of fuels and energy (Article 23 (1) *in fine*), and to support him, provided for a seven-person Consultative Council, appointed and removed by the Prime Minister from among candidates named by the nation-wide energy industry associations and by the nation-wide consumers' organisations (Articles 25 - 26 and Article 64). The legislators also secured transparency of URE's operation, imposing thereon an obligation to publish a bulletin providing, *inter alia*, information on concessions, tariffs and decisions made in cases of conflicts (Article 31). All the remaining matters will depend on the URE President's qualifications and personality.

4.4 The principle of regulation through concessions and tariffs

In the exercise of energy regulation in Poland two instruments can be used: (a) concessions, and (b) tariffs; tariffs mean sets of prices and charges, as well as terms and conditions of application thereof, prepared by the energy enterprises and implemented in compliance with the mode specified under the Act as binding for consumers mentioned therein (Article 3 (18)).

Under the principles and in the mode provided for in the Act under consideration, the URE President will grant concessions to pursue economic activities regarding electricity generation, transmission and distribution, as well as electricity trade, insofar as such activities surpass the volume specified under the Act (Article 32 (1)).

Within a period of 18 months following promulgation of the Act (approximately before the end of 1998), the URE President will grant concessions to the energy enterprises operating or being under construction on the day the Act is promulgated, provided they fulfil the terms and conditions specified thereunder (Article 67).

The Act also accepted a principle stating that energy enterprises holding a concession will determine tariffs for electricity by themselves (Article 47 (1)), following particular principles relating to determination and calculation of tariffs, decided by the Minister of the Economy in the form of an ordinance (Article 46). However, the legislator decided that such tariffs:

(a) should secure the coverage of justified costs of the energy enterprises' operations, including the costs of modernisation, development and environmental protection, as well as the protection of the consumers' interests against an unjustified level of energy prices (Article 45 (1));
(b) may take account of the co-financing costs of undertaking and services aimed at lessening consumers' electricity consumption, borne by the energy enterprises, which costs constitute an economic justification for avoiding the construction of new energy sources (Article 45 (2));
(c) may also take account of the co-financing costs, borne by the energy enterprises in connection with development of unconventional energy sources (Article 45 (3)), by which the Act means the sources where, in course of their processing, no fossil fuels are burned (Article 3 (20)).

Tariffs for electricity may be differentiated for various groups of consumers, only because of justified costs arising out of supplying electricity (Article 45 (4)).

Those tariffs are subject to approval and control by the URE President (Article 47 (1) and (2) and Article 23 (2) (2) of the Act). When the URE President believes that an energy enterprise operates in a competitive market, he may exempt it from the obligation of having to present its tariffs to be approved (Article 49).

However, that provision will not enter into force immediately. In accordance with Article 69 of the Act, for 24 months from the day the Act enters in force tariffs for electricity will be set by the Minister of Finance, under the provision of Article 25 (1) (4) of the Prices Act of 26 February 1988, unless this period is shortened by the Council of Ministers.

4.5 The principle of regulation for competition

The principle of regulation for competition is not explicitly expressed by the Act. However, in order to realise one of the most important objectives of the Act (the creation of conditions to develop competition and to prevent negative consequences of natural monopolies - Article 1 (2)), the URE President should also promote competition (Article 21 (1)) wherever it is possible. That provision should constitute a general principle to implement his powers as a body granting, refusing, changing and withdrawing concessions (Article 23 (2) (1), Article 32 - 33 and

Article 41), approving and controlling electricity tariffs in compliance with the principles decided in Article 45 and 46 of the Act (Article 23 (2) (2)) and solving disputes relating to terms and conditions of provision of transmission services, according to the principle of third party access to networks (Article 8 (1)).

The URE President will be able to enforce, by imposing fines (Article 56 (1), 8)), the keeping of energy enterprises' accounts in such a way as to enable calculation of fixed and variable costs and of revenue, separately for generation, transmission and distribution, for each kind of energy, and with respect to particular tariffs (Article 44). That should prevent cross subsidisation of various kinds of activities, and therefore it should prevent infringement of free and fair competition among enterprises having equal rights.

The URE President's right to require the energy enterprises to provide him with information on their activity (Article 28) also assures transparency of their operation; failure to provide such information may be punished by a fine (Article 56 (1) (7)).

4.6 The principle of ensuring observance of the Act by criminal-administrative means

Failure to observe the provisions of the Act (e.g. unjustified refusal to conclude an agreement to supply electricity or to connect to the network, use of non-approved prices and tariffs or higher than the approved ones, non-observance of obligations resulting from the concession) may be penalised by a fine (Article 56 (1)) up to the amount of 15% of the revenue of the economic entity on which the fine is imposed (Article 56 (3)), paid out of the after-tax income (Article 56 (2)); a fine may also be imposed on the manager of an economic entity, up to the amount of 300% of his monthly remuneration (Article 56 (5)).

The above mentioned fines are imposed by the URE President (Article 56 (1)), who will take into account the harmfulness of the Act, the degree of culpability as well as the conduct of the entity so far and its financial position (Article 56 (6)).

5 Secondary legislation

It is quite universally believed that the new Polish Energy Law is, to a great degree, harmonised with EU law. The only clear departure from EU law (mainly with the Directive 96/92 of 1996 concerning common rules for the internal market in electricity) is that limiting third party access to electricity generated in Poland. However, as is well known, the devil is in the detail and those are to be provided by the secondary legislation (almost thirty ordinances to be issued by various Ministers and the Prime Minister).

As has been shown above, the Energy Act of 10 April 1997 entered into force in December 1997. However, almost no ordinances were ready, and even the deadline of June 1998 was missed for many of them, and it is only in the summer and autumn of 1998 that the process of issuing them really started to gather speed. Article 70 of the Act provides that the secondary legislation issued and remaining in force under the Act of 1984, until now binding, will remain in force until the time they are substituted by new provisions, issued under the new Act of 1997, however not longer than for six months from the day the Act enters in force (i.e. until the beginning of June 1998). Those provisions of secondary legislation which contradict the provision of the new Act will immediately (i.e. on the day the Act enters in force) cease to be in force.

6 The Regulator

As we have already said, the new Energy Law was promulgated on 4 June 1997, and the period of *vacatio legis* was to last 6 months, with the exception of Article 21 creating the Energy Regulatory Office. The President of this new body was to be appointed, and he or she was supposed to use the period before the rest of the Act entered into force to organise the whole regulatory structure. It was not however known whether the Prime Minister was going to appoint the regulator or only the organiser of the Energy Regulatory Body.

The appointment was announced on 23 June 1997. Dr Leszek Juchniewicz became the first president of the URE. It is fair to say that his appointment was a surprise even for people quite close to the government and those from the industry. The new appointee is a 44 years old economist, still keeping his position at the Faculty of Economics at Warsaw University. His academic field of interest is banking and finance, and in recent years he also worked in various banks, most recently in the Polish Investment Bank.

From February 1994 to November 1996 Dr Juchniewicz was deputy minister at the Ministry of Ownership Changes, and is considered to be closely connected with the then Minister of Ownership Changes and the present Minister of the Economy, Wiesaw Kaczmarek.

7 Privatisation in the electricity supply industry

Although most generation and distribution ESI enterprises have been commercialised, not one of them has been privatised, even partially. Although in some cases intensive efforts were undertaken, all that was generated was frustration equally among employees and foreign investors. It is fair to say that until now no clear privatisation strategy in this sector has appeared, not to mention any prospect for a UK-type approach. The only conceptual work was done by PSE S.A., and even when in 1997 the Ministry of the State Treasury created a team (headed by Mirosaw Pczak, deputy minister) supposed to work out such a strategy, very little progress has been achieved. Different versions of the original PSE ideas were presented, but the main problem with them is their caution and lack of comprehensiveness. The PSE strategists wanted to test the ground by encouraging the government to sell one distribution company and one power station, and then to reassess the situation before doing anything more, but so far all that they have achieved is to have their ideas repeated in many fora.

In the new government, in power since October 1997, it is Deputy Treasury Minister Konaszewski who is responsible for the privatisation of the energy sector. The Ministry - in competition with the Economy Ministry, that wanted to merge power stations with coal mines - presented a programme of selling off individual enterprises, starting with a CHP station in Cracow, already sold to Electricité de France, the Patnów - Adamów - Konin complex of lignite power stations (National Power is one of the two shortlisted contenders) and EC Bodzin (another power station). They are to be followed by other companies from all subsectors, and the PSE is supposed to be the last to be (partially) privatised in 2002.

8 Conclusions

The overall conclusion of this chapter is that the picture is still very complicated and not exactly clear. On the one hand, some progress has definitely been achieved: after a four year delay the new Energy Law has been passed and is in the process of entering into force. There is an independent regulatory body, and its first president has been appointed. Many organisational steps towards creating a competitive energy market have been taken, although there are some unorthodox ideas circulating regarding energy prices, investment and the split between local and nation-wide markets.

On the other hand, as far as privatisation is concerned, nothing has really happened and although the decision to go ahead with social security reform will have to have an effect on ownership transformation in the energy sector, nothing is certain to happen at a predetermined date. There will also be budgetary pressure to increase revenues but, while in the energy industry the book value of assets appears quite high, huge investment requirements and low rates of return will reduce their market price. A change of government in itself caused some delays, although the two parties that have formed the new government command a substantial majority; some problems may be caused by the fact that AWS is actually a coalition of some thirty groupings. Be that it as it may, the election resulted in the change of key players: the only one who can be certain to stay in power is the regulator, Dr Juchniewicz.

CHAPTER 5

ELECTRICITY COMPETITION, REGULATION AND THE ENVIRONMENT - AN ASSESSMENT OF THE AUSTRALIAN APPROACH

HUGH OUTHRED

School of Electrical Engineering, University of New South Wales,
Sydney, NSW 2052 Australia
Email: h.outhred@unsw.edu.au

Keywords: Australia; competition; electricity; environment; equity; regulation.

1 Introduction

The electricity industry in Australia has traditionally been state-based. However restructuring is now occurring in the states of New South Wales (NSW), Victoria, South Australia (SA) and Queensland involving[1]:

- functional separation into generation, transmission, distribution and retail supply;
- creation of competing entities in generation and retail supply;
- corporatisation and in some cases privatisation of the new entities;
- staged implementation of the 'National Electricity Market' (NEM).

An interim version of the NEM wholesale electricity market is now in operation (NECA, 1997). One section of the interim NEM serves the interconnected states of NSW, Victoria and SA in a two-region spot market. The other one-region section serves the state of Queensland, which is due to be interconnected to the other states by about the year 2000. Further evolution is expected to occur in late 1998, with formal transfer of market operation to the National Electricity Market Management Company (NEMMCO), the introduction of a market region for SA and the division of NSW into two market regions. In its final form, the NEM spot market may have six or more regions, for which a set of spot prices is simultaneously calculated.

Retail electricity markets are also being developed by progressive elimination of monopoly franchises, with all retail customers supplied via the NEM expected to become contestable early next century. At present, approximately 50% of all electricity sold in NSW and Victoria is in the contestable retail market (customers with consumption above 160 MWH pa).

This chapter is structured as follows. Section 2 summarises the conceptual background to the competitive electricity industry model adopted in Australia. Section 3 discusses initial trading results to date for the National Electricity Market. Section 4 discusses network pricing. Section 5 discusses regulatory arrangements

[1] See NEMMCO (1997), Outhred (1998) and Walsh (1998) for more detailed discussions of electricity restructuring in Australia.

and considers their strengths and weaknesses, particularly with regard to their effectiveness in resolving the multiple objectives of economic efficiency, social equity and sustainability. Section 6 contains conclusions, including a discussion of aspects of the market rules where further development would be desirable.

2 Conceptual background

Successful implementation of a pool-style competitive electricity spot market such as the National Electricity Market relies on the assumption that electrical energy can be treated as a commodity, that is, not differentiated by source of production[2]. Because of the lack of cost-effective storage for electrical energy, it is also assumed that market participants have the ability to consume or to generate at a steady rate within a spot market interval.

It is further assumed that, during each spot market interval, and for each location at which a spot price is determined, the commodity is of uniform quality. For electrical energy, the key attributes of quality are voltage, frequency, waveform purity and phase balance. Collectively, these attributes are given the umbrella label 'quality of supply'.

In practice, quality of supply can vary within and between spot market intervals and from one network node to another. The operating decisions of generators, consumers and network operators, as well as the technical characteristics of their equipment, can all influence the quality of supply achieved at a particular node - for either better or worse. In a competitive electricity industry, this leads to the concept that network operators and market participants are providers or consumers of the 'ancillary services' required to maintain quality of supply, including short-term supply/demand balance.

2.1 A conceptual model for a competitive electricity industry

A conceptual model for an electricity industry that takes account of these issues has the following key elements (Outhred and Kaye 1996):

- A competitive electricity industry may be regarded as a collection of 'nodal electricity markets' joined by an electricity network. The participants in a particular nodal electricity market are the generators and consumers at that location, plus the network elements terminating at that location.
- Each nodal market comprises an ancillary services market, a spot market, and forward markets in financial instruments for price discovery and risk management functions.

---

[2] In practice, each electricity consumer will actually receive a mixture of electrical energy from all operating power stations so that, even in the absence of network effects, no one power station can be held accountable for a consumer's quality of supply.

- 'Network based arbitrage' exploits all opportunities for trade between the nodal markets subject to network topology, impedances, losses and operating constraints. The process of network-based arbitrage can be modelled mathematically in terms of the physical laws that govern network behaviour. In turn, this model can be incorporated into a multi-node auction which solves for the set of nodal prices which maximise the benefits of trade based on the bids and offers submitted by market participants. Such an approach can be used to solve for both spot prices and (short term) forward prices. Studies have shown that modified versions of this approach can be used for some ancillary services.

2.2 The need for aggregation

In an ideal world, the conceptual model described in Section 2.1 would be implemented for each network node at which a market participant was located. With such an implementation, no separate network service pricing would be required, as the network would earn appropriate income through network-based arbitrage in ancillary services, spot and forward markets.

However such an approach may not work well in practice under conditions of network constraint, when problems of market power might emerge at affected nodes. Outhred and Kaye (1996) describes a regionally-aggregated nodal market model in which market regions are defined such that important network constraints are located on regional boundaries. Within each region, there would be a regional spot market and associated ancillary service and forward markets. A regional market operator would be responsible for computing a derived (approximate) nodal spot price for each participant and organising ancillary services and risk management facilities. Mathematical models of the inter-connectors between regions would be incorporated in the spot market auction algorithm. However separate regulated network pricing arrangements would be required within each region.

It is important to realise that the regional nodal market model does not eliminate the underlying market power problems evident in the fully detailed nodal market model, which are due to network flow constraints and the lack of cost effective storage for electrical energy. Thus a trade-off may be required between avoiding over-investment in a regional transmission network and achieving effective competition in the associated bulk electricity market.

2.3 Practical implementation

To achieve the best results in practice, a number of factors should be kept in mind in applying the regional nodal market model:

- region boundaries should be chosen so that all significant network constraints are located on region boundaries, and adjusted if necessary in response to evolving patterns of constraints;
- consistent compromises should be adopted between ancillary services, spot and forward market functions;
- attention should be paid to achieving the greatest possible liquidity in ancillary service, spot and forward markets, including the encouragement of active demand side participation;
- attention should be paid to encouraging the deployment of distributed resources[3] at nodes where they can add value and reduce the monopoly power of network service providers.

3 The National Electricity Market

The regional nodal market model has been adopted for the NEM, with 'notional' inter-connectors between market regions[4].

3.1 Market rules

The rules for the NEM are set out in the National Electricity Code (NEC) and the current version (2.2) is available via the Internet (NECA, 1998). The National Electricity Code Administrator (NECA), owned by the participating jurisdictions, is given responsibility for administering the NEC including code enforcement and development[5]. NECA is also responsible for dispute resolution arrangements, reviewing transmission and distribution pricing arrangements and publishing information on NEM performance. It is required to establish a reliability panel and liaise effectively with other electricity industry regulatory bodies.

Key features of the NEC Version 2.2 include the following:

- all physical trading of electricity is through a multi-region pool-style spot market, with a half-hour trading interval and an embedded model of inter-connector losses and flow constraints;
- the National Electricity Market Management Company (NEMMCO), owned by the participating jurisdictions, operates the spot market, manages ancillary services and undertakes market projection functions including the forecasting of demand other than that bid in as price-sensitive;

---

[3] Small-scale generation, storage and demand side options.

[4] The notional inter-connectors are simplified, representations of the actual transmission links, in which losses are modelled as a function of the power flow, and flow limits are based on thermal or stability constraints.

[5] To provide indemnity from prosecution, the initial NEC and later changes to it, are submitted to the Australian Competition and Consumer Commission for authorisation. This process is discussed in Section 5.

- network service providers are subject to regulatory oversight by either the Australian Competition and Consumer Commission (ACCC) or state regulatory bodies;
- committees of market participants provide advice to NEMMCO and NECA.

3.2 Initial results of spot market trading in the NEM

Pool-style bulk electricity markets were implemented in Victoria in December 1994 and in NSW in May 1996. These markets were subsumed in the interim National Electricity Market 'NEM1' on 4 May 1997. The following diagram, Figure 1 from the NEM1 Monthly report for September 1997 (http://www.tg.nsw.gov.au/sem/data-sheet/monthly/97-09/), shows daily average spot prices for NSW and Victoria for the period 1 January to 30 September 1997. The prices in NSW and Victoria were set independently for the period January - April and jointly from 4 May, and differences in price behaviour for these two periods can clearly be seen. In particular, differences between the regional spot prices were small for most of the period after joint market operation began, and average prices declined.

Very hot weather during the 1997 summer period (January-February 1997) contributed to the high spot prices experienced in Victoria, and the implications of this for market behaviour are currently under review (NECA 1997, p 8). Steps were taken to ensure additional reserve capacity for the 1998 summer. However cooler weather and lower spot prices prevailed.

Spot prices generally declined over the period covered in Figure 5.1, and the average spot prices for the month of September 1997 were 14.5 $/MWh for NSW and 14.3 $/MWh for Victoria, close to the incremental cost of operation for much of the NSW coal-fired generating plant. Various factors appear to have contributed to these low prices including the following:

- to remain committed, operating generators have to be successful in the bidding process for each half-hour spot market interval;
- most generators have substantial forward market cover, so wish to have their generators committed to an equivalent level to insure against high spot prices, and are insulated in the short term from low spot prices, particularly for vesting contracts;
- most coal-fired generators have coal contracts that anticipate a high level of operation and so have a strong incentive to maintain market share (a recent trend towards increasing spot prices may indicate that the battle for market share is waning).

Outhred (1998) discusses the behaviour of the NSW & Victorian spot market in more detail. The Queensland market cannot be discussed here for lack of space, but supply is more constrained and spot prices are higher and more volatile than in the southern states.

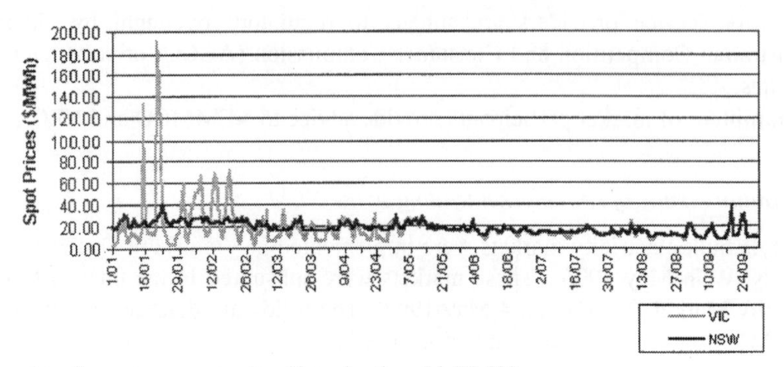

(source: http://www.tg.nsw.gov.au/sem/data-sheet/monthly/97-09/)

Figure 5.1: Daily time weighted average spot prices 1 Jan and 30 Sep 1997

3.3 Forward markets in financial instruments

When the state electricity markets were launched in NSW and Victoria, the state governments instituted vesting contracts with strike prices around $40/MWh. These terminate in accordance with the declining franchise threshold and are being replaced by voluntary trading in financial instruments. Most trading to date has been in form of over-the-counter instruments. However on 29 September 1997, The Sydney Futures Exchange launched two exchange-traded contracts related to monthly average spot prices in NSW and Victoria, with a twelve-month projection.

There is as yet no exchange-based trading in short-term financial instruments, nor has trading commenced in inter-regional hedges. The latter will be discussed in Section 4 because they are linked to the arrangements for inter-connector pricing.

The long-term success of the NEM will depend greatly on effective financial instrument trading, and it is too early to say whether an adequate level of performance will be achieved.

3.4 Ancillary services and projections of system adequacy

The close coupling between the provision of ancillary services, the spot market, and projections of system adequacy was recognised in the NEC by giving NEMMCO responsibility for all of these activities.

Proposals for a market-based approach to acquiring ancillary services are currently under development (NEM1AS, 1997). Challenges must be overcome in defining services, developing contractual arrangements and allocating costs among market participants. One open question is whether adequate competition can be achieved in the provision of ancillary services.

The NEC requires NEMMCO to develop projections of system adequacy (PASA) for periods up to seven days (short term), twelve weeks (medium term) and twenty-four months (long term). The intention is that these projections should inform market behaviour rather than supplant it. However there are provisions for intervention by NEMMCO as a last resort. There are some concerns that PASA may prove to be a cause of market distortion.

4 Network pricing

4.1 NEC provisions for inter-connectors between regions

The treatment of inter-connectors in the NEC can be thought of as a form of network pricing, in which:

- differences in the regional spot prices reflect the effects of (notional) inter-connector losses and constraints;
- markets in inter-regional hedges (yet to be formally implemented) provide risk management and market projection functions[6];
- there are no guarantees of sunk cost recovery for unregulated inter-connectors (pre-existing 'regulated' inter-connectors have been assigned to regional transmission networks for the purpose of cost recovery).

This approach to network pricing is close to ideal in economic efficiency terms, so long as the regional spot markets are sufficiently liquid and efficient risk management facilities for regional price differences are available.

In the absence of inter-connector flow constraints, differences between regional prices are small, as are the operating surpluses resulting from inter-connector trading. With an active inter-connector flow constraint, differences between regional spot prices emerge as each regional market finds its own clearing price. Regularly constrained conditions with differing supply/demand conditions at either end of the inter-connector (or a high probability that these will occur in future) provide the justification and the financial incentive to consider augmenting inter-connector capacity or investing in alternative means to relax the flow constraint.

In a broader context, this behaviour is characteristic of the economic role that a network should play in a competitive electricity industry, given the network's characteristics of high capital cost, economies of scale and low operating cost. Ideally, network augmentation or expansion should not proceed without appropriate risk sharing arrangements involving potential beneficiaries.

[6] Early versions of the NEC required NEMMCO to offer inter-regional hedges consistent with the inter-connector capacity. However it was argued that this function should be left to the market and the requirement was deleted. The relevant issues are canvassed in Putnam, Hayes and Bartlett (1997) and the ensuing discussion (available at: http://www.electricity.net.au/whatsnew.htm).

4.2 *NEC provisions for intra-regional transmission network pricing and access*

The current provisions of the National Electricity Code with regard to transmission pricing and access arrangements are summarised in [ACCC, 1997, p22]:

- Classes of transmission services:
  - *entry services* are those services which can be identified as being provided to market generator participants at their specified connection points;
  - *exit services* are those services which can be identified as being provided to market customer participants (consumers or retailers) at their specified connection points;
  - *transmission use of system services* are those services provided to either generators or customers which can be allocated on a location basis;
  - *common services* are those services provided to *customers* which cannot be allocated on a location basis (eg. services to maintain power system security);
  - *generator access services* relate to risk premiums for generators with connection agreements which include firm access compensation arrangements.
- Prices for transmission services are assessed on a cost-recovery basis, initially by the state regulators and in time by the ACCC as regulator, using a methodology which is outlined in the NEC and which is to be refined by a NECA review into transmission pricing. A mix of fixed charges, demand charges and energy charges may be used.
- Generators are to pay for entry services and may consent to pay for some transmission use of system services. They may also negotiate generator access services. Generators with ratings less than 30 MW are not required to participate in the NEM and are defined as *embedded* generators, that is, embedded in a distribution network.
- Customers pay all other costs of providing transmission services (over 90%), with 50% of cost of the transmission use of system services being assigned to points of connection by 'cost reflective network pricing' (CRNP) and the rest postage stamped.

Of the above transmission services, entry, exit and generator access services are negotiable between the network service provider (NSP) and the network service user. This provides some scope for achieving economically efficient outcomes when there are avoidable network investment costs. However cost-recovery terms dominate transmission pricing, so great reliance is placed on regulatory oversight.

For the transmission use of system service, which covers the bulk of network assets, the following cost-recovery algorithm sets price levels:

- an approved asset base is determined based on the Optimal Deprival Methodology, which optimises network design according to projected future use, and assessed cost of assets;
- an annual revenue requirement is determined for a NSP, based on its approved asset base and allowed rate of return;
- this annual revenue requirement is then allocated to network users according to their assessed (rather than actual) use of the network.

This approach to transmission pricing assigns most transmission network sunk cost recovery to customers, while existing generators pay network prices that approximate short run marginal costs. This creates particular problems in dealing with self-generation facilities that may appear like generators at some times and consumers at others. Also, it provides little encouragement for embedded generators to locate in distribution networks close to load centres. Transmission pricing arrangements are presently under review by NECA.

4.3 NEC provisions for distribution network pricing and access

The current provisions of the National Electricity Code with regard to distribution network pricing and access arrangements are summarised in ACCC (1997), p 40:

- Classes of distribution services:
  - *entry services* are those services which can identified as being provided to (embedded) generators at their specified connection points or, with the agreement of the regulator, by creating entry service cost pools which may differ by voltage level, load class and/or location;
  - *exit services* are those services which can identified as being provided to customers (consumers or retailers) at their specified connection points or, with the agreement of the regulator, by creating exit service cost pools which may differ by voltage level, load class and/or location;
  - *distribution use of system services* are those services provided to either embedded generators or customers which can be allocated on a location basis or, with the agreement of the regulator, by creating distribution use of system service cost pools which may differ by voltage level, load class and/or location;
  - *common services* are those services provided to *customers* which cannot be allocated on a location basis;
  - *firm access services* relate to risk premiums for (embedded) generators with connection agreements which include firm access compensation arrangements.
- Prices for distribution services are assessed on a cost-recovery basis, by state regulators but in accordance with the principles set out in the NEC, which are

subject to review by NECA. A mix of fixed charges, demand charges and energy charges may be used.
- Embedded generators are to pay for entry services and negotiated fractions of distribution use of system services subject to a cap equal to the long run marginal cost of network augmentation applicable to the circumstance. Embedded generators may also negotiate firm access services.
- Customers pay all other costs of providing distribution services, on a 'cost-reflective' basis or other basis determined by the jurisdictional regulator.

As with transmission services, these arrangements focus on recovery of 'sunk' costs rather than economic efficiency and so involve judgements of equity. As presently structured they provide little encouragement to embedded generators. Distribution pricing arrangements are presently under review by NECA.

5 Regulatory arrangements

Regulation of the NEM is undertaken at the federal level, whereas regulation of retail markets occurs at state level. This has resulted in some diversity between the participating states in the timetables for reducing the retail franchise threshold and in the manner in which retailers and distribution wires businesses are regulated. For example, NSW established the Licence Compliance Advisory Board (LCAB) to advise the parliament via the Minister for Energy on licence compliance. It reported for the first time in 1997 (LCAB 1997). The NSW government also established the Sustainable Energy Development Authority (SEDA) to promote market transformation with regard to sustainable energy technologies. SEDA also monitors 'green pricing' schemes introduced by retailers.

The standard conditions for NSW electricity distributor licences include (LCAB 1997):

- an obligation to provide customer connection services under a customer contract;
- demand management strategies must be explored prior to network augmentation/expansion;
- preparation and adherence to plans that protect customers and ensure effective operation;
- reporting and independent auditing requirements.

The standard conditions for retail supplier licences include (LCAB 1997):

- no discrimination against alternative forms of generation or demand reduction;
- development of strategies to reduce greenhouse gas emissions, which are to be audited by the Environment Protection Authority not less than every three years;
- preparation and adherence to plans for complaints management and cost effective service;

- standard form contracts for franchise customers stipulating a minimum standard of supply;
- reporting and independent auditing requirements.

These licence conditions place considerable reliance on self-regulation, emphasising quality management and independent auditing. This is a considerable change from the traditional NSW regulatory model and it is too early to say how successful it will be (the LCAB has recommended further clarification and refinement of the regulatory regime).

5.1 Economic efficiency

Australia has adopted a 'National Competition Policy' which involves a complex reform process intended to engender a climate of competition in the Australian economy as a key means for achieving economic efficiency (Fels, 1997).

One important aspect of this process is to minimise industry-specific regulation. Rather, an economy-wide approach has been adopted with complementary regulatory bodies at state and federal levels, with the ACCC being the relevant federal regulatory body. As previously indicated, a relevant function of the ACCC is to authorise the National Electricity Code, and any modifications to the code, if it judges them to be in the public interest, primarily defined, but not restricted to, a test of economic efficiency. Authorisation by the ACCC provides protection against prosecution under the Trade Practices Act.

5.2 Social equity

Considerations of social equity mainly arise in the context of retail markets, which are the regulatory responsibility of the relevant state regulatory bodies. This responsibility can be exercised in the setting of wires charges for the monopoly wires businesses and, so long as the franchise remains, in setting retail tariffs for franchise consumers. As indicated in Section 4, network charges are largely based on cost recovery and so equity judgements are inevitable. For example, all participating states have averaged both wires charges and franchise tariffs to balance economic efficiency and equity considerations.

5.3 Environmental impact

The most important environmental impacts of the Australian electricity industry are those associated with greenhouse gas emissions. The present Federal government favours voluntary reduction mechanisms such as emissions trading, however the NSW government has taken a more interventionist approach through licence conditions on distributors and retailers.

6 Conclusions

The experience to date of the operation of the National Electricity Market operation is promising, in so far as the revealed level of competition between generators is concerned. Also, the shared regulatory responsibility that is inevitable in a federation may have advantages from diversity that partly counteract divided accountability and complexity.

There are a number of areas where improvements could clearly be made:

- It has yet to be demonstrated that the combination of ancillary service, spot and forward markets will support appropriate investment behaviour, and, in particular, more needs to be done to encourage demand side participation.
- Refinements in ancillary services markets and network service pricing are required to facilitate economically efficient participation by distributed resources such as small scale generation, storage, energy efficiency measures and demand management.
- Satisfactory reconciliation of the objectives of economic efficiency, sustainability and equity has yet to be demonstrated, although some promising initiatives have been taken.

7 References

Australian Competition and Consumer Commission (ACCC, 1997), *Draft Determination, Application for Acceptance, National Electricity Market Access Code*, 22 August.

Fels A. (1997), *National Competition Policy and Director's Duties under the Trade Practices Act 1974*, Australian Competition and Consumer Commission, 1 May 1997 (available at http://www.accc.gov.au/docs).

Licence Compliance Advisory Board (LCAB, 1997), *Annual Report*, October.

National Electricity Code Administrator (NECA, 1997), *Annual Report 1996-97*, http://www.electricity.net.au/neca.htm.

National Electricity Code Administrator (NECA, 1998), *National Electricity Code*, Version 2.2, April 1998, http://electricity.net.au/.

National Electricity Market Management Company (NEMMCO, 1997), *Electricity Market Structure*, http://www.electricity.net.au/mktstruc.htm.

NEM1 Ancillary Services Steering Committee (NEM1AS, 1997), *NEM1 Ancillary Services Project Report*, May 1997, http://www.tg.nsw.gov.au/sem/doc/reports/asrep/.

Outhred H.R. (1998), 'A Review of Electricity Industry Restructuring in Australia', *Electric Power Systems Research,* 44, pp 15-25.

Outhred H.R. and R.J. Kaye (1996), 'Incorporating Network Effects in a Competitive Electricity Industry: an Australian Perspective', Chapter 9 in *Issues in Transmission Pricing and Technology*, Kluwer Academic Publishers, pp 207-228.

Putnam, Hayes and Bartlett (1997), *Inter-Regional Hedging in the Australian National Electricity Market*, 11 September, http://www.electricity.net.au/whatsnew.htm.
Walsh P.J. (1998), *Implementing Competition Reforms in South Australia*, National Competition Policy Conference, Sydney, 1-2 September.

CHAPTER 6

REGULATING ENERGY IN FEDERAL TRANSITION ECONOMIES: THE CASE OF CHINA

PHILIP ANDREWS-SPEED and STEPHEN DOW[1]

Centre for Energy, Petroleum and Mineral Law and Policy,
University of Dundee, Dundee DD1 4HN
Email: c.p.andrewsspeed@dundee.ac.uk

MINYING YANG

Institute of Quantitative and Technical Economics,
China Academy of Social Sciences, 5 Jianguomennei,
Beijing 100732, China

Keywords: China; energy markets; enforcement; political intervention; provincial governments; regulation; transition economies.

1 Introduction

At the heart of transition lies the evolution from the command to the market economy, and during the period of transition the economy comprises a new and growing private sector alongside a static or declining state sector (Lavigne, 1995). One of the central objectives of developing a market economy is to improve the productive and allocative efficiency of the industrial sector. As a result the debate on transition and the advice given to transition countries has been led by the discipline of economics, with law playing a subordinate role. Whilst efforts have been made in most transition countries to draft new legislation, insufficient thought has been given to the process of legal transition from a command system which directly controls, or seeks to control, all the components of the economy to a legal system which regulates markets and players (Ronne, 1997; Waelde and Gunderson, 1997). If the necessary legal reforms are not successfully implemented the economic objectives of transition cannot be realised.

Managing its energy sector through the transition process is one of the greater and more important challenges facing a government, for three reasons: the importance of energy in any industrial economy; the range and strength of vested interests in the domestic energy industry; and the technical difficulties involved in introducing competition and developing markets in the energy sector. In addition to

[1] The support of a Social Science Travel Grant from the Nuffield Foundation is gratefully acknowledged.

these sector-specific issues there exist fundamental legal and political obstacles, which are exacerbated in a federal regime. Prominent among these are the increasing importance of the non-state sector, the institutional weakness of the judicial system and, in a federal country, the rising power of the provincial governments.

China is a federal state in transition. The period of transition has been marked by annual rates of economic growth consistently in the range 8-12% until 1998 when it fell to less than 8%. The government has identified the energy sector as being strategic to the economy and has placed great emphasis on increasing the capacity and output of the main energy industries (coal, oil and electricity) with a considerable degree of success. However the very importance of the energy sector has led the government to attempt to maintain central control over many aspects of energy investment, output and sales.

Central government control over economic activity in China has declined as power has been devolved to the provinces, as direct control of certain state enterprises has been relaxed and as the role of the non-state sector has increased. The attempt of the central government to maintain control over 'strategic' sectors, such as energy, has been against this overall trend of decentralisation. Where limited and gradual liberalisation of the energy sector has been allowed, the government has been unable to regulate. Where central controls are still in place, they are increasingly ignored.

This chapter describes and analyses some of the regulatory challenges facing the Chinese government in the energy sector, focusing on the increasing power of the provinces and the evolving regulatory framework. This account is based on our understanding of the administrative structure in place before the reforms announced in March 1998. It does not attempt to evaluate the likely impact of these reforms, as the restructuring of government had not been completed at the time of writing. The concluding section discusses how features of other federal regimes might be applied in China and identifies some fundamental obstacles to regulatory reform.

2 The balance of power between central government and the provinces

2.1 The increasing power of the provinces

The Chinese government has long been characterised by a degree of decentralisation which is high relative to that in communist Russia of the past (Goodman, 1994). On the one hand the last twenty years has seen a substantial decline in the role of China's central state administration in capital investment (Yang, 1994). On the other hand, local government leaders have wide-ranging incentives to invest heavily and raise local economic output (Huang, 1996). Combined with the national preference for reaching decisions by consensus, this has allowed local administrations to become a powerful force in economic reform and growth, while at the same time retaining an effective working relationship with central government (Oi, 1992; Yusuf, 1994).

The key to the increasing power of the provinces has been their active support of the main government policy of rapid and sustained economic growth (Lieberthal, 1995) and the success of local enterprises in generating wealth and revenue (Solnick, 1996). This has led to the provinces gaining influence in both administrative and financial activities. The provinces now play a greater role in formulating and administering policy, and are an integral part of the planning cycle (Yusuf, 1994; Huang, 1996). They have achieved a high level of control over resources, funds and the workforce (Oi, 1992); in the case of funds for investment, this has been aided by the influence of provincial governments over local banks (Solnick, 1996). Provincial governments have gained stronger control over tax revenues, in part through formal contracts with central government, and in part through the manipulation of financial data and the raising of illegal fees (Wong et al, 1995 ; Solnick, 1996).

2.2 The key to central government influence

In any federal country the central government requires a certain number of instruments by which it can maintain control over the provinces in selected activities. In a federal democracy these controls are likely to be rooted in the constitution and in the judicial system. In the case of China, these controls are more widely dispersed among a range of mechanisms. At one end of the spectrum is the continued role of the state in appointing and dismissing senior provincial leaders (Lieberthal and Oksenburg, 1988; Huang, 1996) behind which lies the potential threat that the central government still has the power to send in the army to oust a recalcitrant provincial government. Less direct influence derives from the central government's control over key resources such as energy and transport, its expertise in technical and development issues, and its ability to send in special teams to investigate misdemeanours (Lieberthal and Oksenberg, 1988; Lieberthal, 1995).

Such instruments for central control may be found in other transition countries. In addition, Chinese officials show a propensity and an ability to seek consensus through negotiation (Lieberthal and Oksenberg, 1988; Solnick, 1996). What might be described as a national or cultural characteristic is reinforced by the Communist Party of China which has maintained its central role in the country's politics and administration during transition, unlike the Communist Party in Russia (Nolan, 1995).

3 The weakness of the evolving regulatory framework

The progressive shift of power from the centre to the provinces has fundamentally affected the lines of command and control for state enterprises. At the same time the old system of regulation has been unable to react to the growth of the non-state sector.

The overall hierarchy of control over state enterprises has remained largely unchanged. At the top lie the National People's Congress (parliament), the State Council (cabinet) and various commissions such as the State Planning Commission[2]. Three types of organisation report into this core: the provincial governments; the industrial ministries and state companies; and the specialist ministries and agencies which have a predominantly regulatory role. The Planning Commission has local offices, and the industrial and specialist ministries all have bureaux at provincial (and lower) levels. State enterprises at local level report to these bureaux (Figure 6.1).

What has changed during transition is not the lines of command and control but their relative strength. Provincial (and lower level) governments have increased their influence over the bureaux of the industrial and specialist ministries and thus have reduced the ability of the central government to control or regulate the state enterprises. At the same time, an effective apparatus for regulating non-state enterprises has not been put in place. The overall weakness of local specialist bureaux and regulatory agencies is exacerbated by the low rank of their officials, their dependence on the local government for funds, a shortage of resources, and a lack of incentive to regulate effectively (Huang, 1996; Lieberthal, 1995; Jahiel, 1997).

Despite the continuing enactment of laws designed to regulate this transition economy it is now clear that these laws cannot be implemented effectively within the existing command and control structure of regulation. The legal system offers little protection of contractual arrangements, though improvements can be seen as public awareness grows. Further, the power of local governments prevents the courts from providing substantial control over official arbitrariness (Lubman, 1995). A fundamental change is required if China is to have a regulatory system attuned to its evolving, mixed economy (Shao, 1997; Montinola, 1995; Johnson, 1995).

---

[2] Renamed the State Development Planning Commission in March 1998.

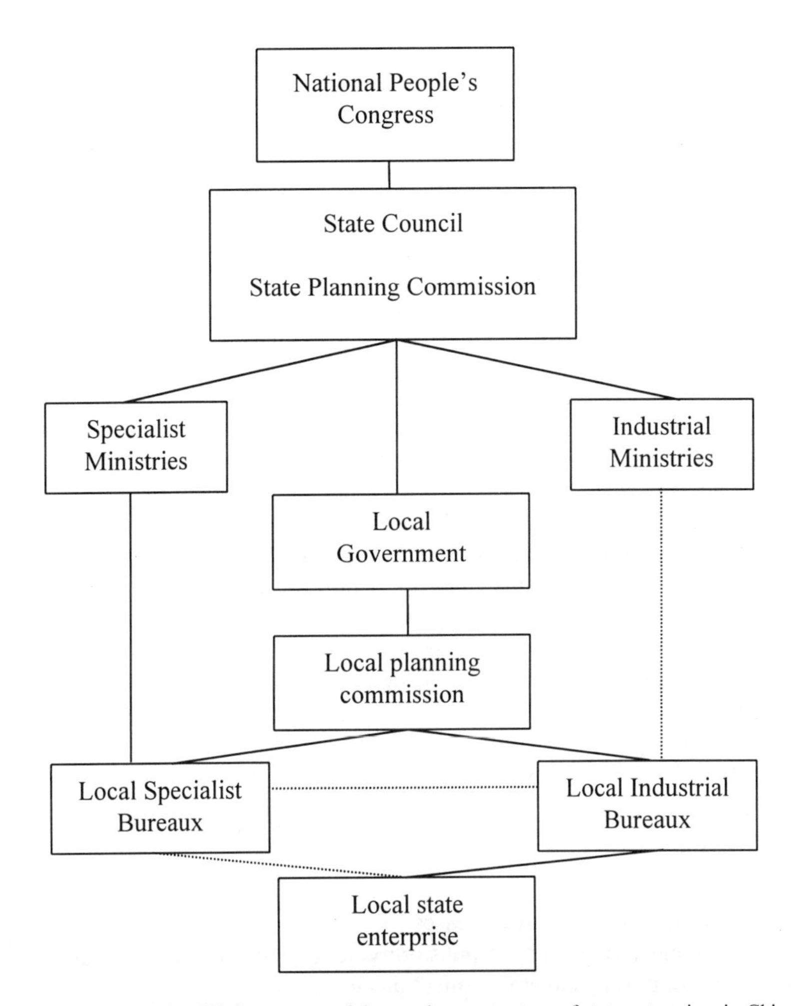

Figure 6.1: Schematic and simplified summary of the regulatory structure of state enterprises in China before the reforms announced in March 1998 (modified from Lu, 1996). This diagram ignores the role of the Communist Party and the Military. Solid lines indicate a stronger relationship and dashed lines a weaker relationship.

4 The impact on regulation of the energy industry

The regulatory structure of China's energy industries before March 1998 approximated to that outlined in Figure 6.1. The key industrial ministries and state companies were as follows:

- Ministry of Electric Power (MOEP);
- Ministry of Coal Industries (MCI);
- China National Petroleum Corporation (CNPC);
- China National Offshore Oil Corporation (CNOOC);
- China National Petrochemical Corporation (Sinopec).

Each of these institutions acted as both regulator and operator within its defined sphere of interest. The only exception was the power industry. A State Power Company was created in 1997 to act as a holding company for all state assets in the power sector.

With the exception of the coal industry, each of these enterprises owned and controlled a majority of the industry within its sector. Other minor participants include direct or indirect foreign investors, other state enterprises, township and village enterprises (TVEs) and domestic private investors (Andrews-Speed, 1998). The coal industry is exceptional in that some 40% of output comes from TVEs, which are owned and controlled by the lowest levels of local government, and from private mines (Thomson, 1996).

In addition to the lack of or inadequacy of the separation of regulation and enterprise management, the regulatory system suffered from a number of other deficiencies:

- highly centralised nature of approval processes;
- complexity of approval processes;
- lack of clear division of responsibility between central and local government;
- poor monitoring and enforcement mechanisms.

These characteristics are well illustrated by the electricity industry. The State Planning Commission was required to approve any new power station with a capacity of more than 50 MW and the State Council approved projects in excess of US $100 million if foreign funds are involved. Such approval processes would also involve a variety of institutions at local and central levels: the Ministries of Electrical Power, Water Resources and Railways; the local bureaux of these Ministries, the local bureaux of the Planning Commission and Finance Ministry, and the local governments themselves.

These low thresholds for central approval combined with the complexity of the approval process resulted in either projects being delayed or in the deliberate avoidance of these procedures by local governments keen to increase the power generation capacity in their districts. This latter reaction was recognised by central government officials but sanctions were rarely imposed on those officials who circumnavigated the official approval process. Thus the division of responsibility between central and local government was becoming increasingly blurred.

A lack of clear allocation of responsibility between the federal and provincial governments has also been evident in both the coal and petroleum industries. Lieberthal and Oksenberg (1988) have described in some detail the nature of the protracted negotiations between the Ministry of Coal Industries and the Shanxi provincial authorities concerning the division of rights and responsibilities in the development of major coal reserves in the province during the 1980s. Although agreement was reached at that time, the province has clearly failed to discharge its responsibilities in the field of regulating local coal miners (see below).

The regulatory structure of the upstream petroleum industry has been weakened by two developments. As CNPC, the onshore oil company, has devolved responsibility to individual oil-field managers, so the respective provincial governments have sought to increase their influence over the management of the fields and the issuing of exploration or production licences to other state enterprises. A further complication was introduced early in 1997 by the creation of a third state oil exploration and development company, China National Star, which has its roots in the Ministry of Geology and Mineral Resources and which operates both onshore and offshore. In the absence of a single Petroleum Law and a clear licensing system, the scope for confusion has only been increased (Gao, 1998).

The lack of the ability of the specialist and industrial ministries to enforce regulations has been discussed in Section 3 and these arguments apply as much to energy as to other sectors of the economy. The aspect of energy regulation which suffers from the greatest deficiencies is environmental protection. Though the importance of environmental protection is being increasingly emphasised by the central government, policy is only slowly being transformed into action. At the local level the priority given to economic growth far outweighs that placed on environmental protection, and the National Environmental Protection Agency lacks the resources and political power to enforce the law (Lotspiech and Chen, 1997; Jahiel, 1997).

The symptoms of this deficient regulatory system are wide ranging, and a selection are listed in Table 6.1. The great majority of the examples relate to production of primary natural resources. The coal industry is the worst offender. Poor regulation at local levels results in large numbers of small miners, both legal and illegal, exploiting surface or near-surface deposits in a technically-inefficient, polluting and unsafe manner. This has resulted in the destruction of reserves, the pollution of rivers, and loss of life (Thomson, 1996; Yang, 1995). A new Coal Law was passed in 1996 and recent press reports indicate that the Ministry of Coal

Industries has embarked on a campaign to bring order to the small coal mines (China Daily, 1997), though it is not clear that systems have been put in place to sustain any improvements implemented. Illegal operations are also reported from the oil industry. These include the theft of oil from wells and pipelines to feed illegal and polluting refineries in Hebei Province (Yang, 1995). Illegal or poorly regulated activity in any of these sectors results in unreported production as well as a loss of tax revenue.

The examples given in Table 6.1 refer mainly to the regulation of operational activity rather than to the regulation of competitive investment or of a market. The state enterprises hold strong monopoly and monopsony powers in the energy sector, and entry barriers for new investors (foreign or domestic) are high both for investment and for sales in the domestic market (Andrews-Speed, 1998). Within the major state corporations themselves, recent years have seen an increase in the intensity of vertical rivalry between different levels of management. This is particularly strongly developed in the power industry where state, provincial and county enterprises all seek to protect their own interests (Andrews-Speed and Dow, 1998).

A further deficiency relates to the lack of transparency in the regulatory process which affects both investment approvals and pricing. For example, no formal tendering process exists for construction and operation of new power plants, with the exception of the two build-operate-transfer (B.O.T) rounds which have taken place in 1996 and 1997. Further, production from all new power plants is covered by power purchase agreements (PPAs), but the Pricing Bureaux have the authority to overrule the pricing agreement within PPAs, with the exception of the recent BOT projects. A more fundamental flaw is the absence of any systematic and transparent mechanism to transfer to the customer any benefits obtained from increased productive efficiency in the power industry.

In a market economy the role of enforcing regulations commonly falls on a combination of independent regulators, parliament, the judicial system, the citizens and the press. In the absence of independent regulators, judiciary and press, the National People's Congress (NPC) has taken on the task of overseeing the enforcement of regulations. Whilst clearly keen to wield this newly-found authority, NPC Deputies are constrained by a shortage of resources and limited access to information. Only the largest and most flagrant breaches are likely to be worthy of their investigation. Further, it is evident that Deputies to the NPC face a wide range of conflicting pressures. Amongst these, loyalty to the Party and the Government are more likely to be of primary importance than the concerns of constituents (O'Brien, 1994). This is unlikely to change substantially until the independence of the judiciary is enhanced.

Table 6.1: Selected symptoms of deficient regulation in China's energy industries

Activity	Industry	Location	Symptom
Production	Coal	nation-wide	>50,000 illegal mines
	Coal	Inner Mongolia	530 entities in 1400 km^2
	Coal	nation-wide	air, land and water pollution
	Coal	nation-wide	unsafe operations
	Crude oil	Shaanxi	illegal operations
	Crude oil	Jilin	unreported production
	Power	Guangdong	illegal power plants
Processing	Oil	Hebei	200 illegal refiners
	Oil	nation-wide	air, land and water pollution
Pricing	Power	nation-wide	no contractual sanctity

Sources: Yang (1995), Thomson (1996) and unpublished investigations of the authors.

At a lower level, individuals or groups of individuals are increasingly taking their cases to the courts or to local government (e.g. Lotspeich and Chen, 1997). The courts are largely still controlled by local party and government cadres, and thus the judgements lack predictability (Lubman, 1995). However, anecdotal evidence suggests that complaints by citizens are increasingly drawing a response from local governments which is consistent with the law. In the absence of effective courts, firm control or guidance from the central government is still required.

5 Issues to be addressed

The sections above have briefly demonstrated the fundamental inability of the current legal and regulatory system in China's energy sector to cope with the transition to a market economy. A change is needed in the government's understanding of the nature of regulation. Direct, pervasive and non-transparent control should evolve into limited and transparent regulation (Shao et al., 1997). Regulation by vertical command should be replaced by horizontal contractual relations (Waelde and Gunderson, 1997).

The separation of the institutions of regulation and enterprise management is an important first step. This appears to have been started in 1997 in the case of the power industry and in 1998 in the case of the coal and petroleum industries. The new regulatory bodies should have responsibility for overseeing both the state and the private sectors.

A further step should involve the clear allocation of regulatory responsibilities between central and local governments. Agreement has to be reached as to which tasks are best regulated at which level. Some principles for this allocation might be drawn from the experience of federal, market economies such as the U.S.A. and Canada (Fox, 1983; Farrell and Forshay, 1994; Hancher, 1997). Federal governments tend to retain powers over issues which are of truly national importance, and delegate other powers. For example the production and processing of energy is regulated mainly at the provincial level, with the exception of offshore seas and federal lands. However, the federal government may draw up guidelines or minimum standards, as it does in India for the power industry, for example (Bath, 1997). The regulation of inter-provincial energy transport networks and international sales may also lie with the federal government, as may the regulation of wholesale prices. Environmental protection should lie firmly in the remit of the federal government because of the long-term and widespread environmental impact of the energy industries. As a result of such a reallocation of responsibilities, the presently over-stretched central government of China should find itself able to focus on and provide resources for a more tightly defined list of priorities in the field of energy regulation. These priorities should be limited to market failure, policy formulation and implementation, and monitoring functions (Waelde and Gunderson, 1997).

Whilst some aspects of the American federal systems of regulation may be transferable to China, Shao et al (1997) have argued that, in the case of the power industry, the provincial regulators should be answerable to the federal regulator rather than to provincial governments as in the USA. In the absence of democratic processes, such an arrangement should counteract the ease with which local regulators can be captured by local governments.

Such amendments to the Chinese regulatory system, though desirable, are not necessarily sufficient for effective regulation of the energy sector. In the field of law, greater independence for the judiciary and the clarification of ownership rights are prerequisites for effective regulation. In society as a whole, a range of requirements may be identified: for example, a professional, well-paid civil service; a well-informed and independent press; and a respect for the rule of law (Waelde and Gunderson, 1997).

6 Conclusions

As China evolves towards a market economy the long-standing, command and control apparatus of regulation is becoming increasingly ineffective. This is the case in many sectors of the economy, but especially in the energy sector where the central government seeks to maintain a high degree of control. One of the principal reasons for this regulatory deficiency is the failure of the regulatory system to adapt to the increasing *de facto* devolution of power away from the centre to the provinces. This deficiency is compounded by the overall weakness of the judicial system.

The regulation of China's energy sector is of great importance to the national economy. The effective use of energy natural resources, a sustained level of investment in the energy industry and the protection of the environment all depend on a credible and transparent system of regulation. At present, the long-term national interest is subordinated to short-term local interests, and the central government lacks the mechanisms to impose its priorities in the operation of the energy industries. A key step to improving this situation is to devolve to the provinces powers over those issues which are primarily of local importance, and retain for the central government powers in those fields which are truly of national importance. Until this is achieved, regulation in all parts of the energy sector is likely to remain ineffective.

References

Andrews-Speed P. (1998), 'Reform of China's energy sector: slow progress to an uncertain goal', in Zhuang J., Cook S. and S. Yao (eds.), *China's Transitional Economy*, London: Macmillan.

Andrews-Speed P. and S. Dow (1998), 'Reform of China's electric power industry. Challenges facing the government', submitted to *Energy Policy.*

Bath D.S. (1997), 'India's power sector: tariffs and electricity pricing', *Oil and Gas Law and Taxation Review*, 15 (9), pp. 325-334.

China Daily (1997), 'Coal mines targeted for upgrade', *China Daily CBNet*, 2nd September.

Farrell J.H. and P.F. Forshay (1994), 'Competition versus regulation: reform of energy regulation in North America', *Journal of Energy and Natural Resources Law and Policy*, 12 (4), pp. 385-405.

Fox W.F. Jr. (1983), *Federal Regulation of Energy*, Colorado Springs: McGraw-Hill.

Gao Z. (1998), 'A new star among state oil companies: China Star and its implications for the development of the industry', *Oil and Gas Law and Taxation Review*, 16 (4), pp.131-139.

Goodman D.S.G. (1994), 'The politics of regionalism: economic development, conflict and negotiation', in Goodman D.S.G. and G. Segal (eds.), *China Deconstructs. Politics, Trade and Regionalism*, London: Routledge, pp.1-20.

Hancher L. (1997), 'Energy regulation and competition in Canada', *Journal of Energy and Natural Resources Law and Policy,* 15 (4), pp.338-365.

Huang Y. (1996), *Inflation and Investment Controls in China. The Political Economy of Central-Local Relations During the Reform Era*, Cambridge: Cambridge University Press.

Jahiel A.R. (1997), 'The contradictory impact of reform on environmental protection in China', *The China Quarterly,* 149, pp. 81-103.

Johnson T. (1995), 'Development of China's energy sector: reform, efficiency and environmental impacts', *Oxford Review of Economic Policy,* 11 (4), pp.118-132.

Lavigne M. (1995), *The Economics of Transition. From Socialist Economy to Market Economy*, London: Macmillan.
Lieberthal K. (1995), *Governing China. From Revolution Through Reform*, New York: Norton.
Lieberthal K. and M. Oksenberg (1988), *Policy Making in China. Leaders, Structures and Processes*, Princeton: Princeton University Press.
Lotspeich R. and A. Chen (1997), 'Environmental protection in the People's Republic of China', *Journal of Contemporary China,* 6, pp.33-59.
Lu Y. (1996), *Management Decision-Making in Chinese Enterprises*, London: Macmillan.
Lubman S. (1995), 'The future of Chinese law', *The China Quarterly*, 141, pp.1-21.
Montinola G., Qian Y. and B.R. Weingast (1995), 'Federalism Chinese style. The political basis for economic success in China', *World Politics*, 48, pp.50-81.
Nolan P. (1995), *China's Rise, Russia's Fall. Politics, Economics and Planning in the Transition from Stalinism*, London: Macmillan.
O'Brien K.J. (1994), 'Agents and remonstrators: role accumulation by Chinese People's Congress Deputies', *The China Quarterly*, 137, pp. 359-380.
Oi J.C. (1992), 'Fiscal reform and the economic foundations of local state corporatism in China', *World Politics*, 45, pp.99-126.
Ronne A. (1997), 'Alternative approaches to regulatory agency structures and powers: eastern and western Europe', *Journal of Energy and Natural Resources Law and Policy*, 15 (1), pp. 41-50.
Shao S et al. (1997), *China. Power Sector Regulation in a Socialist Market Economy*, World Bank Discussion Paper No. 361, Washington: World Bank.
Solnick S.L. (1996), 'The breakdown of hierarchies in the Soviet Union and China. A neoinstitutional perspective', *World Politics*, 48, pp.209-238.
Thomson E. (1996), 'Reforming China's coal industry', *The China Quarterly*, 147, pp. 726-750.
Wong C.P.W., Heady C. and W.T. Woo (1995), *Fiscal Management and Economic Reform in the People's Republic of China*, Hong Kong: Oxford University Press.
Yang M. (1995), 'The cost of natural resources and economic benefit', *Quantitative and Technical Economics*, 12 (7), pp. 39-53.
Yusuf S. (1994), 'China's macroeconomic performance and management during transition', *Journal of Economic Perspectives*, 8, pp.71-92.
Waelde T.W. and J. Gunderson (1997), 'Same name, different content. Institutional aspects of regulatory reform in post-Soviet transformation', *D.I.W. German Institute for Economic Research*, Discussion Paper 155.

SECTION 3

ELECTRICITY AND GAS: MARKETS AND REGULATION

105-17

CHAPTER 7

WHOLESALE TRADING ARRANGEMENTS: COMPETING OPTIONS FOR EUROPE

L94
L98

MICHAEL MORRISON
Managing Director, Caminus Energy Limited

ILESH PATEL
Consultant, Caminus Energy Limited

Caminus Energy Limited
Caminus House, Castle Park, Cambridge CB3 0RA, UK
Email: Energy@Caminus.co.uk

Keywords: contracts; efficiency; electricity; Europe; spot markets; wholesale markets.

1 Introduction

The England and Wales electricity Pool has been in operation since 1990. During the 1990s other European energy markets have established competitive markets for balancing supply with demand and setting wholesale prices. These include the gas market in the UK and electricity markets in Norway, Sweden, Finland and Spain. In the same period, but outside Europe, the introduction of competitive electricity trading arrangements has been much more widespread. Prominent examples include national and state trading arrangements in Australia, New Zealand's spot market, South America[1] and North America[2].

In the rest of Europe, traditional centrally planned vertically integrated supply industries with municipally based local distribution to captive customers still persist. With much of Scandinavia, England and Wales and Spain having set up competitive wholesale markets, and the EU Electricity Directive now signed, other European countries now need to contemplate how they will establish wholesale trading arrangements in their own electricity markets.

The EU Electricity Directive on liberalisation of the internal electricity market finally came into force in February 1997 after almost a decade of consultation and debate. The majority of EU Member States have to transpose the Directive into national law by February 1999. The Directive sets an agenda for change, and as

[1] A number of South American countries have followed the lead provided by Chile's reforms in the 1980s.

[2] In North America, individual states in the US including California and 'PJM' (incorporating Pennsylvania jointly with five other states), and provinces in Canada including Alberta have established competitive trading arrangements so far. Other states and provinces are in the process of establishing new competitive trading arrangements.

such represents an important step towards the establishment of competitive electricity markets and the further development of international trade in electricity.

The Directive lays out some general principles and aims, some structural options, and sets a timetable for the opening of Member States' electricity supply markets.

- Member states can choose between an authorisation (free entry) and a tendering (centrally planned) procedure for new entry in generation.
- System operation should be managed separately from transmission and generating plant despatch should be non-discriminatory.
- Access to the market can be Regulated, Negotiated, or via a Single Buyer.
- The Single Buyer model would involve the purchase of electricity on behalf of eligible customers at a price offered by the single buyer.

However, the Directive is silent on the specific form that wholesale trading arrangements may take. In this chapter, we discuss the need for wholesale trading arrangements in competitive electricity markets. In particular, we consider the key principles, objectives and issues associated with the establishment of electricity trading arrangements. We then explore the main competing options for wholesale trading. In particular, we examine the trade-offs that need to be made in practice and how national market characteristics may influence the choice of trading arrangements. Finally, we look at the implications for Europe-wide wholesale electricity trading.

2 Why are wholesale electricity trading arrangements needed?

Electricity is a unique commodity. There is a need to match supply and demand instantaneously because electricity is not easily or cheaply stored. An electricity system, therefore, needs to match generator output and customer load.

Historically, electricity systems have tended to match supply and demand through the role of a central planning and despatch authority (for example the Central Electricity Generating Board in England and Wales prior to 1990). Such central authorities would despatch plant, based on least cost, to meet the demand of customers who did not have a choice of supplier.

In a competitive market structure, suppliers could establish a portfolio of bilateral contractual arrangements with generators in order to secure supplies to meet their customers' needs. However, in a market solely consisting of bilateral contracts between generators and suppliers, inevitably mismatches will occur, both between generators' outputs and their contractual commitments, and between suppliers' contract cover and their customers' consumption patterns. The transaction costs of resolving a large number of such contractual mismatches are likely to be high. A wholesale trading market offers the possibility of solving the problem of mismatches efficiently and at much lower transaction costs.

Competitive wholesale trading arrangements have the same function as a central authority - that of matching supply and customer demand at least cost. However, competitive trading arrangements involve more than just a move away from a central planning and despatch approach. They incorporate the idea of customer choice. Furthermore, wholesale trading arrangements offer a move away from an approach in which electricity is thought of and sold as a single indivisible product. Instead, electricity can be differentiated into a number of distinct products that have different costs and can be valued separately.

These products and services might include different delivery periods, flexible generation, ancillary services, reserve and transmission constraint services (including minimisation of system losses). The development of markets which are able to provide these services at a level that customers are willing to pay for them and the encouragement of cost efficient investment in the long run will help to promote allocative and dynamic efficiency. For these reasons, market based mechanisms for electricity trading can be superior to a centrally planned despatch mechanism in terms of economic efficiency.

The mechanism through which the need for products and services is communicated is through price signals and the price discovery process. This is an important feature of competitive electricity trading arrangements and in itself can vary between different forms of wholesale trading arrangements[3].

A move from centrally planned authorities with captive customers to competitive markets also transfers risks, typically from customers to other market participants. These risks might include:

- Market based risks where the risk lies in the movement of market prices against an unhedged position.
- Credit risks where a counterparty fails to perform their obligations.
- Operational risks and any associated exposure to market prices.

However, in response to such risks, and the needs of market participants more generally, competitive trading arrangements should encourage the emergence of financial products and services, as opposed to the physical products and services discussed above, that provide the opportunity to manage the risks faced effectively.

3 Wholesale trading arrangements – principles, objectives and issues

A number of key principles underlie the operation of competitive and efficient trading arrangements:

- A trading system should be transparent so as to pose few barriers to entry.
- The market should result in efficient and competitive price setting and charging mechanisms.

[3] Price discovery can take place via an auction process or as a consequence of continuous trading.

- The mechanisms for scheduling and despatch should be efficient, non-discriminatory and transparent.
- A trading system should be implementable and reflect true market conditions.

There are a number of economic objectives which new trading arrangements should attempt to meet. These include:

- A new trading system should result in choice for, and competitive prices to, end-users.
- By establishing a basis for entry, a new trading system should encourage incumbent market participants to reduce costs, thereby increasing productive efficiency.
- Competitive trading arrangements should result in long-run investment decisions that are made at least cost, thus achieving dynamic efficiency.
- Market based trading systems should attempt to increase allocative efficiency. Allocative efficiency would be a feature of a trading system in which distinct products and services are provided to consumers based on the willingness of those consumers to buy and pay for those services.
- Conditions should be established under which appropriate markets and/or tools for the effective management of market based risks might emerge.

The establishment of new trading arrangements faces a number of complex challenges and issues:

- Market power of incumbent generators needs to be addressed. In particular, the initial market structure chosen will be critical.
- Transmission constraints will be more prevalent in some countries than in others. This may influence the choice of wholesale trading system to be adopted.
- The plant mix and level of fuel diversity is an important determinant of the type of trading arrangements which may be implemented, since the type of plant will influence, in part, the type of scheduling and despatch process adopted.
- New trading systems will face transitional issues that they must be able to address. Amongst others, support for indigenous fuels and the elimination of cross-subsidies are likely to be important in many countries.
- The extent of demand-side participation is a fundamental issue. Market mechanisms can incorporate the concept of price-elastic demand.
- Environmental concerns and emissions limits are an important determinant of what investments may be made in the long run.

4 Characterisation of wholesale trading arrangements – three generic models

It is not surprising that wholesale trading arrangements have differed in liberalised markets, given differences in market features and starting points. There are a number of critical issues that have faced the design of electricity trading arrangements:

- Whether electricity should be a single price commodity, or consist of multi-priced products, or vary by location.
- Whether scheduling, despatch and price setting should be based on engineering principles and practice or market based principles and practice.
- Whether price setting should be *ex ante*, in real time, or *ex post*.
- Whether participation in the trading market should be mandatory or optional.
- Whether the system should be based on self-despatch or central despatch of generating plant.
- Whether generators should be rewarded separately for making their plant available, as well as for delivering energy.
- Whether and how the demand side of the market should participate in the price setting mechanism.
- The inclusion or exclusion of transmission constraints from the price setting mechanism.
- How the existing market structure and features should be incorporated within new trading arrangements (e.g. subsidies for coal producers and renewable forms of electricity generation).
- How price discovery should be facilitated.

Countries embarking on the establishment of new trading arrangements need to address *all* of these key issues. In this section, we consider three main types of wholesale trading arrangements that have been implemented around the world:

- Mandatory Pooling[4].
- Bilateral contracting and residual trading.
- Nodal pricing.

4.1 Mandatory pooling

A Mandatory Pool, in broad terms, is likely to be based more on engineering principles than market principles. Such systems or markets emphasise efficient price signalling and economic scheduling and despatch in generating electricity

[4] Whilst we use the term Mandatory Pool in this chapter, the concept underlying this term relates to systems in which generation is pooled into a single despatch or merit order. The term used for a system based on this idea, however, can vary from system to system.

rather than direct competition between generators to provide distinct products and services.

A single price is set for the whole market and participation is likely to be mandatory. Mandatory participation means that all generators (above a relatively low level of capacity) must sell all their output into the wholesale trading market and all customers must purchase their requirements from the market. Examples of such systems include the England and Wales Pool and the Pool in Victoria, Australia up until March 1998, and its successor the National Electricity Market (NEM).

A mandatory market is more likely to be necessary if a system is dominated by large thermal plant. As such, the costs at different levels of output of individual plant and their dynamic constraints can be incorporated explicitly within the bids that generators place into the Pool.

A crucial distinction between Mandatory Pool systems and other types of trading arrangements that we describe below is that contracts do not play a direct part in the despatch of plant - financial contracts are used primarily for hedging the price of power not physical contracts for the guaranteed delivery of power from a generator to a supplier. This is because in a Mandatory Pool there is central merit order despatch and generally there are no financial penalties over and above the lost revenues of not generating. No generator can guarantee that they will be despatched.

The England and Wales Pool, which has been in operation since 1990, is a good example of a Mandatory Pool based market:

- The England and Wales Pool is based on a central scheduling and despatch model. All electricity generated is scheduled and despatched by the grid operator.
- The Pool generates a single price which generators are paid for energy and capacity, while consumers pay a single price for energy, capacity and services to support the reliability of the grid.
- Scheduling and despatch is based on a least cost merit order to meet demand, given the bids offered by generators which include no-load costs, start-up costs and incremental bids. Plant dynamics are taken into account in the scheduling algorithm.
- There is only limited demand side participation and no self-despatch.
- The Pool does not operate a distinct market for services such as ancillary services, although the National Grid Company (NGC) does contract separately for such services.
- Financial hedging instruments known as Contracts for Differences (CfDs) are available to enable generators and suppliers to lock in the price of power.

The Victoria Power Pool (or VicPool), that operated between 1994 and 1997, was another example of a Mandatory Pool. Whilst it was similar in operation to the England and Wales Pool there were a number of important differences:

- Prior to 1994, the electricity industry in Victoria, as in England and Wales, was operated by a single vertically integrated and state-owned company – the State Electricity Commission of Victoria (SEC).
- After 1994, the functional businesses of the SEC were separated and privatised, and a wholesale trading market – VicPool – was created.
- VicPool had a single price for the whole system and there was mandatory participation. However, there were no capacity payments and the System Marginal Price (SMP) was determined half-hourly after the event (after constraints and losses were taken into account).
- The structure of bids placed into VicPool, which generators submitted, were much simpler than in England and Wales. Generators were not required to submit no-load or start-up costs, but were allowed to submit more incremental bids reflecting their willingness to, and the costs of, generating at incremental levels of output.
- Generators were able to self-commit and synchronise with the network at their minimum stable generation level. The self-commitment level of output was included in the daily generator offers.
- In most other respects the operation of VicPool was similar to the England and Wales Pool.

In May 1998, the NEM in Australia began operation and VicPool simultaneously stopped operating as an independent entity. Like VicPool the NEM is a mandatory market. The major differences between NEM and VicPool are:

- In the NEM, prices are set on an *ex-ante* five minute basis with the trading period for settlement purposes being half an hour, the settlement price being calculated simply as the time-weighted average of the five minute prices over the half-hour.
- The NEM uses zonal pricing to cope with transmission constraints. Since major transmission constraints occur only on flows between states, if transmission constraints arise across interconnectors, market prices will diverge between the regions. The marginal generator in each region, allowing for the maximum power flow on the interconnector, will set the clearing price.

A special case of the Mandatory Pool can be observed in many South American countries including Chile, Argentina and Peru. The model in these countries might be termed a Generator-only Pool. In such a system all electricity is despatched through a spot market using a merit order approach. Generators sign contracts for the supply of electricity directly with large customers and distribution companies. Generation companies meet their contractual obligations with despatched electricity, whether produced by them or purchased by them in the spot market. Generators do not submit their own offer prices: instead the market operator in each system calculates marginal costs using predefined algorithms and fuel prices. Based on these parameters, *ex-ante* energy prices are determined. The contracts that generators sign, although nominally contracts to supply, are financial contracts. Contract volumes are deducted from the metered generation and consumption that have to be settled by the parties concerned. Residual quantities are settled by the market operator at spot prices.

The other significant example of a Mandatory Pool in Europe is in Spain. On 1 January 1998, the Spanish market liberalisation law came into effect and the Spanish mandatory day-ahead power market began operating. Generators are able to sign bilateral contracts with large customers. These contracts are, as in other Mandatory Pools, financial in nature since generators cannot guarantee despatch. An hour-ahead market began operating on 1 April 1998. Hence, although a Mandatory Pool, the Spanish model, by including significantly more demand side participation and the ability to participate in a market that operates much closer to real time, has built on rather than duplicated the England and Wales mandatory Pool model.

In 1998 the Department of Trade and industry (DTI) and the Office of Electricity Regulation (OFFER) in the UK jointly conducted a review of the current electricity trading arrangements in England and Wales. In the proposals that came out of this review, OFFER envisages the abolition of the Pool and its replacement by a series of voluntary markets supplemented by a mandatory settlements process. Instead of operating in one compulsory market (the Pool) and receiving the Pool Purchase Price, generating plant could choose to sell power into several markets:

- Forwards markets that could be exchange-based and/or over the counter.
- An organised short-term exchange.
- A balancing market.
- A settlement process for imbalances.

These proposals, if implemented, would represent a move to a bilateral contracting and residual trading type market.

4.2 Bilateral contracting and residual trading

Bilateral trading markets are, broadly speaking, based more on market related principles than Mandatory Pooling systems. The principles underlying this type of wholesale trading arrangement are that it is possible to identify distinct products and that distinct markets can be set up to trade these products. In such a market not only is energy priced in a market but also transmission constraints and location, different delivery periods and flexibility. The creation of real-time balancing markets and zonal prices are also features of such markets. Participation in any organised power exchanges or spot markets is voluntary.

In bilateral trading markets, plant dynamics are unlikely to be as important as in Mandatory Pool systems. Such markets are also characterised by greater involvement from the demand-side. Bilateral trading markets have tended to be implemented in systems that have previously been based on bilateral contracts between generators and suppliers.

Bilateral trading markets are, in general, exchange based systems with sellers interacting directly with buyers, perhaps via a trading screen in which offers to sell and bids to buy products might be placed. A market-clearing price can be determined and trades are done when an offer and bid can be matched. The price discovery process in such a market is based not on a merit order as in a Mandatory Pool, but on the willingness of sellers to be paid an amount for their product and buyers to pay an amount to purchase a product.

Examples of such markets include Norway, Sweden and Finland (NordPool), California, and the gas market in Great Britain.

The NordPool was established in 1992. In 1996 it expanded to include Sweden, and in June 1998 Finland has also joined the NordPool. The electricity system in Norway and Sweden, prior to the establishing of NordPool, was based on bilateral contracts between generators and suppliers (and customers).

- In Norway and Sweden generators offer simple price-quantity bids. These bids are used to derive supply and demand curves and thus derive prices.
- NordPool has three constituent markets:
  - An ex-ante spot market, Elspot, is an auction based wholesale trading market.
  - A real-time balancing market is used to deal with real-time constraints and changes in bids between the ex-ante market and real-time operation.
  - A financial futures market (Eltermin) also exists to provide price and volume risk management tools. Contracts in the financial futures market are not physically cleared but used only to hedge purchases in the spot market.
- Important differences exist between the three member countries of NordPool in the mechanism through which real time balancing takes place and how imbalance charges or prices are calculated.
  - In Norway, for example, offers and bids for real-time balancing power are submitted up to 3 hours ahead and imbalance prices are calculated for each hour based upon the direction in which most balancing actions were taken. Imbalances are cleared at this marginal price. Actions taken to deal with constraints cannot set the imbalance market price (constraints are dealt with through the spot market).
  - In Sweden, offers and bids can be submitted much closer to real time and marginal balancing prices are calculated for both increment and decrement actions. Transmission constraint costs are recovered through grid charges. Svenska Kraftnett, the Swedish system operator, also runs an hour-ahead market to try and 'forecast' balancing market prices.
- Around 200 players are now registered to trade in the market including international players such as Enron and Eastern Power and Energy Trading (a subsidiary of Energy Group).

- Plant dynamics are not included in bids placed into NordPool. If transmission constraints exist then zonal prices are determined.
- NordPool is a voluntary market with bilateral contracts. This means that there is no obligation to sell electricity into the spot market or to buy electricity from the spot market.
- Because there are physical bilateral contracts there is some self-despatch based on these bilateral contracts.
- There is significant demand-side participation as generators and suppliers are able to submit price-elastic demand curves.

The emphasis within the NordPool is very much on distinguishing products and services. The range of products includes energy, the location of delivery and the delivery period. NordPool relates the pricing and provision of these products and services to the willingness of customers to consume and pay for them.

The Review of Electricity Trading Arrangements in England and Wales, carried out in 1998, has produced proposals that if implemented would abolish the Pool and replace it with a series of voluntary markets. The proposals have a number of key characteristics. These include:

- A market for imbalances and a mandatory settlements process.
- Prices will be determined by the interactions of buyers and sellers.
- Bids and offers will be firm and will be made in a simple format.
- Prices in all the markets will be determined by one party accepting another party's offer, i.e., via a "pay as bid" mechanism.
- Greater freedom of contractual form, with market participants having the ability to sign contracts for physical delivery as well as financial contracts for differences (CfDs).
- Participants who take or supply more or less than they contracted for will be paid or charged individually.
- Prices will be comprehensive, i.e., there will be no separate capacity payment.
- NGC will continue to act as System Operator (SO). A separate Market Operator (MO) will be appointed for the short-term exchange. Other MOs may emerge as longer term forwards and futures markets develop.
- Governance arrangements for the SO and MO should provide incentives for efficiency and flexibility.
- Central despatch by NGC will be largely replaced by self-despatch, with NGC using a balancing market to secure the system.

Clearly these proposals bring England and Wales much closer to the trading arrangements in NordPool. It is envisaged that the new arrangements should be in place by April 2000.

Recently, plans have been announced for an international spot market based in the Netherlands to be called the Amsterdam Power Exchange (APX). APX is due to start operating at the beginning of 1999 and will be based on the NordPool model.

The exchange will have contract delivery in the Dutch transmission grid but will incorporate trading by participants in the Benelux countries and Germany. Separate prices will be quoted for these countries. This development represents a significant step forward in electricity trading in continental Europe and the design of APX itself builds very much on the ideas behind a bilateral trading market as described above.

In summary, the existing bilateral trading markets have started from a different point to Mandatory Pool systems, and have developed to address different issues and achieve different objectives. The principles and practice of markets such as NordPool, the GB gas market and California are quite distinct from the England and Wales Pool and VicPool.

4.3 Nodal pricing

In nodal pricing systems, the market based pricing of products and services is set very much to the fore. Nodal pricing systems price generation and transmission simultaneously. Hence, market prices should make clear whether it is more efficient to invest in the transmission network (to reduce losses and constraints) or in new generation capacity. This results in economic and efficient decision making by existing players and new entrants. Features of nodal price systems might include dispersed demand centres and generation capacity, an underdeveloped transmission network, and strong demand growth.

In nodal systems a different price is set for each node or grid point, and participation is mandatory, i.e., all generators are scheduled and despatched through a single market. Examples of such systems include the New Zealand spot market, and to a lesser extent the generator only pool in Chile and the pool in Argentina.

The New Zealand spot market began operation in October 1996 after a process of structural change in the New Zealand market. A substantial amount of hydro capacity exists in New Zealand dispersed mainly around the South Island. Major demand centres exist in the North and South Islands.

- The most significant feature of the New Zealand spot market is nodal pricing.
- Generators and purchasers make simple price-quantity bids and offers (supply and demand curves) at a particular grid point.
- Since, the system is over 66% hydro-based, plant dynamics were not seen as a major issue and therefore are not included as part of the bids of generators.
- Prices are generated for 477 points or nodes.
- There is significant demand-side participation.

Hence, nodal pricing systems are distinct from the previous two models discussed. Whilst Chile and Argentina do set nodal prices, the nodal prices are not set directly through a market based mechanism. Nodal prices in these countries reflect the underdeveloped transmission system and the dispersed nature of generation capacity. Nodal pricing in these markets is intended to signal the need, in particular areas, for either new generation capacity or investment in the transmission network, and to encourage new and existing players to make these much needed investments.

5 Competing wholesale trading options for Europe – which to choose?

The three broad characterisations we have made reflect how the principles of competitive wholesale trading arrangements can be interpreted and applied in practice. We have described three competing options which have been implemented in liberalised markets around the world, and which European countries need to consider as a basis for wholesale trading in electricity in the near future.

Countries are unlikely, of course, to adopt identical trading systems or directly copy existing models. Indeed, countries should learn from and improve upon these models. Furthermore, the adoption of a particular type of trading arrangement should also reflect the particular features of an electricity system as they exist if the market is to operate efficiently. The fact that the Mandatory Pool model and the bilateral trading model are exemplified in at least four European countries, demonstrates that these options are very relevant in a European context and need to be carefully considered by other European countries. The success of NordPool in two countries that began from a municipal system like those seen in other continental countries suggests that competitive trading arrangements could bring real benefits to industry and other customers in continental Europe.

Although bilateral trading markets and Mandatory Pools are very much competing options for wholesale trading arrangements in European countries, recent developments in NordPool, the proposals for new trading arrangements in England and Wales and the development of a Power Exchange in the Netherlands suggest that advantages of direct participation and contracting between generator and customers/suppliers in bilateral trading markets have begun to be seen for European countries.

Furthermore, we would expect that price differences and price volatility (demonstrated in the graph below) particularly between interconnected systems should result in greater pressure for a convergence in the design of trading arrangements. NordPool and the experience of the NEM in Australia show that efficient trading arrangements between interconnected systems can be developed. Moreover, as systems become increasingly integrated we would expect international price indices to develop at 'hub' countries (the Swiss electricity price index is one example already developed) and the emergence of independent commercial market operators in response to demand and trading activity.

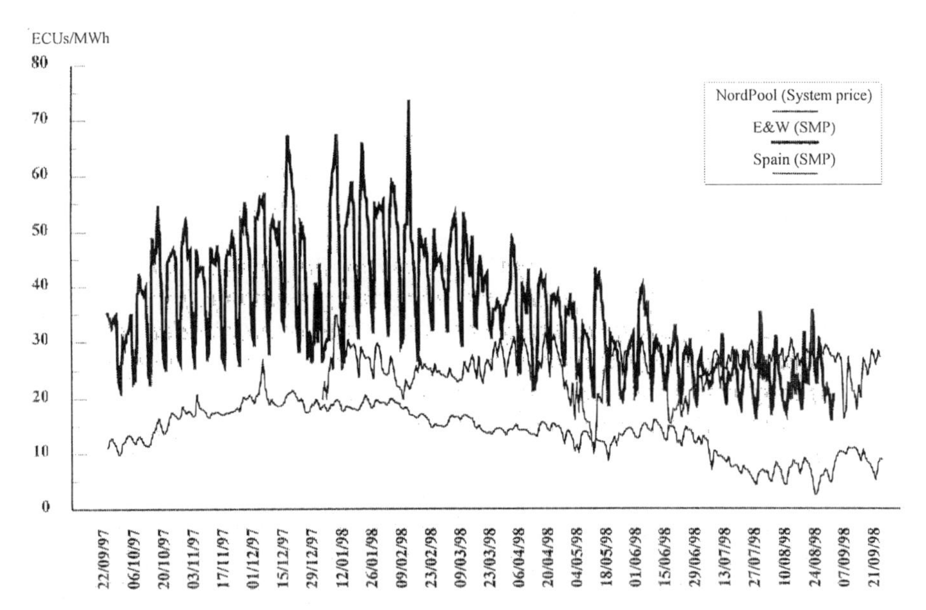

Note: Exchange rates as of 23/9/98

Figure 7.1: Daily average primary electricity market prices: 22/9/97 - 23/9/98

The EU Electricity Directive attempts to ensure that differences in approach do not prevent the emergence of an internal market in electricity. If different models for wholesale trading are adopted by a broad range of countries, reflecting their specific circumstances, this does raise interesting issues about the interaction between different markets and price setting mechanisms.

However, we do not believe that this raises any fundamental obstacles to the development of Europe-wide competition and trade in electricity. Price disparities will likely be competed away by the activity of traders, and with the possible development of a European futures market in electricity, we may even see convergence in trading arrangements in the longer run.

[illegible]

[illegible] markets and price setting [illegible].

[illegible] believe that this [illegible] development of [illegible] competition [illegible] will likely be [illegible] development of [illegible] European [illegible] markets [illegible] trading arrangements in the longer term.

CHAPTER 8

REGULATION POLICY AND COMPETITIVE PROCESS IN THE UK CONTRACT GAS MARKET: A THEORETICAL ANALYSIS[1]

HUW D. DIXON

Department of Economics, University of York, Heslington, York YO10 5DD
Email: hddl@york.ac.uk

JOSHY Z. EASAW

Economics Group, Middlesex University Business School

Keywords: competition; gas; network access; OFGAS; pricing behaviour; regulation.

1 Introduction

The Gas Act of 1986 which led to the privatisation of British Gas, also created the gas regulator, Ofgas, with an unequivocal mandate to introduce competition and, subsequently, deregulate the industry. Together with the Office of Fair Trading (OFT) and the Monopolies and Mergers Commission (MMC), Ofgas embarked on a series of regulatory and competitive policies to increase competition in the contract gas market. The purpose of the present chapter is to analyse the ensuing interaction and strategic behaviour that arose as a result of Ofgas' policies, given the nature of both the incumbent and entrants in the contract gas market.

When analysing the strategic behaviour of both the incumbent and entrant, it is important to distinguish between pre-entry and post-entry advantage. British Gas (BG), as the incumbent shipper, has pre-entry advantage. Furthermore, BG operated as a natural monopoly provider of the transmission network in a vertically integrated industry. Conversely, the main entrant shippers, which are North Sea gas producers, operate as "upstream" firms in the contract gas market. This enables them to make quantity pre-commitments in the final market by opting to sell directly to end users rather than BG. The upstream firms have an intrinsic first mover advantage during post-entry.

The chapter is organised as follows; Section 2 considers the regulatory policies adopted in the retail or final goods market and the ensuing BG's pricing behaviour. Section 3 reviews the network access regulation in a vertically integrated gas industry and the strategic reactions to these policies and, finally, a summary and main conclusions are drawn up in Section 4.

[1] We are grateful to John Hall Associates for making available their dataset on the competitive gas market which is the basis of the analysis in this chapter.

2 Regulation and pricing behaviour in the final goods market

2.1 Ofgas' policies

Soon after the privatisation of BG in 1986, Ofgas embarked on trying to pursue pro-competitive policies and increasing entry. In a report; "Competition in Gas Supply" (Dec 1987), Ofgas put forward the rationale for promoting greater competition. First and foremost, it would bring about general efficiency gains. Secondly, it would give BG greater incentives and effectiveness to buy and sell gas at efficient prices. Thirdly and most importantly, it would change the structure of the industry so that BG's market powers were considerably diminished.

In the same report, Ofgas identified the contract gas market as the best potential for developing competition. Ofgas felt that gas producers operating in the North Sea were most likely to enter the gas contract market for three main reasons. Firstly, they could expect to sell gas at a higher price than they would receive from BG. Secondly, selling directly to large industrial users would prove more attractive. Industrial users' demand for gas has less seasonal variation than that of residential users and the smaller "swing factor" implies lower supply cost. Lastly, gas producers might be able to bring a given oil field on stream more quickly by supplying directly to end users.

Ofgas together with the MMC and OFT embarked on trying to reduce BG's pre-entry strategic advantage as part of their pro-competitive policy. The MMC Report (1988) recommended three regulatory objectives; prevention of abuse, promotion of competition and the management of the market. In the same report the MMC highlighted the possibility that BG may practise first degree price discrimination. There was a suspicion that the prices charged to customers depended on how easily alternative sources of fuel were available to them. Ofgas responded by requiring BG to publish its price schedules on 1 May 1989. BG could not alter any particular schedule more than once in any 28 day period without the consent of Ofgas. Furthermore, Ofgas must be given 21 days notice of any change to the classes or description of schedules. This was intended to prevent abuse and the requirement was suspended in October 1994. In the Ofgas Review of 1994, Ofgas was satisfied with the level of competition in the contract gas market and it also increased focus on network access price regulation.

The OFT Gas Review of October 1991 reiterated the MMC's call to promote competition and endorsed BG's requirement to publish its pricing schedules. BG responded by committing to give up 60% of the competitive market to its competitors. In addition, BG also gave an undertaking to the OFT under the "Gas Release Programme" to release at least 500 million therms for the years 92/3, 93/4 and 94/5 and 250 million therms in 95/6 to competing shippers.

2.2 BG's pricing behaviour

The present sub-section investigates BG's pricing policy, which was clearly a reaction to Ofgas' regulatory and competition policy. BG was under regulatory pressure to make entry easier, whilst BG's pricing behaviour came under scrutiny as Ofgas required BG to publish their pricing schedules to eliminate any possibility of the practise of first degree price discrimination.

We will briefly consider two theoretical perspectives on the possible reaction of BG in the face of the regulation measures adopted and the increased competition from upstream firms: (i) contestability and (ii) inter-temporal pricing strategy. Whilst the Cournot model is appropriate for determining the general level of gas prices, the *structure* of prices charged to different customers at different times is better modelled using the Bertrand framework. Gas can usually be switched between different customers with ease, whilst some types of customers have relatively low switching costs between suppliers. One of the main aims of the Ofgas regulations was to promote price competition between suppliers in particular submarkets, and the range of markets has been expanding over time.

The switch from a situation in which there is little price competition to one in which there is effective competition would lead us to expect a change in pricing policy. The theory of contestable markets, in which incumbent firms can be undercut by actual or potential competitors from the related markets, predicts that the resultant prices will be Ramsey-optimal. Ramsey-optimality requires price equal to marginal and average cost when there are decreasing returns, and price equal to average cost with increasing returns. Ramsey-optimality implies zero supernormal profits.

The increase in price competition could lead to two effects. First, a reduction in the general *level* of prices, as the possibility of being undercut forced the incumbent to lower prices. Second, the *structure* of pricing would have to reflect the structure of cost. In particular, cross-subsidisation would become impossible. If all of the various submarkets were contestable, all prices would be Ramsey-optimal. One of the main victims of price competition of this nature would be the use of demand based inter-temporal variations in tariffs, third degree price discrimination. In particular, from the time of nationalisation, gas has had a seasonal tariff, being higher in the winter months when demand is inelastic. Whilst there is some cost based justification for this, in terms of providing peak capacity, the commercial reason has always been based on the demand side (the inelastic demand). However, in the case where some markets remain effectively sheltered (for example the domestic gas users) these markets could still face a price well in excess of the Ramsey-optimal.

The conditions for a contestable market are very stringent. In the case where demand is less price sensitive there is the possibility that the incumbent, anticipating intense competition and loss of market share, will adopt an inter-temporal pricing strategy. In a Gaskin-type (1971) dynamic limit pricing model, an incumbent or dominant firm may choose a monopolistic price behaviour now in the face of inevitable entry into the market and sacrifice future profits through the loss of market share. Indeed, BG voluntarily agreed to give up its market share.

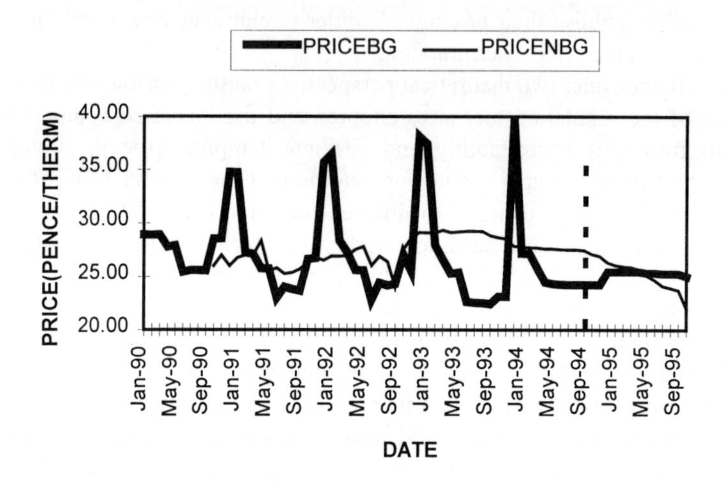

Figure 8.1: Price of British Gas and non-British Gas

Figure 8.1 above depicts BG and non-BG volume weighted prices. While Ofgas attempted to eliminate the possibility of first degree price discrimination, BG evidently practised intertemporal or third degree price discrimination as the graph depicts distinct seasonal variations. More importantly, BG ceased this practise when the requirement to publish their pricing schedules was suspended in October 1994, as depicted by the dotted lines in Figure 8.1. The breakdowns of the individual volume weighted average prices of the main gas shippers are given in Figure 8.2. It indicates that non-BG prices do not display any distinct seasonal patterns, and the entrants into the contract gas market do not follow the incumbent's prices. The main competing shippers are; Mobil Gas, Alliance Gas, Kinectica, AGAS, Amerada Hess and Quadrant and are either wholly or partially owned by gas producers and traders.

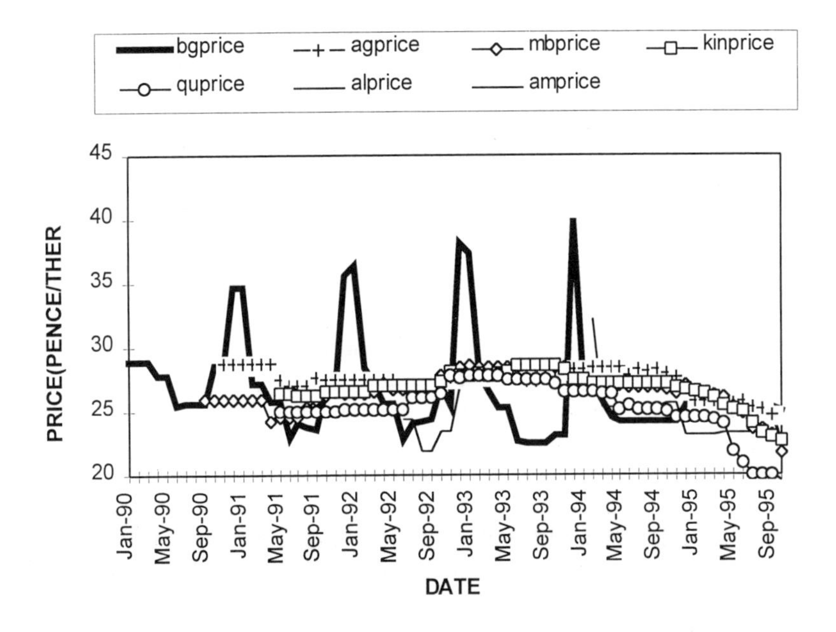

Figure 8.2: Average prices of 7 main gas suppliers 1990-96

When BG was required to publish its pricing schedule and public pricing policy was introduced, Ofgas justifiably anticipated contestable outcomes and BG's pricing to be Ramsey-optimal. However, from Figures 8.1 and 8.2, BG persisted with inter-temporal price discriminations with seasonal variations. Though, Ofgas was keen to eliminate any first degree price discrimination but would appear to tolerate third degree price discrimination. A possible answer could be found in Ofgas (1994). Ofgas highlighted the possibility that increased competition could lead to erosion of cross-subsidises and the ability of BG to provide Universal Service Obligations (USOs).

Taking into account seasonal variations in gas supply, even the low seasonal variation to be found in the contract gas market, indicates that BG's prices are considerably higher than those of the entrant shippers in the contract gas market. This, however, reverses as the requirement to publish their price schedules was suspended and subsequently BG's prices closely followed non-BG's[2]. This suggests that BG was using its strategic advantage as an incumbent to practise an inter-temporal pricing strategy.

[2] It must be noted that BG's weighted average prices were never really lower than those of the entrants.

3 Network access regulation and British Gas's strategic behaviour

3.1 Network access regulation

Although network access is particularly important to vertically integrated industries such as the gas industry, Ofgas was slow in introducing any formal regulation. Only after the MMC (1988) recommendations were "third-party carriages" formally introduced in October 1989. Nevertheless, the regulation of gas transportation soon became an integral part of Ofgas' pro-competitive policies as stated in Ofgas (1993):

> "The gas transportation system is a significant part of the infrastructure of the UK economy. It is important therefore that the structure of prices adopted for the system encourages, to the maximum extent possible, the efficient use of resources, to the benefit of gas customers and *the economy as a whole*."

Ofgas eased the regulation in the final goods market, while concentrating on network access. In December 1993, the President of the Board of Trade announced that Transco, the transportation and storage wing of BG's operations and Business Gas were to "separate to the satisfaction of Ofgas". BG was required to maintain "accounting separation" between Transco and the rest of BG's operations. This decision was undertaken by the government, despite the second MMC report in 1993 strongly recommending vertical separation. Increasingly, Ofgas focused on network access regulation and the Ofgas Review 1995 highlighted certain barriers of entry into the competitive segment. Transco was required to compensate entrant shippers financially for unsatisfactory service.

Prior to 1989 BG was only required to provide indicative access, or transportation, charges and they were negotiated individually. The first formal "third party carriage" was introduced in October 1989. Basically, the charges are based on the geographical distance of gas transported for third party shippers and load factor, where the latter refers to the proportion of gas transported at peak times. In the first three years when Ofgas was keenly promoting pro-competitive policies, access prices fell sharply, as much as 50% in real terms. BG complied to Ofgas' wishes as they hoped that it would ease regulation in other areas such as the requirement to publish price schedules.

Access price was based on the rate of return on capital assets, set at 4.5%. This form of regulation has two effects; as the rate of return was set above the cost of capital it encourages over-capitalisation (see Averch and Johnson (1962)) and it allows rebalancing within the cap to suit the incumbent's own intent. Over-capitalisation provides incentives for the undercharging of capital intensive demand, thereby encouraging its expansion. BG's transportation charges indicated these imbalances as the distance and peak-related elements of access charges were relatively low. These issues were clearly highlighted when BG and Ofgas engaged in an extensive debate on access charges in 1994. The *capacity charge* element, which is payable on the peak daily capacity required by the shipper, and independent of actual usage, and the *commodity charge* element, which is paid on each therm of gas transported, were split 90:10 in the access charges. However, Ofgas wanted to revise this split to 50:50. This implied that off-peak gas supply would become relatively more expensive and would suit BG's profit maximising strategy. However, as pointed out by Waddams Price (1997), a more likely motivation is to increase peak demand and expand the capital base which would determine the allowed return in the future. Sherman (1989) maintains that such behaviour is consistent with rate-of-return regulation.

BG also proposed charges that depended on the point of entry of the gas and its point of exit, in the form of a matrix. BG suggested that the new method would be more reflective of long-run marginal cost. It would take account of flows and congestion within the system, which is monitored by Transco, rather than the distance travelled. Though the sophisticated formulae reflected long-run marginal cost, they had the disadvantage of being more obscure, increasing the uncertainty of potential entrants. Ofgas, despite reservations, relented to BG's position and the new access charge regime was introduced in September 1994.

In addition, the new charges increased the proportion of charges for the use of medium and low pressure systems while reducing the distance related component. This made the supply of gas to smaller consumers relatively more expensive. More recently, there has been a change of attitude; in 1997 Transco was moving away from the 50:50 split. As the transportation operations and the rest of BG's operations demerged and as lower peak charges do not benefit Transco, this implied that undercharging directly benefited BG's own supply arm.

3.2 British Gas's strategic reaction: vertical integration and capacity pre-commitment

We showed in Section 2 that BG used its pre-entry advantage as an incumbent to exercise an inter-temporal pricing strategy. BG maintained its status as a vertically integrated operator in the gas industry. Hence, the regulation of network access is an integral element in effectively reducing BG's pre-entry strategic advantage.

This section examines the effectiveness of Ofgas' policy of maintaining BG as a vertically integrated entity and the use of rate-of-return regulation as a basis of network access regulation. The incumbent in a vertically integrated industry has pre-entry advantage and is able to make capacity pre-commitment; this could pre-determine the outcomes in post-entry. Indeed, even when the entrant is a Stackelberg leader, the entrant, who is an upstream firm, as in the case with the contract gas market where the main entrants are North Sea gas producers, has post-entry Stackelberg leadership advantage. It may chose to pre-commit quantity directly to the end user rather than BG.

The form of access price regulation, where the allowed rate of return on capital base was higher than the cost of capital, gives incentives for: (i) increased capitalisation and (ii) large capacity precommitment by the incumbent. Both implications can affect the reaction functions of both the incumbent and entrants. The former can potentially shift the entrant's reaction function further to the left, due to higher access charges, as a result of rebalancing and a larger capital base. The Averch-Johnson effect on Transco's operations and the consequent rebalancing of access charges has been much discussed (see Price (1994) and Waddams Price (1997)).

The essential part of the present analysis relates to the costs of production, where each firm is assumed to have: (a) constant variable cost of output, (b) constant unit cost of capacity expansion and (c) set-up cost. In the case of the contract gas market, we assume a duopoly scenario between BG and a non-BG gas shipper, Mobil, which is the largest non-BG supplier at present, and also a North Sea gas producer, and therefore operates as an "upstream" firm in the contract gas market. The capacity or transmission network is vertically integrated. Prior to entry we have:

$$n = n_I + x \tag{1}$$

where n denotes the transmission network $n_I = q_I$ represents the usage of the network (i.e., output) of the network by the incumbent, x the excess capacity which may be present if the network is under utilised (as in Spence (1977)), $x \geq 0$ .After entry (1) becomes:

$$n = n_I + n_E + x \tag{2}$$

Though the network is shared by both the incumbent and entrant, BG retained control and the cost relating to the network appears in its cost function and subsequently affects its reaction function. More importantly, BG is able to precommit the transmission network, which is not a binding constraint. The entrant shipper's cost function is:

$$C_E = f_E + (b_E + a_E)q_E \tag{3}$$

where a_E refers to access price, b_E is the marginal cost of production, f_E the fixed cost. The incumbent's cost function is as follows:

- when $q_I + q_E > \bar{n}$, the network has to expand beyond the precommitted level:

$$C_I = f_I + (b_I + a_I)q_I + a_I q_E \tag{4}$$

where a_I is the constant unit cost of capacity

- when $q_I + q_E \leq \bar{n}$, there may be a possibility of excess capacity:

$$C_I = f_I + b_I q_I + a_I \bar{n} \tag{5}$$

The entrant's marginal cost of production is:

$$mc_E = b_E + a_E \quad (6)$$

whilst the incumbent's marginal cost of production, when $x < 0$ is:

$$mc_I = b_I \quad (7)$$

and when $x < 0$ and additional capacity is needed marginal cost is:

$$mc_I = b_I + a_I \quad (8)$$

When capacity expansion does not matter, capacity cost is independent of output and there is a possibility of excess capacity ($x < 0$) capacity cost becomes part of fixed cost. When capacity expansion takes place, capacity expansion is dependent on output and the constant unit cost of capacity becomes part of marginal cost.

As a vertically integrated firm, BG is able to precommit the capacity; in the present case, transmission network. It also uses the access price charges to manipulate the entrant's usage of the transmission network, thereby limiting the extent to which it can exercise its post-entry advantage. These issues relating to the contract gas market are examined in detail in the following sub-section.

The form of rate of return regulation introduced by Ofgas for transportation charges also gives the incumbent an incentive to make large capacity pre-commitments[3]. There is every likelihood that both the incumbent and entrant's output would be within the precommitted capacity ($n_I + n_E \leq \bar{n}$) and there is the possibility of excess capacity ($x \geq 0$). Consequently, BG has lower marginal cost and operates on the *JJ'* reaction function in Figure 8.3. *GG'* and *JJ'* are the incumbent's reaction functions when marginal costs are given by (8) and (7) respectively (whether additional capacity needs to be made). In addition, the higher access price as a result of over-capitalisation, can shift the entrant's reaction function from RR' to *SS'*. Using the standard Dixit (1980) analysis, the possibility of capacity precommitment *on its own* would lead a range of possible post-entry outcomes in the segment *TV* indicated in Figure 8.3. However, by shifting the entrant's reaction function by use of the access price, the more favourable range of outcomes *T'V'* is possible: the incumbent has a higher market share, output is lower and the price higher.

[3] As the rate of return regulation is set higher than the cost of capital, it would be relatively "costless" for BG to pre-commit large transmission capacity.

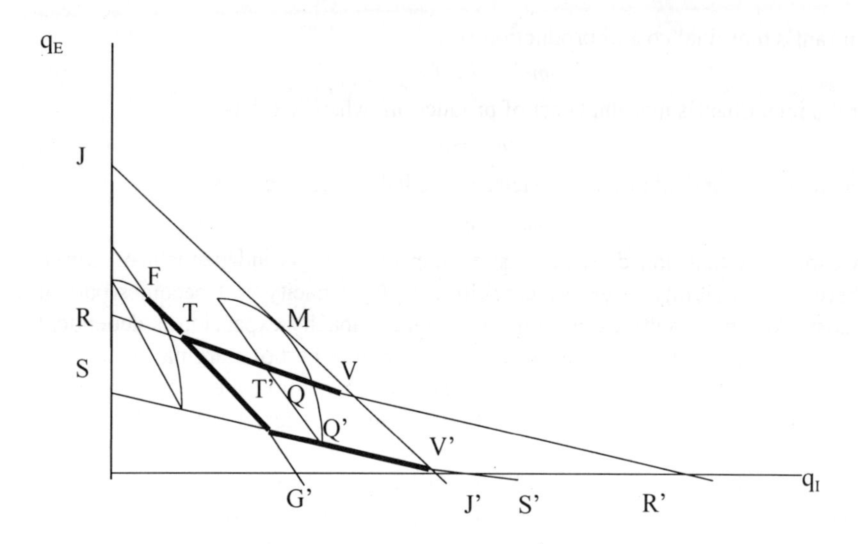

Figure 8.3: Incumbent's and entrant's reaction functions

Large pre-commitment of capacity by an incumbent implies that there is excess capacity (see Spence (1977) and Dixit (1980)) and the incumbent has to secure gas supply. In Ofgas' review of transportation charges for 1997 (Ofgas (1996)), it proposed to reduce the valuation of Transco's asset base by 8%, and one of the main reasons for such a move is its under utilisation. Furthermore, in the 80s and early/mid 90s BG attempted to secure fixed gas supply, and engaged in "take-or-pay" contracts. They anticipated only a small loss of market shares in the long run despite their participation in the "Gas Release Programme" and agreeing to give up a fair size of the market. Capacity pre-commitment would have, in general, a negative effect on the possibility of bypass from the competitors; such a scenario may not be directly relevant to the UK gas industry, where the network is generally deemed a natural monopoly. Therefore, setting the rate of return regulation higher than the cost of capital for network access charge, as in the case with the gas industry, gives incentives to:

(i) over-capitalisation and increase the marginal cost of the entrant,
(ii) large precommitment of capacity by the incumbent, reducing its marginal cost,
(iii) in a vertically integrated industry, large precommitment of capacity means that the incumbent has to secure high levels of gas supply, and
(iv) large capacity precommitment, reduces the possibility of bypass, which may be optimal as entrants reduce their long-term investments.

Figure 8.4 depicts BG and non-BG outputs respectively, while Figure 8.5 gives the individual outputs of the main seven gas shippers in the contract gas market.

Non-BG shippers' output expands and at an increasing rate in mid/late 1992 and early in 1993. This is consistent with the publication of the OFT 1991 report and the consequent BG commitment to market share targets and participation in the gas release programme.

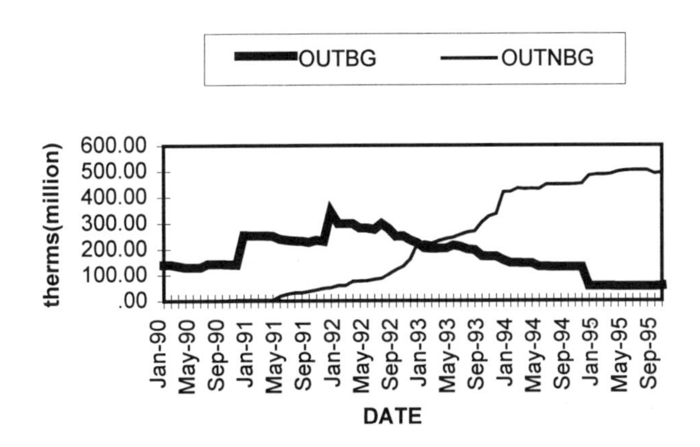

Figure 8.4: Outputs of British Gas (BG) and non-British Gas (NBG)

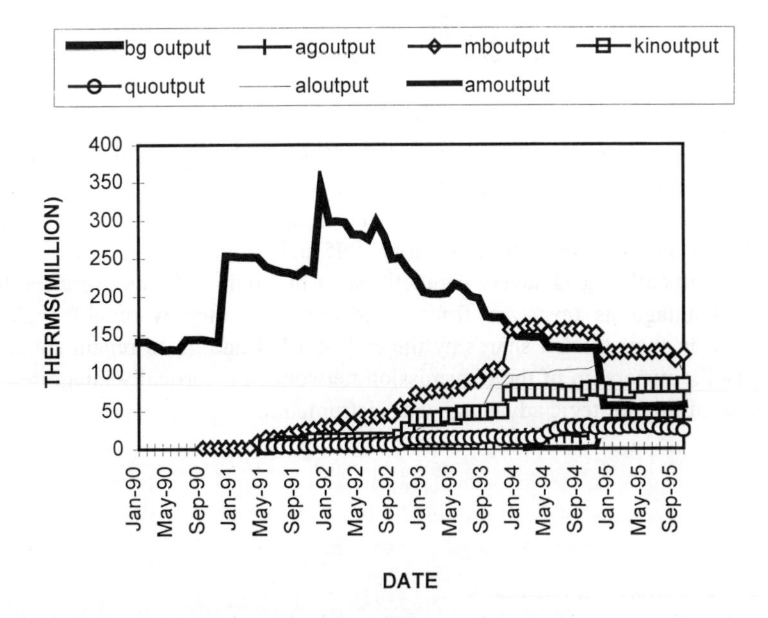

Figure 8.5: Output of 7 main gas suppliers 1990 - 1996

Mobil would appear to be the leading non-BG shipper; by the end of 1994 their output is twice that of BG. Alliance Gas and Kinectica follow closely behind as these upstream firms take advantage of their post-entry advantage.

Figure 8.6 shows the dramatic fall in BG's market share in the contract gas market. This halted in early 1995 and held constant around the 10% mark.

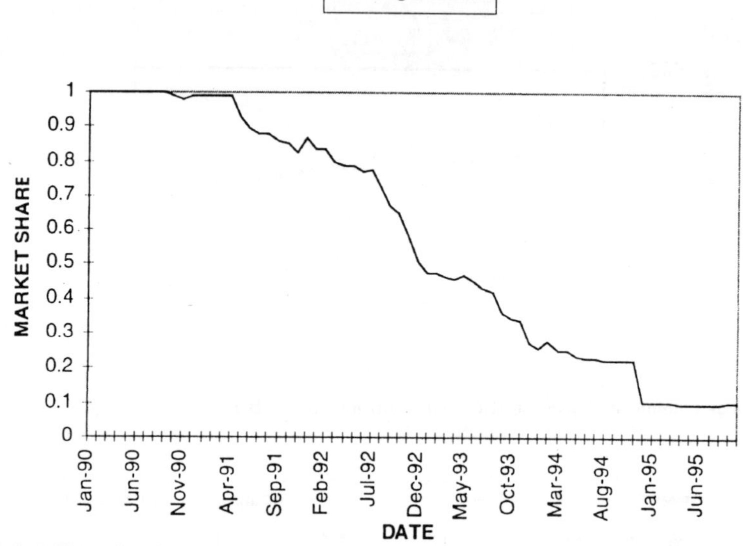

Figure 8.6: BG's market share 1990-1996

Table 8.1 gives BG's market share in the entire competitive market[4]. Since the lowering of the threshold in 1992, competition was extended beyond the contract gas market. This again indicates that BG's market share falls sharply and the fall is halted while some lost ground is regained in 1995 and 1996.

BG undoubtedly faced severe competition as the main entrants exercised their strategic advantage as upstream firms. Nevertheless, they were able halt the dramatic fall in their market shares by the end of 1994 and go to regain some lost ground. BG's ownership of the transmission network in a vertically integrated and the exercise of this strategic advantage was instrumental.

---

4 John Hall Assoc.'s database relating to the competitive market beyond the contract gas market is fairly limited. The figures given in Table 8.1 are obtained from Ofgas (1994 and 1996).

Table 8.1: BG's share in the competitive market, 1990-96

DATE	MARKET SHARE(%)
Oct 90	93
Oct 91	80
Oct 92	57
Oct 93	32
Mar 94	20
Dec 94	9
Apr 95	10
Jun 96	19

4 Conclusion

Concerted attempts were made by the policy-makers to introduce competition and deregulate the UK Contract gas market. However, Ofgas' public pricing policy only prompted BG to invoked its strategic advantage, giving credence to Newbery's (1997) dictum that regulation is essentially inefficient. While the public pricing policy may have a transitory role, the decision to privatise and maintain BG as a vertically integrated entity could have more long-term repercussions on pro-competitive and deregulation policies.

Ofgas between 1995 and 1997 engaged a very active policy towards BG's transportation operations. They adopted a very stringent policy, such as reducing BG Transco's allowable asset base. These policies eventually led to the demerger of BG's operations in 1997 and the creation of separate entities; Centrica Plc which controlled gas supply, while BG Plc concentrated on the transportation and storage operations. This effectively meant that the gas industry became vertically separated. A more comprehensive review of BG's strategic behaviour and non-BG entrant shippers' behaviour is analysed in Dixon and Easaw (1998).

References

Averch H. and L. Johnson (1962), 'Behaviour of firms under regulatory constraint', *American Economic Review*, 52.

British Gas (1987), *Contract Gas After the MMC: A Commentary by BG*, London: British Gas.

___ (1995), *Gas Transportation Charges 1995/96: Revised Prices*, London: British Gas.

Dixit A. (1980), 'The Role of Investment in Entry-Deterrence', *Economic Journal*, 90.

Dixon H. and J. Easaw (1998), 'Strategic Responses to Regulatory Policy: What Lessons Can Be Learned From The UK Contract Gas Market, 1986-1996', *Centre for Economic Policy Research Discussion Papers, forthcoming.*
Gas Act (1986).
Gas Act (1995).
Gaskin D. (1971), 'Dynamic Limit Pricing: Optimal Pricing under Threat of Entry', *Journal of Economic Theory*, 3.
Monopolies and Merges Commission (1988), *Gas*, Cmnd 500, HMSO.
____ (1993), *Gas and British Gas plc*, Cmnd 2314-2317, HMSO.
Newbery D.M. (1997), 'Privatisation and Liberalisation of Network Utilities', *European Economic Review, Vol. 41, pp 357-383, 1997 Presidential Address.*
Office of Fair Trading (1991), *The Gas Review*, OFT.
Ofgas (1987*), Competition in Gas Supply*, London, Ofgas.
___(1989), *Direction on Common Carriage*, London, Ofgas.
___(1992a), *Separation of BG' Transportation and Storage*, London, Ofgas.
___(1992b), *Gas Transportation and Storage: A Discussion Document,* London, Ofgas.
___(1993), *A Pricing Structure for Gas Transportation and Storage: A Consultation Document,* London, Ofgas.
___(1994), *Regulation of the Competitive Gas Market: The Way Forward,* London, Ofgas.
___(1995a), *The Competitive Market Review*, London, Ofgas.
___(1995b), *Price Control Review, BG' Transportation and Storage: A Consultation Document*, London, Ofgas.
___(1995c), *The Competitive Market Review: Decision to Suspend the Requirement to Price According to Published Schedules*, London, Ofgas.
___ (1996a), *Review of the Competition Gas Supply Market Above 2,500 therms a Year*, London, Ofgas.
___(1996b), *1997 Price Control Review British Gas' Transportation and Storage*, London, Ofgas.
Price, C. (1994), 'Transportation Charges in the gas industry' *Utilities Policy* 4(3) pp 191-197.
___(1997), 'Competition and Regulation in the UK Gas Industry', *Oxford Review of Economic Policy*, Vol. 13, No. 1, pp 47-63.

CHAPTER 9

"REGULATORY SPARKS ABOUT TO FLY?" THE ELECTRICITY GENERATION INDUSTRY

MELINDA ACUTT

Department of Economics and Accounting, University of Liverpool, L69 7ZA
Email: M.Acutt@liv.ac.uk

CAROLINE ELLIOTT[1]

Department of Economics, The Management School, Lancaster University, LA1 4YX
Email: c.elliott@lancaster.ac.uk

Keywords: economic regulation; electricity; environmental regulation; game theory; market power; modelling.

1 Introduction

The privatisation of the electricity industry in England and Wales has been contrasted with previous privatisations as it was accompanied by the vertical and horizontal break up of the industry. The aim was to increase competition in both the generation and supply of electricity. The earlier privatisation of British Gas, for example, was not accompanied by a comparable restructuring of the industry, and its consequent post privatisation monopoly power was criticised by many commentators (see, for example, Waddams Price 1998). Nevertheless, despite the break up of the electricity industry, significant market power has remained both in the generation and supply sectors, which has necessitated the use of economic regulation.[2]

The generation sector is also subject to environmental regulation relating to pollutants produced as by-products of generation. The electricity generation sector is a significant contributor to UK emissions of a range of pollutants including sulphur dioxide, nitrogen oxides and carbon dioxide.[3] The electricity generation

[1]Support from the Lancaster University Research Committee is gratefully acknowledged. The authors would like to thank Bob Barker, Martin Hall, Mick Howard, Ronan Palmer, John Whittaker and participants at the 1997 BIEE conference for their very helpful comments, suggestions and time. We would also like to say thank you to Grace Lee for excellent research assistance.

[2] While the generation of electricity has not been continuously subject to regulation, the economic regulator (OFFER) imposed pool price regulation for the financial years 1994/5 and 1995/6, and the generators have also agreed to some divestment of plant. The generators remain constantly monitored by the new Energy Regulator.

[3]Electricity generation contributed 92 per cent of UK sulphur dioxide, 28 per cent of UK nitrogen oxides and 34 per cent of UK carbon dioxide emissions in 1990; as well as contributing to black smoke including particulates (Newbery and Pollitt, 1997).

sector is, therefore, subject to both economic and environmental regulation. However, the aims of the regulators may conflict as the principal economic regulator (the Energy Regulator, previously OFFER) may desire a higher output of generated electricity than that of the Environment Agency (EA). The Environment Agency will be concerned with the positive relationship between industry output and emissions levels, while the Energy Regulator (ER) may associate a larger amount of electricity generated with potentially lower prices for consumers.

In this chapter we examine the implications of a move from separate, simultaneous regulation by the two regulators to a co-operative regulatory regime. We begin by reviewing the relevant literature and outlining our theoretical expectations. We then present our modelling assumptions and results, and conclude with a summary of our findings.

2 A theoretical framework

2.1 Literature review

There is an extensive literature on the optimal methods of regulating firms with market power (See Armstrong *et al.* (1994) for a summary of many of the significant models). Similarly, existing literature describes the implications of differing methods of reducing market failures associated with the pollution that can emerge as part of a production process. This research spans the production of pollution by competitive, oligopolistic and monopolistic firms.[4] However, relatively little work has been undertaken regarding the simultaneous impact of economic and environmental regulation on firms with market power. Baron (1985) describes the impact of economic and environmental regulation on a monopolistic producer of a non-localised pollution externality, studying the implications of both co-operative and non-co-operative policy making by the two regulatory bodies concerned. Fullerton *et al.* (1997) model the implications of alternative economic regulation regimes on the costs of sulphur dioxide compliance under the US Clean Air Act Amendments of 1990 which provide for a system of tradable emissions permits. However, these analyses are not directly applicable to the English and Welsh electricity generating industry. Post privatisation, this industry is oligopolistic, with the largest producers until now being identified as market leaders. Further, Baron and Fullerton *et al.* discuss possible forms of pollution regulation that differ from those applied in the UK.

[4]See, for example, Perman *et al.* (1996) for a review of economic policies for pollution reduction. See Buchanan (1969), Endres (1978) and Dnes (1981) for discussion of pollution control with monopoly production; and Levin (1985) for discussion of pollution control with oligopolistic production.

2.2 *Theoretical expectations*

In this section we outline our theoretical expectations based on the literature to date. In order to simplify the analysis we discuss the regulatory problems at the industry level, and so refer to one generator representing the aggregation of the industry generators.

Figure 9.1 depicts the industry situation faced by both the generator and the two regulators.[5] It is assumed that the marginal costs of production are linear and increasing with output. In the absence of regulation, the generator will choose to produce at the profit maximising level Qg, where marginal revenue is equal to marginal private costs. Due to the imperfectly competitive nature of the industry the generator can charge Pg, making positive profits.

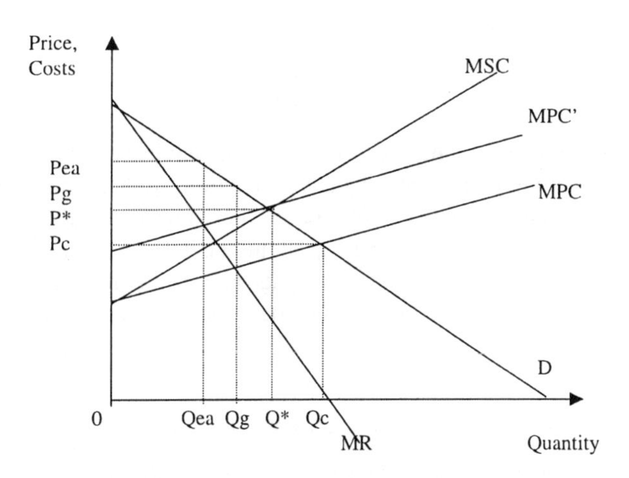

Figure 9.1: The interaction of economic and environmental regulation

An economic regulator is required to protect the interests of the consumers of electricity. If it wishes to maximise the economic welfare of these consumers, ignoring both producer interests and the costs of pollution, then its preferred outcome is at Qc, Pc, where a perfectly competitive generation market is mimicked and consumer surplus is maximised. However, this level of output is higher than the generator's preferred level, with the consequence that pollution levels and the associated external costs of pollution are also higher than under the generator's unregulated level of output.

The Environment Agency (EA) is required to prevent, minimise, remedy or mitigate the effects of pollution of the environment. Pollution abatement can be achieved by regulations requiring the installation of abatement technology, as is

[5] Note that the diagram is for illustrative purposes only and does not fully reflect the theoretical model later developed.

currently adopted in the UK. This policy will lead to a rise in the marginal private costs of generation faced by the generator to, for example, MPC'. If the EA is concerned with achieving the optimal level of pollution, then the EA's preferred outcome is at the socially optimal point where marginal social costs of electricity generation equal the marginal consumer benefits at Q*, P*. However, as Buchanan first noted with regard to a monopoly in 1969, due to the imperfectly competitive nature of the industry, this intervention will lead not to Q*,P*, but to Qea,Pea. This is because the profit maximising generator (in the absence of intervention by the economic regulator) will maximise profits by setting MPC' equal to marginal revenue, not demand. This outcome will produce a lower level of output and a higher price than both the socially optimal level and the generator's original private optimum.

Therefore, we would expect a situation where, with both regulators acting independently with different aims and remits, the outcomes generated by each regulator will be conflicting, pushing the generator in opposite directions.

The following analysis describes a theoretical model that can be used to illustrate the implications of the economic and environmental regulators acting together to maximise a joint objective function, rather than acting non-co-operatively and simultaneously. We model the environmental regulator setting a technology standard, as is currently used in the UK. The model continues to examine regulation at an industry level, its aim being to highlight the welfare implications of a move to Cupertino between environmental and economic regulators.

2.3 *The UK electricity generation industry - modelling assumptions*

A stylised, theoretical model is developed to capture the principal elements of the post privatisation English and Welsh electricity generation industry. As the model examines regulation at an industry level it is equally applicable to jointly profit maximising oligopolists and a monopolist producer. Typically, economic regulation of the largest electricity generators is believed to be limited. However, the imposition of pool price regulation for the financial years 1994/5 and 1995/6 suggests that it is appropriate to include economic regulation of the dominant electricity generators. Further, the Energy Regulator (ER) continues to monitor pool prices and it has been suggested that the generators remain subject to an implicit price cap (Acutt and Elliott 1998). Consequently, the model characterises the ER as interested in maximising a weighted average of consumer surplus and producer surplus.

Given the nature of the output the firm(s) also face environmental regulation. Electricity is generated from a range of fuels including coal, gas, nuclear and a small proportion of renewables. Despite the 'dash for gas' a significant proportion of generation capacity in England and Wales remains dependent on coal. In order to simplify our exposition, we assume that the Environment Agency is concerned only with the level of sulphur dioxide emissions produced with electricity. Emissions are modelled as synonymous with the environmental damage from electricity generation. As in the model of Baron (1985) it is appropriate to model emissions as a non localised pollution externality. The Environment Agency regulates using the BATNEEC rule - best available technology not entailing excessive costs and currently specifies a technology standard (such as the fitting of desulphurisation equipment or the use of clean coal technology) rather than using an emissions tax. In the analysis outlined below we assume that the generators do not have the ability to switch generation capacity to additional gas or nuclear plants as these options are available in the long run only.

The consequences of the game are infinitely lived, and both regulators are assumed to have full information or act as if they had full information. There are two further simplifying assumptions of the model: the model focuses exclusively on the pool market for electricity[6,7]. Also, no distinction is made between business and household consumers. The model aims to compare the consequences of the regulators' co-operating and acting non-co-operatively.

2.4 Model

It is initially assumed that the regulators act non-co-operatively. This is an appropriate assumption to make as (Chesshire 1997, p.12):

> "...the House of Commons Trade and Industry Committee heard during its recent inquiry into energy regulation, there is no effective dialogue between the economic regulators (OFGAS and OFFER) and the environmental regulators (Environment Agency)."

2.4.1 Environmental regulation

When the EA sets a technology standard its objective function can be stated as:

$$Min.Z = X(Q,t) + \delta[A(Q,t) + C(Q,t)] \quad (1)$$

where:
X = sulphur dioxide emissions;
Q = industry output of generated electricity;
t = an index of the 'cleanliness' of electricity generation technology;
A = total abatement costs that derive from fitting a plant with a cleaner technology;

---

[6]The long term contract market is not considered.
[7]Note that the industry is modelled as comprising the generating companies and the regional electricity companies (RECs), but these are modelled as distinct.

C = industry total costs of production;
δ = the proportion of the burden of additional abatement and production costs faced by generators that the EA takes into account, $0 < \delta < 1$.

This objective function is based on BATNEEC, which requires the best available technology not entailing excessive cost. The EA is minimising emission levels and a proportion (δ) of abatement and additional production costs to avoid the firm incurring excessive costs as a result of the installation and use of the abatement equipment. For analytical tractability we assume a single continuous abatement technology, as do Coggins and Smith (1993). Fullerton *et al.* (1997) have modelled discrete technology choices of: fitting fluegas-desulphurisation units; switching to low sulphur coal; or switching production between plants, in their model of the costs of compliance with the US Clean Air Act Amendments of 1990. However, as they themselves note, this means that their model must be solved by numerical rather than analytical methods.

It is assumed that: $\frac{\partial X}{\partial Q} > 0, \frac{\partial X}{\partial t} < 0, \ \frac{\partial A}{\partial t} > 0, \frac{\partial C}{\partial t} > 0$

Minimisation of the Environment Agency's objective function with respect to t results in:

$$\delta\left(\frac{\partial A}{\partial t} + \frac{\partial C}{\partial t}\right) = -\frac{\partial X}{\partial t} \tag{2}$$

2.4.2 Economic regulation

We model the ER as aiming to maximise a weighted sum of a) consumer surplus plus profits, and b) profits, taking account of the costs of abatement incurred by the industry in order to satisfy the EA. The ER then achieves these objectives by imposing an explicit price cap on electricity generation prices. These objectives enable the ER to maximise both current benefits to consumers in terms of consumer surplus, but also allows firms at least to cover costs. If firms are permitted to make profits, this may encourage investment and aid the achievement of dynamic efficiency.

The ER's objective function is, therefore, defined as:

$$Max.V = (1-\beta)\left[\int_0^Q P(Q)dQ - C(Q,t) - A(Q,t)\right] + \beta[P(Q)Q - C(Q,t) - A(Q,t)] \tag{3}$$

where:

P = the industry price faced by the regional electricity companies[8], who then charge final consumers of electricity a price equal to P plus their (regulated) increment to generators' prices;
β = a parameter to reflect the relative weight the regulator places on consumer surplus and producer profits, $0 \leq \beta \leq 1$.

It is assumed that both production and abatement costs rise with output, that is:

$$\frac{\partial C}{\partial Q} > 0 \quad \frac{\partial A}{\partial Q} > 0$$

Maximisation with respect to Q gives rise to equation 4) and the ER will set a price ceiling accordingly:

$$P - \beta bQ = \frac{\partial C}{\partial Q} + \frac{\partial A}{\partial Q} \tag{4}$$

The resulting optimal price ceiling will be set where the generators' price is equal to the sum of the marginal costs of production and abatement when β, the weight placed on producer profits, is zero. As β increases towards unity the price ceiling rises, enabling generators to increase profits. Ultimately, when β takes its maximum value of one, the price ceiling is set at the level where profits are maximised.

When $\beta = 0$:

$$P = \frac{\partial C}{\partial Q} + \frac{\partial A}{\partial Q} \tag{5}$$

and, when $\beta = 1$:

$$\frac{d(PQ)}{dQ} = \frac{\partial C}{\partial Q} + \frac{\partial A}{\partial Q} \tag{6}$$

2.4.3 Implications of the non-co-operative regulatory regime

Minimisation of the EA's objective function is achieved when the technology standard imposed is higher than the optimal level of abatement technology. The optimal level would be achieved by setting the marginal reduction in emissions resulting from the installation and use of 'cleaner' technology equal to the marginal costs of installing and using the 'cleaner' technology. However, the EA will set the technology standard where the marginal benefits of the standard are equal to only a

[8] It is assumed that demand for generated electricity can be represented by a linear inverse demand schedule, $P = a - bQ$.

proportion of marginal abatement and production costs. This is because the EA is only taking a proportion of the costs involved into account, in order to avoid excessive costs.

The generators will, as a result of the technology standard, now face increased marginal costs and so will reduce output. The price at which the generators' output can be sold is constrained by the economic regulator's price ceiling. Assuming that the economic regulator correctly estimates the technology standard imposed by the EA (and therefore its associated costs), when $\beta = 0$ the generators will only be able to cover their costs. Output is lower than in the absence of regulation, and price is greater than in the absence of environmental regulation because costs have increased as a result of the imposition of the technology standard.

2.4.4 Cupertino

Let us now assume that the two regulators co-operate. Their combined objectives can be modelled as maximising a weighted average of a) consumer surplus plus profits, and b) profits, minus emissions. The proportion of generators' profits taken into account must now be agreed between the two regulators. As noted above, the ER may prefer generators to make some positive level of profits. However, the EA would only wish to take account of increased costs resulting from technology regulation if they are 'excessive'.

The joint objective function given a technology standard can be defined as:

$$\begin{aligned} Max.U &= (1-\gamma)\left[\int_0^Q P(Q)dQ - C(Q,t) - A(Q,t)\right] \\ &+ \gamma[P(Q)Q - C(Q,t) - A(Q,t)] - X(Q,t) \end{aligned} \tag{7}$$

where:
γ = a parameter reflecting the relative weight the two regulators place on consumer surplus and producer profits, $0 \le \gamma \le 1$.

Optimal conditions for the maximisation of the joint objective function are:

$$\begin{aligned} P - \gamma bQ &= \frac{\partial C}{\partial Q} + \frac{\partial A}{\partial Q} + \frac{\partial X}{\partial Q} \\ \frac{\partial C}{\partial t} + \frac{\partial A}{\partial t} &= -\frac{\partial X}{\partial t} \end{aligned} \tag{8}$$

Therefore, when $\gamma = 0$:

$$P = \frac{\partial C}{\partial Q} + \frac{\partial A}{\partial Q} + \frac{\partial X}{\partial Q} \tag{9}$$

and, when $\gamma = 1$:

$$\frac{d(PQ)}{dQ} = \frac{\partial C}{\partial Q} + \frac{\partial A}{\partial Q} + \frac{\partial X}{\partial Q} \tag{10}$$

Welfare gains can be shown to derive from co-operative, as opposed to non-co-operative, regulation. From 8) it can be seen that the technology standard imposed will no longer be as severe once the regulators co-operate, as the full marginal costs of abatement and production are now taken into account.[9] This results in increased emissions and so a reduction in welfare. However, the price ceiling imposed on generators can be expected to be less stringent than when regulation was non-co-operative. Consumers benefit from a greater output being produced, and producers will always enjoy positive profits. These welfare benefits counteract the reduction in welfare that results from the higher level of pollution. Overall welfare gains from Cupertino compared to non-Cupertino depend on the size of the increase in consumer surplus plus producer profits compared to the change in the environmental costs resulting from an increase in pollution. However, as pollution levels remain below the theoretically optimal level, there will be an overall welfare gain from Cupertino.

The model results of less than optimal pollution levels may at first appear at odds with 'observed' outcomes of excess pollution. This is due in large part to problems associated with valuing the marginal damage costs of the emissions. Damage valuations, by their very nature, can be estimated only after pollution has been emitted. It is only recently that investment has been put into the valuation of external costs - a whole area of research in its own right (see, for example, Johansson 1993). Hence, it is likely that actual emission levels are higher than the optimal level. In our theoretical model we assume that emissions are directly analogous to damage costs and so valuation is not a problem, and there is no time lag between emission production and valuation. Further, actual pollution outcomes are constrained by jurisdictional boundaries, which can lead a national environmental regulator to take account only of the damage costs incurred in the home country, rather than the full damage costs.

3 Conclusions

A model has been suggested to delineate the basic characteristics of the electricity generation industry in England and Wales. The chapter analyses and compares the implications of the regulators co-operating and acting non-co-operatively. This analysis is timely given that in June 1997 the new UK government announced its

[9] Provided that $\delta < 1$ when the regulators did not cooperate.

intention to review the regulation of the privatised utility companies (Chesshire 1997), as a result of which a 'Green Paper' has recently been published (Department of Trade and Industry 1998). Our modelling work suggests that a potential welfare improvement is available as a result of a move to regulator Cupertino. Under a co-operative regime, whilst welfare is reduced as a result of less investment in cleaner technology by firms, welfare is simultaneously increased as industry output and consumer plus producer surplus will be greater than under non-Cupertino. The overall impact on welfare depends on the relative size of the two welfare changes. However, as the co-operative outcome results in lower levels of production and pollution than the optimal level of both of these outputs, the increase in consumer plus producer surplus will outweigh the increase in pollution costs, making Cupertino welfare improving.

References

Acutt M. and C. Elliott (1998), *Hit and Run Regulation: Regulatory Contestability*, Department of Economics discussion paper EC9/98, Lancaster University.

Armstrong M., Cowan S. and J. Vickers (1994), *Regulatory Reform: Economic Analysis and British Experience*, MIT Press.

Baron D.P. (1985), 'Non-co-operative regulation of a nonlocalized externality', *RAND Journal of Economics*, Vol. 16, pp. 553-568.

Buchanan J.M. (1969), 'External diseconomies, corrective taxes and market structure', *American Economic Review*, Vol. 59, pp. 174-177.

Chesshire J. (1997), 'Regulating the energy sector', *ESRC Global Environmental Change Special Briefing No.1*, pp. 12-13.

Coggins J.S. and V.H. Smith (1993), 'Some Welfare Effects of Emissions Allowance Trading in a Twice-Regulated Industry', *Journal of Environmental Economics and Management*, Vol. 25, pp. 275-297.

Department of Trade and Industry (1998), *A Fair Deal for Consumers: Modernising the Framework for Utility Regulation*, Green Paper CM3898, HMSO.

Dnes A.W. (1981), 'The case of monopoly and pollution', *Journal of Industrial Economics*, Vol. 30, No. 2, pp. 213-216.

Endres A. (1978), 'Monopoly-power as a means for pollution control?', *Journal of Industrial Economics*, Vol. 27, No. 2, pp. 185-187.

Fullerton D., McDermott S.P. and J.P. Caulkin (1997), 'Sulphur dioxide compliance of a regulated utility', *Journal of Environmental Economics and Management*, Vol. 34, pp. 32-53.

Johansson P-O. (1993), *Cost-Benefit Analysis of Environmental Change*, Cambridge University Press.

Levin D. (1985), 'Taxation within Cournot oligopoly', *Journal of Public Economics*, Vol. 27, pp. 281-290.

Newbery D.M. and M.G. Pollitt (1997), 'The restructuring and privatisation of Britain's CEGB - was it worth it?', *Journal of Industrial Economics*, Vol. XLV, No. 3, pp. 269-303.
Perman R., Y. Ma and J. McGilvray (1996), *Natural Resource and Environmental Economics*, Longman.
Waddams Price C. (1998), 'The UK Gas Industry' in Helm D. and T. Jenkinson (eds.), *Competition in Regulated Industries*, Oxford University Press.

CHAPTER 10

HOW WILL ELECTRICITY PRICES IN DEREGULATED MARKETS DEVELOP IN THE LONG RUN? ARGUMENTS WHY THERE WON'T BE ANY REALLY CHEAP ELECTRICITY

REINHARD HAAS, HANS AUER, CLAUS HUBER and WOLFGANG ORASCH

Gusshausstrasse 27-29/E357, A-1040 Vienna, Austria
e-mail: haas@risc.iew.tuwien.ac.at

Keywords: competition; deregulation; electricity; electricity prices; regulation; utilities.

A major intention for guidelines on deregulation of electricity markets is to provide electricity at more competitive (= lower) prices than under monopolies with franchised service areas. Despite little doubt that there will be short term decreases in electricity prices – at least for large customers – it is not clear how long these price reductions will last. This is due to the fact that nobody knows how the structure of the European power industry will develop in the long run. Since there is no final stage of an equilibrium market it is likely that mergers will take place and that small utilities will disappear. The major conclusion of this chapter is that without a new strong regulation – probably on an EC level – electricity prices in the long run may be at the same or an even higher level as under the past regimes of monopolies with franchised service areas.

1 Introduction

The major purpose of world-wide observable efforts to introduce deregulation and competition (D&C) in electricity markets is to provide electricity at more competitive prices than under regulated monopolies with franchised service areas. More precisely, it is expected that, especially for large customers, substantial short term as well as long term price decreases will take place.

This expectation is based on the observation that as a result of the regulation of the power industry, various distortions exist with respect to both signals to consumers regarding the value and the costs of electricity generated at different times and signals regarding the need for new power stations. Moreover, due to the Averch-Johnson-Effect (see Humer (1997)) the investment costs in the past have been higher than they would have been under competition, and, hence, also strongly regulated prices (as e.g., in the U.S.) turned out to be higher than the social optimum.

Other major arguments for lower electricity prices under D&C are:

- Utilities will become leaner (e.g., reduction of bureaucracy). As a result costs will drop.
- Free choice of supplier (for all customers in the long run) will bring down prices close to (short run) marginal costs.

- Prices for different groups of consumers will be more justified and – at least for large customers – drop.
- Unwanted side-effects due to cross-subsidisation of other sectors (e.g., transport, district heating) will disappear.

Yet, despite little doubt that there will be short term decreases in electricity prices – at least for large customers – it is not clear of what magnitude these price reductions will be and how long they will last. Furthermore, most D&C approaches are half-hearted:

- E.g., a rigorous privatisation may lead to a disappearance of the nuclear industry in Western Europe and a strong decentralisation of power generation.
- Since consumer groups will be divided into "eligible" and "captured" customers, where only "eligible" customers (mainly large industrial companies) are allowed to freely choose their supplier, new distortions have to be expected.

Historically, over the past 50 years the development of electricity generation costs has been shaped by one issue: increasing "Economies of Scale". Power plants were constructed in increasing sizes and electricity demand increased hand in hand. Apparently, cheap fossil fuel and subsidised nuclear power stations peaked at about 1400 MW (see Grubb (1997)).

Moreover, until the early 1980s, in most countries investments in new power stations were made under fairly stable regulatory conditions and allowed long term planning horizons.

Yet, about ten to fifteen years ago this pattern began to break up (Grubb (1997)). Besides altered trends in demand, environmental issues, and input costs, the perception that smaller capacities were economically justified (Banks (1996)) led to an end of "Economies of Scale" as well as to an end of the assumption that the power industry is a natural monopoly. Farther on, we will see that this argument must not necessarily hold.

Additionally, these reflections led to a discussion about D&C where the following words were often mixed up:

- privatisation (describing ownership);
- liberalisation (concerning the free choice of supplier);
- deregulation (concerning mainly electricity prices and investment recovery);
- competition (is only seriously possible if there exists a power pool and a separate grid company).

In general, up to now, only very limited knowledge on the effect of D&C on the development of prices in electricity markets is available. Moreover, strategic issues (mergers!) have been largely neglected. Furthermore, various issues of international differences in supply and demand structures have been completely ignored.

Some of these caveats are described in a recent paper by Banks (1996). He refers to the experience in the UK and Norway and concludes that "the market is a wonderful thing, and it should be exploited as far as possible; but it also has its limits." Hunt (1996) mentions that restructuring and privatisation must be kept apart since they are two different dimensions of change. Restructuring is about commercial arrangements for selling electricity, whereas privatisation is a change of ownership. In Jaccard (1995) the changes in the two historic rationales, natural monopoly and public good, are detailed. The information is synthesised to probe the implications for policy with the result that there will remain opportunities and justifications for significant differences between regions and countries in the future.

2 How can competition be brought about? Side-effects and open questions

Beyond the confusion in the discussion, there are some other vital issues with respect to D&C which still have not been clarified up to now. The guideline launched by the EC set the conditions for the degree of competition to be reached by 2003. Furthermore, it provides suggestions on how to introduce competition. Yet, there are a lot of issues which are not discussed in this guideline.

Moreover, the promises of D&C are accepted widely. Yet, surprisingly no corresponding comprehensive underlying analysis has been done so far. Neither rigorous investigations on the actual distortion effects and on the losses of various types of existing regulation have been conducted, nor have comprehensive investigations on the possible side-effects (e.g., on the environment) of D&C been conducted. Some examples for unconsidered side-effects include: the effects on Combined Heat and Power (CHP), on the technical development of the industry such as centralisation versus decentralisation of power generation, and on the role of renewables in electricity generation. Moreover, important components like strategic issues and environmental aspects are largely neglected in this discussion. We are convinced that there will be substantial backlashes with respect to environmental issues and the role of renewables. With liberalisation, it becomes very difficult to reflect external costs other than by direct resource and pollution taxation.

Moreover, the possible effect on "captured" customers respective of remaining supply obligation has not yet been analysed. Without strong regulatory price caps it is obvious that they will pay the entire bill of a liberalised market.

Besides these side-effects open questions with respect to D&C are:

- How will the wheeling fees be set and what impact will they have on the absolute magnitude of traded electricity? The current discussion in Austria takes place within a range of 0.03 ATS/kWh to 0.15 ATS/kWh which is up to 30% of the generation costs.

- How to cope with stranded investments (if serious D&C takes place)? This is especially important for recently constructed hydro power stations.
- Are regulation policies necessary to avoid new distortions with respect to pricing and service for different consumer groups and to protect "captured" customers (e.g., to avoid cross-subsidies from private households to industry)?
- How will differences between countries be taken into account? E.g., will roughly the same number of new power stations be constructed in Austria as in the UK after the liberalisation took place? Or will it be much different due to the different generation structure by fuel?
- How will the structure of the power industry (size and number of utilities, ownership, and control of distribution networks) and the structure of the customers develop in the long run? Will municipal utilities have a chance to survive? To what extent will mergers be economically justified? Furthermore, will mechanisms be introduced allowing associations of electricity consuming households to act like a large electricity consumer with favourable contracts?

3 Why there won't be any really cheap electricity in the long run

Besides the confusing discussion on deregulation, mentioned above, there are some fundamental contradictions which clearly show the limits of a free market. The basic assumption of a liberalised and competitive market is that a customer may choose between different suppliers. Moreover, it is assumed that only investor-owned companies will be able to generate electricity at the lowest possible costs. Hence, for real competition a large number of investor-owned utilities is necessary to bring the electricity prices down to the marginal costs of generation. Yet, this is an illusion. Experiences with deregulation and competition so far tell us that there is no real successful example (see Figure 10.1).

What can be derived from Figure 10.1? Either there is a reasonably high number of utilities which are largely public-owned (e.g., in Norway). In this case it can be seen that Liberalisation (the unrestricted free choice of suppliers) does not necessarily lead to real competition. Or utilities are mainly investor-owned (e.g., UK, U.S.), but if something like deregulation is under discussion they start to merge immediately and ultimately head towards "virtual" monopolies charging monopoly prices, see Figure 10.2.

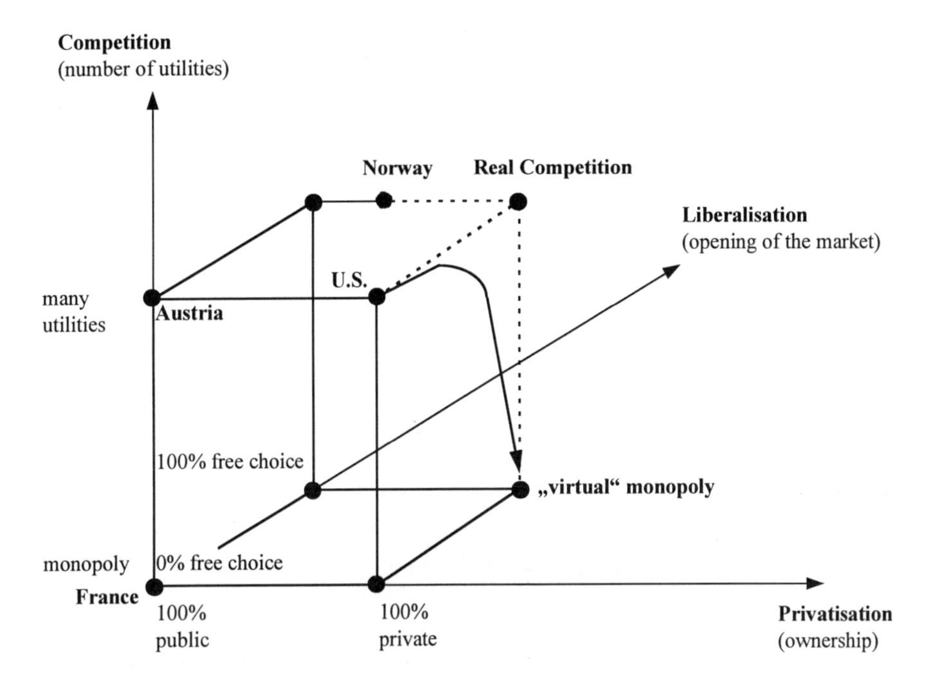

Figure 10.1: No way to real competition without strong regulation?

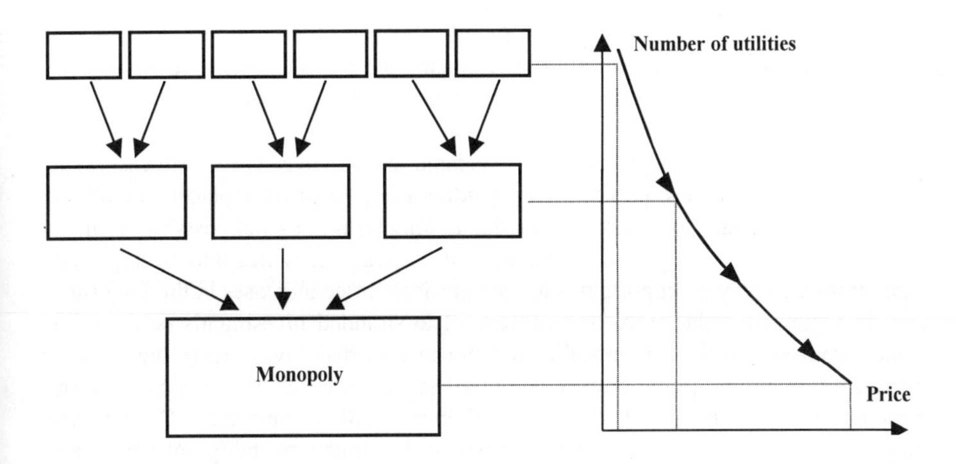

Figure 10.2: Utilities merge and electricity prices increase

Now, the major question for all types of customers in liberalised markets is: who is really expected to sell cheap electricity and at what price? The major candidates are:

- Large utilities with a lot of excess capacity and equity (e.g., EdF)
- Small IPPs.

Summarising the arguments so far, in Figure 10.3 the development of electricity prices over time is shown in principle. The main impact factors in the short run as well as in the long run are indicated (see also Gundersen (1997)).

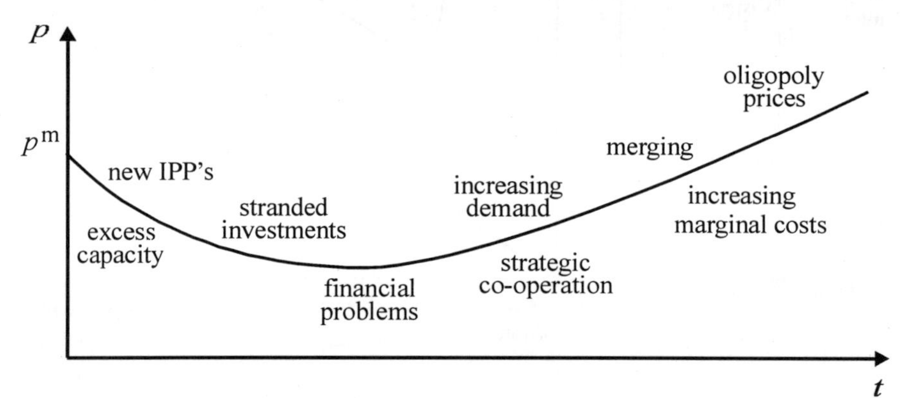

Figure 10.3: Evolution of electricity prices over time (in principle)

In the short run electricity prices will decrease. Not only does excess capacity which will be reduced in the near future exist, but new IPPs will also emerge and contest the market. Since stranded investments have to be recovered during the transition to competition, price decreases will be delayed. But, in the long run, utilities will have financial problems by setting low electricity prices. Furthermore, in order to meet the growing electricity demand and to cover costs with respect to the construction of new power plants (additional as well as replacement of old plants) it is necessary to assess charges due to long term marginal costs (investment costs as well as running costs). Additionally, merger activities and strategic co-operations will play an important role for significant price increases in the long run.

In Figure 10.4 the problem with respect to stranded investments is shown in detail from the utilities point-of-view. Since stranded investments have to be recovered, electricity prices have to be set so that the "true" short run marginal costs will never be reached, and, the price minimum will be delayed. Only in the hypothetical case of a "real revolution" would the short term marginal costs have been reached immediately.

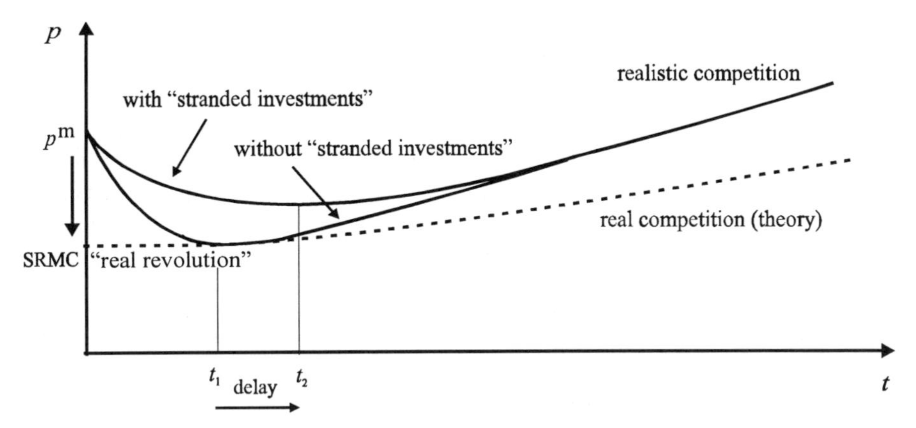

Figure 10.4: Evolution of electricity prices over time taking stranded investments into account

Different scenarios of future price developments are compared in Figure 10.5. Continuing the old price-cap-regulation no price decreases can be expected. By way of contrast, if the electricity market developed towards perfect competition, the short term marginal costs would be reached after a certain transition time and would also hold true in the long run. The development of electricity prices describing actual competition can be expected between these two borderlines. The curve in the middle in Figure 10.5 corresponds to the case with stranded investments in Figure 10.4 until t_3. Whenever the prices increase again after a minimum in t_2 the utilities have to take into account the threat of a new price-cap-regulation after t_3 which will restrict the price level in the long run.

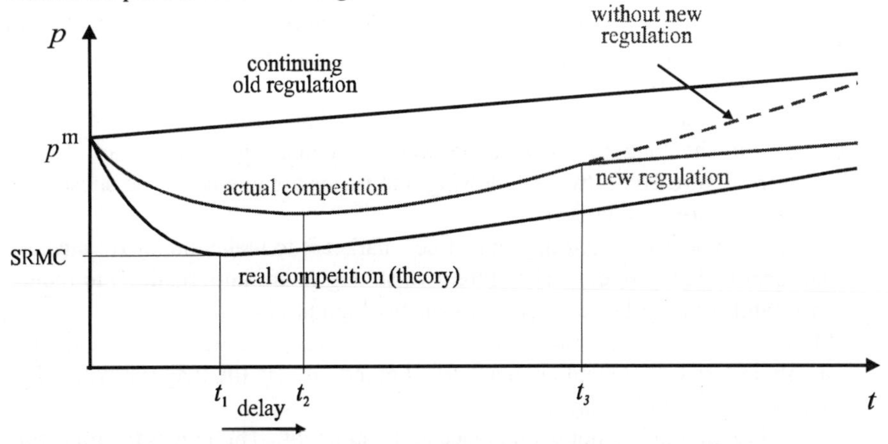

Figure 10.5: Evolution of electricity prices under regulation, competition, and private monopolies

From the utilities point-of-view a short term as well as a long term effect are responsible for the actual price development. On the one hand, in the near future the utilities are forced to decrease electricity prices – at least for large customers – in order to be competitive on the market. As shown in Figure 10.6, compared to the status quo the long term utility losses due to deregulation are indicated by the light area. On the other hand, due to their long term strategic behaviour, the utilities neither have any incentive to reveal their "true" short run marginal costs nor to offer such low prices to their customers. As mentioned above, whenever the prices of electricity increase again (after a – theoretical – minimum at short term marginal costs) there is an increase in the probability that the threat of price-cap-regulation

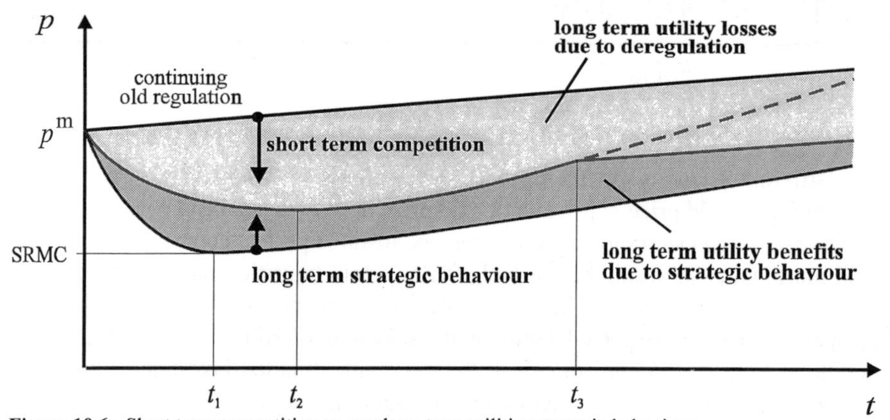

Figure 10.6: Short term competition versus long term utilities strategic behaviour

will take place. Since the regulatory body is incapable of forcing the utility to operate at a specified combination of output, prices, and costs (see also Upadhyaya et al (1997)), this strategic aspect is most important for long term utility benefits (dark area in Figure 10.6).

The basic arguments why electricity prices in a liberalised market won't really be cheaper, in the long run, than under current regulatory regimes can also be derived from Figure 10.6. In the following, the major arguments for no substantial price reductions are summarised:

1. The basic objective of a utility in a "free" market is to make money (to increase the "shareholder value"). Therefore, two strategies are important: (i) to reduce costs, and (ii) to sell electricity at reasonably high prices.
2. Merger activities and strategic alliances in the European electricity- (and gas-) supply industry could end up in a "virtual monopoly" in the long run with a few remaining influential players.
3. A "real revolution" is not taking place at the moment. The longer the transition time towards real competition (see also delay in Figures 10.4 - 10.6) the lower the probability of substantial price decreases. Examples in this context are: stranded investments which have to be recovered; inefficient power plants

which will remain in operation as long as possible; "securitication bonds" which will be introduced.

4. In order to meet the threat of a new price-cap-regulation in the future, long term strategic aspects – as shown above in detail – are very important from the utilities point-of-view, especially since it is also possible that the EC will introduce its own regulatory authority. Regardless of whether regulation is really an effective tool to reduce the price of electricity, these strategic aspects underline the argument that regulation is an endogenous function of the price of electricity.
5. Small flexible IPPs will, of course, contribute to short term competition. But obviously, there is no interest of IPPs to sell very cheap electricity at all as long as other generators are not similarly cheap and as long as there is a high demand for electricity. Hence, electricity may be available at by and large lower prices than today but only if many IPPs really compete will prices go down substantially. Otherwise, it will not be that much cheaper than electricity from a local supplier. This is also a question of the different individual situations with respect to equity and planning horizons as far as investments in new generation capacities are taken into account.
6. The major objective of large utilities with respect to IPPs is to keep these from expanding. Furthermore, the survival of small IPPs in the long run is unlikely, because it is almost always the same story: after a short time successful IPPs sell their assets, of course with high profits, to the large utilities.
7. The magnitude of the transmission fee – compared to the generation price of electricity – will play a considerable role in deciding which amount of electricity will be traded and which degree of competition will be reached. In the case of an expected "high" (political influence, recovery of stranded investments, etc.) transmission fee, real competition – and as a result low electricity prices – seem to be an illusion.
8. The basic idea of unbundling is to break up vertically integrated utilities – which in earlier days acted as monopolies in their franchised service areas in the fields of generation, transmission, and distribution – in order to introduce competition. But it is still unknown, if unbundling will really work in the long run. E.g., in the UK* it is "planned to allow PowerGen and National Power, England's two main non-nuclear generators, to move into electricity supply. [...] Insiders are aware that the decision to allow vertical integration again will reduce competition and impede price cuts for consumers."

* The Sunday Times: "Labour plans to allow generators to buy into Recs", 7 December 1997.

4 Conclusions and outlook

Will the currently discussed D&C approaches fulfil the promises of cheaper electricity? The answer is a little bit ambiguous. On the one hand, in the near future various customers may have the opportunity to freely choose their supplier and to reduce their electricity bills. Naturally, utilities will also have to become less bureaucratic and more customer service oriented. On the other hand, there are a lot of mainly strategic issues still left unconsidered which may curtail the success of D&C considerably in the long run. In the very long run, these issues lead to even higher electricity prices than would develop under the old rules of regulation.

This chapter concludes that, in a completely liberalised market, cheap electricity won't be available in the long run. Hence, primarily regulatory effects will have to take place to ensure real competition and to avoid mergers and strategic alliances. Moreover, from a government's or even the EC's point-of-view a strategic introduction of a price-cap regulation is very important. That is to say, the market in every country has to be observed carefully and if it is obvious that the minimum of the electricity price is reached, a rigorous price-cap-regulation (EC-wide) has to be introduced. Furthermore, to take into account environmental issues and renewables, a strong regulator is vital. This is supported by the fact that since deregulation in the UK was introduced, the staff of the regulatory office OFFER has increased considerably.

The sobering conclusion is that, in the long run, only strong regulation can ensure cheap electricity. The only thing that will change might be the way of regulation.

References

Banks F. (1996), Economics of Electricity Deregulation and Privatisation. An Introduction Survey, *Energy - The International Journal* **21**(1).

Grubb M. (1997), *Renewable Energy Strategies for Europe, Volume II: Electricity Systems and Primary Electricity Sources*, The Royal Institute of International Affairs, Earthscan.

Gundersen B.D. (1997), 'Chancen für die E-Wirtschaft im deregulierten Markt am Beispiel Norwegens', in *Proceedings: Trends und Perspektiven für Energiewirtschaft und Industrie*, ENERCON97-Konferenz, 14.-15. April, Wien.

Haas R., Orasch W., Huber C. and H. Auer (1997), 'Competition versus Regulation in European Electricity Markets', in *Proceedings: European Energy Markets - The Integration of Central European, Baltic and Balkan Countries in the European Energy Economy*, pp. 162-171, July 2-4, Vienna.

Humer H. (1997), 'The Austrian Electricity Sector and the Liberalisation of the European Electricity Market', in *Proceedings: European Energy Markets - The Integration of Central European, Baltic and Balkan Countries in the European Energy Economy*, pp. 138-145, July 2-4, Vienna.

Hunt S. and G. Shuttleworth (1996), *Competition and Choice in Electricity*, ISBN 0-471-95782-8, Wiley.

Jaccard M. (1995), 'Oscillating currents - The changing rationale for government intervention in the electricity industry', *Energy Policy* **23**(7), pp. 579-592.

Upadhyaya K.P., Raymond J.E. and F.G. Moixon (1997), 'The economic theory of regulation versus alternative theories for the electric utilities industry: A simultaneous probit model', *Resource and Energy Economics* **19**(3), pp. 191-202.

SECTION 4

OIL: MARKETS AND REGULATION

CHAPTER 11

WINDOWS ON EXPLORATION: THE ESTIMATION OF OIL SUPPLY FUNCTIONS[1]

G.C. WATKINS

Law & Economics Consulting Group Inc., Emeryville, CA, USA
Email:Campbell_Watkins@lecg.com

Keywords: data; econometrics; modelling; oil supply; OPEC; prices.

1 Purpose

Published analyses of the economics of oil supply in producing countries are generally sparse. The reason is straightforward - lack of consistent and, often even very basic, data. Moreover, the deficiencies are not confined to countries lacking strong data collection agencies, and may be getting worse. It seems US reserve data will now be published only for alternate years.

Yet the issue of whether crude oil supply functions are shifting and if so, in what direction, is at the crux of any assessment of the outlook for the world oil industry. Expectations of more stringent supply, especially in non-OPEC countries, often lie behind the view of those who foresee a return to OPEC dominance and strong price increases.

This chapter attempts to shed light on the issue by estimating supply functions for 41 countries, using publicly available data. These countries cover a wide range of locations including all the major oil producing regions of the world, except the former Soviet Union. They range from established oil producers, mature oil producers, and more recent entries on the production scene.

2 Model specification

The crux of the model framework is to relate reserve additions to the *in situ* price of discovered but undeveloped reserves, and to the passage of time. The price of such potential reserves represents a window within which exploration efforts must pay off. The time variable acts as a surrogate for the latter was intended to measure the net impact of changes in 'prospectivity', resource depletion, cost efficiency and technology. It enables a crucial distinction to be made between shifts along the supply function and shifts in the position of the supply function.

---

[1] This chapter is a much shortened version of a study undertaken jointly with Shane Streifel and published by the World Bank in 1997 (Watkins and Streifel 1997).

Two basic models were specified. The first (Model I) was a straightforward linear function relating reserve additions to the price of undeveloped resources and to time. The second (Model 2) estimated the slope of a notional non-linear supply function over time and then expressed the slope coefficient series as a function of time.

The simple reduced form supply curve given by Model 1 is:

$$RA = a + b(V - I) + ct \tag{1}$$

where:

RA = reserves additions in the given period;
t = time;
V = value of barrel of developed reserves in the ground;
I = development investment per unit of proved reserve.

The sign of the coefficient of time variable, t, is critical. A positive "c" indicates a rightward shift in the supply curve over time: the remaining reserve endowment is expanding. A negative "c" indicates a leftward shift: the remaining reserve endowment is less generous.

A priori, the "b" coefficient of equation (1) is expected to be positive: the greater the spread between *in situ* values and the cost of placing reserves on production, the greater would be potential reserve additions. This spread is an estimate of the price of *undeveloped* reserves, what was referred to earlier as the *window of opportunity for exploration.* Higher prices of undeveloped reserves encourage exploration activity. To test for possible lagged relationships, equation (1) can also be expressed with a one year lag $(V\text{-}I)_{-1}$ and a two year lag $(V\text{-}I)_{-2}$.

The data available for estimating this model are not what one would prefer. There are, of necessity, no reliable data describing the additions to potential, or undeveloped, reserves. Therefore trends in the level of additions to *proved* reserves for a given value of such reserves are assumed to reflect the success over time of all methods for replenishing the stock of potential reserves. The assumption implies a consistent proportionate relation between potential and proved reserves.

Regular patterns of proved reserve appreciation, for which there is evidence[2], provide some support for this notion. But equally one might expect additions to potential reserves to be shrinking faster than additions in proved reserves. If so, the model would tend to understate any underlying resource depletion. There is no solution to this problem without a finer distinction among the reserve and cost data.

For a limited sample of countries, tests were made of the impact of splitting the estimation period between earlier and later intervals. The model was also modified to see whether there was any evidence that the level of reserve prices themselves would affect shifts in supply functions.

---

[2] See ERCB (1969) and Attanasi and Root (1994).

Model 2 is grounded in and adapted from the hypothetical supply curves sketched in Adelman (1990, p29). Again the notion is of a pristine supply function - reserve additions plotted against the (*in situ*) price of reserves. The function is forced through the origin: zero price, zero reserve additions. Also, the function is assumed to be concave upwards, not a straight line, implying diminishing returns for reserve additions as the price increases. A logarithmic transformation for the price term would be a simple expression of this feature. Call the slope of the function 'x'. It is the ratio of the log of the *in situ* price to reserve additions. That is:

$$x = \ln(V\text{-}I+1)/RA \qquad (2)$$

where RA is reserve additions, V is the (real) *in situ* price of reserves, and I is development cost. The '+1' element in (2) is to ensure that when V-I = 0, RA = 0.

The key concern is what happens to 'x' over time. If in this model costs were decreasing via shifts in the *slope* of the curve rather than through movements along the curve, then 'x' will be declining, that is the curve will be shifting downwards to the right. If costs are increasing, then 'x' will be increasing and the curve will be becoming steeper.

For any country 'x' can be calculated as given by Expression (2) for each year. To capture lagged relationships two sets of values for 'x' can be computed, one where (V-I) and RA are contemporaneous, and one where there is a two year lag on (V-I).

Any one set of country values of 'x' by year $\{x_t\}$ generated by equation (2) were simply regressed on a time counter:

$$x_t = b + ct. \qquad (3)$$

Interest focuses on the sign of the 'c' coefficient attached to the time variable. If it were negative, that would indicate more generous supply; if positive, more constrained supply.

Note the difference in interpretation for the time coefficient (t) between the first and second models. In the first model, a positive 't' coefficient indicates expanding supply, in the second model, contraction.

Equations (1) and (2) can also be redefined to reflect the fact for many regions - especially as they mature - reserve additions mainly accrue from reservoir development activity (reservoir extensions and enhanced recovery), that is from reserve appreciation. Here the price variable would be the price of developed reserves, V (not V-I). Results reported in Watkins and Streifel (1997) for the revised specification differ little from those of the main models and are not discussed further here.

3 The data

Model estimation required an extensive effort to gather and assemble data on reserves and reserve additions, production, drilling activity, well costs, development expenditures, operating costs, wellhead prices and other elements for an initial total of 45 countries over a period of time from the mid 1950s to 1994. Insurmountable problems for a few countries led to their deletion from the list, reducing it to 41 - still a very considerable number.

Much of the data (see Adelman and Shahi 1981) gathering for the earlier part of the period of analysis relied on previous efforts by M.A. Adelman. In large measure their series was forward extended, and included some revisions to old data. New data was also developed as dictated by the analysis pursued. It became obvious that the reserve information for many countries contained a lot of anomalies. Adjustments made to eliminate these were quite frequent and relied heavily on judgement.

I emphasise the inadequate coverage and poor quality of some of the basic data. It is true that for some countries the data reliability met good standards - for example the US and Canada. But the degree of adjustment for other countries underlined poor collection procedures and even strong political influences on the numbers, such as the booking of reserves. Moreover, an absence of detail for components of reserve additions and corresponding costs necessitated reliance on total reserve additions as a surrogate for potential reserves - the preferred focus.

In this light, the typically cautionary tenor of researchers' comments about data problems and quality become more emphatic. This situation both dictates and underlines the need for simplicity in model specification. The quality of the data simply will not support any sophisticated supply modelling techniques, and few simple ones. Hence the quite rudimentary nature of the two types of models outlined above.

4 Estimation results

The results were broadly similar for the two versions of the simple models specified. There were disappointments, including the appearance of seemingly perverse effects with higher reserve prices discouraging rather than encouraging reserve additions. This was a suitable reminder that the models doubtless suffer from the omission of important variables needed to adequately explain the supply behaviour of some countries.

Nowhere is this more apparent than in the case of those OPEC countries with very high reserve-production ratios, mainly the Middle Eastern producers. However, paradoxically the poor results for these countries are reassuring. It would be a matter of concern had a model based on competitive responses with countries acting as price takers worked well for OPEC producers marching to a different beat.

For any given Model run, the critical variable expressing shifts in supply functions over time was statistically insignificant for the majority of countries. This means that in most instances there is no evidence of a distinctive shift in oil supply functions, either towards more constrained conditions or towards greater abundance. However, if the preferred results for all Models estimated were combined, 26 countries displayed statistically significant shifts in supply functions. These were pretty evenly split between those in an apparent expansionary phase and those suffering contraction. Some were countries with an especially long production history, such as Burma, Trinidad and the US. Some were OPEC producers - again countries where the model specification of price responses is suspect. Table 11.1 brings together these results.

Tests on a limited sample of countries for differences within the estimation period displayed some evidence of a shift in a more expansionary direction for the latter part of the period of analysis - roughly from 1980 onwards. The same limited sample also revealed some modest evidence that technological innovation, cost efficiency and exploration productivity were stimulated by lower oil prices.

Table 11.1: Summary: countries with evidence of contractionary or expansionary supply conditions

Evidence of Contraction	Evidence of Expansion
Abu Dhabi	Argentina
Algeria	Brazil
Brunei	Brunei/Malaysia
Burma	Cameroon
Canada	Congo
Iran	Egypt
Kuwait	Malaysia
Libya	Norway
Mexico	Oman
Neutral Zone	Venezuela
Nigeria	Syria
Trinidad	United Kingdom
Tunisia	
United States	

5 Implications for oil supply

A gloomy outlook for non-OPEC production is not warranted. Several countries are still in an underlying expansionary phase. Others show no evidence of entering a period of decline. Moreover, there is evidence - albeit based on a limited sample -

that contrary shifts in supply functions may have been mitigated or arrested over the past 15 years or so.

The same limited sample of countries yielded evidence that the lower the price of oil, the greater the stimulus for cost reduction. If so, recent technological enhancements in the upstream petroleum sector - albeit not well measured as yet - are no surprise. It follows that a sustained period of flat prices may not be associated with a steady deterioration in supply from non-OPEC countries.

I emphasise that a leftward shift in the supply function does not mean a country will not continue to add reserves. In a country endowed with substantial proved reserves, significantly more oil resources await finding and development. Reserves accruing from development investment can continue over a long period, as has happened in the US. What leftward movement indicates is that returns from further exploration have started to diminish, and are not offset by continuing technological improvement, or by the opportunity to exploit new plays.

Generalisations all too often gain currency as precise statements. Nevertheless, I suggest the overall results can be characterised in the following broad way. Outside of North America, on balance non-OPEC countries have a rightward (expanding) shifting supply function. Reserves additions will increase even with constant prices. North America is probably moving in the contrary direction - less will be found at a given price. Supply conditions in OPEC countries cannot be depicted by the interaction of conventional supply functions with price; other factors intrude.

References

Adelman M.A. (1992), 'Finding and Developing Costs in The United States, 1945-1986', in Moroney (ed.), *Advances in the Economics of Energy and Resources*, Vol 7, JAI Press Inc.

Adelman M.A. and M. Shahi (1981), 'Oil Development Operating Cost Estimates, 1955-1985', *Energy Economics*, Vol. 11 No 1, pp 2-10, January.

Alberta, Energy Resources Conservation Board (1969), *Reserves of Crude Oil, Gas, Gas Liquids and Sulphur*, Calgary.

Attanasi E.D. and D.H. Root (1994), 'The Enigma of Oil and Gas Field Growth', *AAPG Bulletin* 78 (3), pp 321-332, March.

Bradley P.G. and G.C. Watkins (1994), 'Detecting Resource Scarcity: The Case of Petroleum', IAEE 17th Annual International Conference *Proceedings,* Stavanger, Norway, May.

Watkins G.C. and S. Streifel (1997), 'World Crude Oil Reserves: Evidence from Estimating Supply Function for 41 Countries', *Policy Research Working Paper #1756*, Washington DC: World Bank.

CHAPTER 12

AUCTIONS VS. DISCRETION IN THE LICENSING OF OIL AND GAS ACREAGE

GEOFF FREWER[1]

Chief Economist, Amerada Hess Ltd, 33 Grosvenor Place, London SW1X 7HY
Email: geoff.frewer@hess.com

Keywords: auction; discretion; exploration; oil; production; UK.

1 Introduction

This chapter discusses the relative merits of alternative systems for allocating licenses for oil and gas exploration and production. This subject has been extensively studied in the past, especially in the US context, but it remains of current interest as countries are reviewing and changing their existing systems. The general trend, as with other areas of regulation, is towards more market-based instruments, especially in countries which are privatising and opening their oil and gas industries to international competition. If the alternative licensing systems are viewed on a continuum ranging form direct government control at one extreme, through various degrees of government discretion, to the purely market-based mechanism of auctions at the other extreme, the problem faced by governments is how far to move along the continuum and how much control needs to remain with the state.

In its most extreme form, direct government control consists of assigning all rights and responsibilities for the industry to a government agency, the National Oil Company (NOC). The establishment of NOCs, typically by nationalisation of a previously private structure dominated by multinationals, was a response to a complex set of problems: nationalism, the political power of unions and land owners, the desire to control the rate development of the industry, and of course, the desire to take a greater share of the revenue. However, many of the countries which had taken control of their oil and gas industries in this way have been dissatisfied with the performance of their industries and made moves back towards private sector participation. Countries have turned to market-based systems as a means of improving efficiency, alleviating investment constraints and accelerating development of the industry. Rolling back the frontiers of the NOCs therefore means establishment of new systems for regulating the private sector in these countries.

---

[1] Chief Economist, Amerada Hess Ltd.

Under discretionary systems, such as that operated in the UK, the primary criterion for allocation of the acreage is the level of commitment on the part of the bidder to undertake an exploration programme, typically consisting of seismic surveys and a specified number of exploration wells. Under these systems, technical merit and financial viability are also important criteria since the government would want to be assured that the bidder has the capability to carry out the proposed programme of work, and that the proposal is the best one for developing the industry. The discretionary system gives the government some control over the investment in the industry - they choose which of the proposed exploration programmes are actually carried out - and along with this potential to control comes the need for the government to carry sufficient resources to be in a position to evaluate the proposals. Brazil, for example, is adopting a discretionary approach in opening its industry to private sector competition. Although at the time of writing the Brazillian system has not been finalised, the general approach is for consortia to submit technical assessments of the acreage and fields to Petrobras which will then enter into joint ventures with the selected consortia.

The auction system can potentially be an almost purely market-based system where the government exercises a minimum of discretion or control over the activity of the industry. In auctions the licence is allocated to the highest bidder. Usually the primary criterion is an upfront payment by the bidder (bonus bids), and this early revenue may be a strong attraction to governments, however other variants would take bids for a share of the well-head revenue (royalty bids), or a share of the profits (profit bids). Venezuela, for example, adopted the bonus bid system to auction 18 fields in the 1997 bidding round, raising a total of $2.2bn. In terms of the revenue raised by the government this exercise was regarded as a great success, and industry sources suggest that the winning bids were at a substantial premium above the straight economic values of the assets concerned. In some systems additional criteria may also apply, and in some instances there are special mechanisms to promote the interests of the national oil company, or to encourage the involvement of indigenous companies. Therefore, within this category, the potential for government intervention to promote other policy objectives is not ruled out and in practice the market is allowed freer reign in some countries' systems than in others.

The UK and the US have at different times used a number of methods for allocating acreage, and the objective of this chapter is to draw lessons from the experiences of these two countries. Section 2 sets out the general economic arguments, section 3 examines the data from the UK experience, and section 4 reviews the US experience which has been studied by a number of authors.

2 The economic rationale

Auctions are a theoretically efficient means of capturing economic rent. Given certain assumptions about the competitiveness of the auction, firms will bid their valuation of the acreage. Each block will be allocated to the company with the best exploration plan and therefore the highest valuation; the government will take the economic rent, leaving the firm with the going rate of return on their capital and risk. However, some of the key assumptions required for the auction system to work efficiently are questionable in the context of oil and gas exploration.

Incomplete information is an essential characteristic of the exploration problem - explorers do not know where the oil is, although they have some knowledge indicating where it is likely to be. This is not necessarily an obstacle to the efficiency of the auctions; bidders can form a view about the risk and bid their expected value for the licence. In some instances, bidders will have paid for acreage which turns out to be worthless or have negative value, but if the bidder has a large enough portfolio of licences these failures will be offset by successes where there are larger then expected discoveries and the value of the licence exceeds the bid. If the bidders are risk averse, then they will subtract a risk premium from their bids and the government will fail to extract the optimal level of rent. Moreover, smaller companies with limited financial resources and a small number of interests are more likely to be risk averse, and would therefore tend to be at a disadvantage when bidding against big established players.

A more worrying problem arises where different consortia have different views about the value of the acreage. Clearly, if a company has private information that the block is highly valuable, from experience in operating the adjacent block for example, then it may still be able to secure the licence with a low bid, and the auction system will fail to capture the economic rent efficiently.

Different views about the value do not necessarily arise from private information. Consider for example a situation where all bidders have access to the same noisy data-set, but there are random errors in extracting a signal from the data. Under these conditions there would be a danger that the highest bidder would not be the one with the most accurate estimate of the value of the acreage but instead it would be the bidder who had made the largest overestimate of the value (the winner's curse). In the short run this might inflate the revenues received by the government but in the longer term it could depress industry returns and lead to sub-optimal or cyclical investment levels.

In areas where there has been little or no exploration activity, the lack of information takes an extreme form. It is not that the risks are larger, it is that there is no reasonable basis on which to be able to estimate the risks. Under these circumstance an auction would not work efficiently.

Competition between bidders is essential for the efficient operation of an auction. Ideally, there should be a large number of independent bidders. However, in oil and gas exploration, the investments are large, and the risks are high, and therefore the industry has evolved systems of joint ventures to spread the risks between companies. The relationships between the companies are complex: in some situations a given group of companies will be partners with common interests, in others the same group may be competitors, or alternatively some companies within the group may be customers for services provided by others. Moreover, ownership of infrastructure provides consortia with strategic assets giving significant market power in the vicinity of the infrastructure. As a consequence, the number of consortia bidding for acreage may be quite small even though there are a larger number of companies involved, and this gives rise to concerns about collusion between bidders reducing the level of bids.

With a small number of participants, and significant entry barriers to specific geographical areas, financial constraints may impact on the efficiency of auctions in regulating the industry. If companies are financially constrained, expenditure on licence bids is deducted from exploration and development programmes and therefore tends to reduce activity levels. Clearly, this would not be a consideration if the markets were perfectly competitive - capital would find a way to economically viable projects. However, entry barriers, and control over strategic assets in a core area may not be bankable assets, and the companies holding these assets may have to phase their activity to meet short term financial constraints.

Aside from issues of efficiency, there are a number of other issues affecting the merits of auctions and discretionary systems. Auctions have the advantage of transparency, meaning that it is relatively easy for the government to set up the system in such a way that it would be difficult unfairly to favour any specific group. Where there are risks of corruption, or where industrial lobbying groups can exert powerful influences over the government, or where companies are litigious this transparency is an important consideration.

However, transparency comes at the expense of sacrificing one means of government control over the development of the industry. Exercise of discretion allows governments to choose applicants whose investments are most likely to fit with overall government policy, such as growth of the industry and environmental objectives. In some instances, government discretion over future licence awards provides a powerful means of exerting influence over the activities of companies on an on-going basis.

Auctions also have the advantage, from the government's point of view, of bringing forward revenues. The government still needs to ensure that the overall level of government take is compatible with the attractiveness of the acreage, and is competitive with tax regimes in other countries, so revenue from auctions cannot be considered as an additional source of revenue over and above profits taxes, royalties and so on. However, timing of the receipts may be important, especially when government finances are under pressure.

For companies, these up-front payments can be an entry barrier. This may occur if potential entrants are financially constrained. More significantly, since licence payments may be offset against tax, the auction system would favour companies which have on-going business in the country, and hence a tax shelter, and new entrants would be at a competitive disadvantage.

3 The UK experience

In the UK, licences are applied for in response to invitations (licensing rounds) from the Secretary of State for Trade and Industry. There is no legislated requirement on the timing of licensing rounds; there have been 17 since the first in 1964. In exceptional circumstances, companies may make applications for licenses "out of round". Such circumstances may arise when a discovered field is found to cross the line into unlicensed territory.

The Brown Book describes the criteria for awarding licences as follows: "Licenses are granted at the Secretary of State's discretion. The background against which all applicants are judged is that they meet fully the general objective of encouraging expeditious, thorough and efficient exploration to identify the oil and gas resources of the UKCS. The criteria used to make this judgement are primarily based on the technical capability displayed by the applicants and their plans to explore the blocks applied for, along with environmental considerations".

In addition to the licences awarded under the discretionary system, there have been three auctions. In each of the 4th, 8th, and 9th rounds, fifteen blocks were offered on the basis of cash bids. The auctioned blocks were located in more mature areas, which at the time of the auctions were thought to contain less risky exploration prospects which would command a premium. In addition, the 7th round in 1980/1, allowed allocation of blocks for a flat fee of £5mm per block.

Revenue raised from the auctions is detailed in Table 12.1. The 4th round took place prior to the oil price increase of 1973, and prior to the rapid growth in North Sea oil production, and therefore its proceeds made up a high share of total revenue, 74%. In the 8th and 9th rounds oil prices were high and oil and gas revenues from taxation and royalties were booming, so proceeds from auctions made up a small percentage of total revenue. In proportion to the 1995/6 level of oil and gas tax revenue, the auction proceeds are a small but significant amount. However, it would be incorrect to infer that bids for highly attractive acreage, made under much higher oil prices, would be representative of what might be forthcoming under any future auction.

Table 12.1: Impact of past UK auctions on oil and gas tax revenue

ROUND		MOD £MM	% OF TOTAL REVENUE(1)	REAL 1996 £MM	% OF 1995/6 REVENUE(1)
1971/2	4TH	37	74.0%	274	11.6%
1982/3	8TH	33	0.4%	61	2.6%
1984/5	9TH	121	1.0%	205	8.7%
"SELF-CHOSEN PREMIUM BLOCKS"					
1980/1	7TH	210	5.3%	466	19.8%

(1) TOTAL TAXES AND ROYALTIES ATTRIBUTABLE TO UK OIL AND GAS PRODUCTION

Source: Brown Book

The industry demand for the auction blocks was good: in the 4th round all fifteen blocks were allocated, in the 7th the allocation was seven out of the fifteen offered, and in the 8th it was thirteen out of fifteen. These ratios are much higher than for the discretionary blocks where typically less than half of those offered were allocated. Clearly, the industry shared the government's perception about the attractiveness of the blocks offered for auction. However, the number of parties bidding for the licences was small, between two and three bids per block on average (see Table 12.2), and far fewer companies were involved in the auctions than in the discretionary allocations, raising concerns about the competitiveness of the auctions and their disincentive effect on some groups of companies.

Table 12.2: Bids and outcomes of auction licence rounds

ROUND	DATE	BLOCKS ON OFFER	LOCATION	NO OF APPLICATIONS	NO OF COMPANIES IN CONSORTIA	NO OF BLOCKS APPLIED FOR	LICENCES AWARDED		
							NO OF LICENCES	NO OF BLOCKS	NO OF COMPANIES
4TH	1971/2	421 DISCRETIONARY	NORTH SEA IRISH SEA CELTIC SEA ORKNEY/SHETLAND BASIN	92	228	271		267	
		15 AUCTION	NORTH SEA	31	73	15		15	
		436 TOTAL		123	301	286	118	282	213
8TH	1982/3	169 DISCRETIONARY	NORTH SEA WEST ORKNEY BASIN EAST SHETLAND BASIN UNST FAIR ISLE FORTH APPROACHES BRISTOL CHANNEL	40	94	76	48	63	65
		15 AUCTION	NORTH SEA	20	47	8	7	7	16
		184 TOTAL		60	141	84	55	70	81
9TH	1984/5	180 DISCRETIONARY	NORTH SEA WEST SHETLAND BASIN ROCALL TROUGH FAEROES TROUGH MORECAMBE BAY CELTIC SEA ENGLISH CHANNEL	117	134	107	76	80	75
		15 AUCTION	NORTH SEA	32	52	13	13	13	28
		195 TOTAL		149	186	120	89	93	103

Source: Brown Book

The auctions were effective in generating exploration activity on the blocks, and a number of significant discoveries were made, some of which are currently producing fields (see Table 12.3). However, in addition to the successes there were some notable failures. In the 8th round Amerada paid £10.1 million for 21/15b where there has been no significant discovery, and in the 9th round BP paid £25.5 million for 15/18b which has yielded no significant discoveries. Clearly, the risks and uncertainties associated with these blocks were still substantial even though they were relatively mature areas.

Table 12.3: Significant discoveries on auction blocks

FIELD	WELL	OPERATOR	ROUND	DATE
GALLEY	15/23-1Z	TEXACO	4	1974
	15/24A-4	HAMILTON	4	1990
N CORMORANT	211/21-2	SHELL	4	1974
BERYL A	9/13-1	MOBIL	4	1972
	9/13-18	MOBIL	4	1977
NEVIS	9/13-4	MOBIL	4	1974
BERYL B	9/13-7	MOBIL	4	1975
	9/13A-23	MOBIL	4	1984
	15/25B-3	CONOCO	8	1990
	21/20B-4	SHELL	8	1991
FORTH	9/23B-7	BRITOIL	9	1988

Source: Arthur Andersen Petroview

One important difference between the auctioned and discretionary blocks has been the frequency with which second wells have been drilled on the acreage. In the 4th round nearly all blocks both discretionary and auctioned have had two or more wells. In the 8th round 86% of discretionary blocks and 60% of auctioned blocks have had second wells, and in the 9th round the corresponding numbers are 63% and 33%. Obligations undertaken by consortia under the discretionary system appear to have led explorers to seek out additional prospects on the block when the first well failed. By contrast, in the auction blocks initial failure has tended to shift management's attention to more favourable area or troublesome obligations on other acreage.

In other respects, auctioned and discretionary blocks have performed in a similar fashion. It is not true that companies work faster on blocks which they have paid for (Appendix Table 12.1 compares the elapsed time to first and second wells). There is no clear difference between the number and size of discoveries on auctioned and discretionary blocks. This is surprising since a more mature a block would be expected to yield smaller discoveries but with a higher success rate. (Appendix Table 12.2 shows that the 4th round auction blocks yielded more and slightly larger discoveries, the 8th round auction and discretionary blocks yielded similar numbers and sizes of discoveries, and in the 9th round only one discovery has been made on an auction block).

Comparison between the auction and discretionary blocks in terms of activity shows that in the 4th round there was more exploration drilling per block on the auction acreage than the discretionary acreage, but in the 8th and 9th rounds there was little difference. In all three rounds there was more appraisal drill on the auction blocks, as would be expected in areas which had already been explored (see Appendix Table 12.3). In terms of the proportion of fallow acreage (which has not been drilled for six years), there is little difference between the auction and discretionary blocks (see Appendix Table 12.4).

In summary, the UK experience has been mixed. Auctions were shown to be an effective method for allocating acreage, companies were willing to participate in the auctions, revenue was raised for the government and there has been significant activity and discoveries on the blocks. However, the number of consortia bidding was small enough to raise concerns about the competitiveness of the process. Also, the benefit of following through an exploration programme which the obligations in the discretionary system has effectively promoted, have been lost under the auction system where the frequency of second wells has been lower. Moreover, the risks and uncertainties on the "mature" acreage have still been high as witnessed by the outstanding failures.

4 Lessons from the US

The US has a much longer history of licence allocation than the UK. By the mid 1970s the bonus bid system which was in general use in the US was subject to growing criticism on the grounds that it required too much money up front, thereby reducing the number of companies participating, and increasing the probability of bid-fixing. To some extent these and related concerns were reflected in the academic literature. Hughart (1975) found the system of sale to the highest bidder non-optimal in a game-theory model where one bidder has superior information. Reece (1978) showed the greater the uncertainty, and the smaller the number of bidders, then the lower would be the bids. Ramsey (1980) concluded that if the average number of bidders is less than three or four then lack of competition would have a significant effect on revenue, and the government should slow the rate of issue of licences.

Partly in response to these kinds of criticisms Congress passed the Outer Continental Shelf Lands Act Amendment in 1978 requiring the Department of the Interior to experiment with other systems. The main alternatives were work programme commitments, royalty bids, and profit bids. The experiments in the US yielded a large volume of data reflected in the following studies. Mead et al (1984) and (1986) concluded that the bonus bidding system gives rise to outcomes which are very close to what would be expected by profit maximising behaviour under competitive conditions[2]. However, although the government was raising revenue efficiently, they also noted informational asymmetry where the owners of adjacent acreage had superior information[3], and the tendency for smaller firms to be unable to compete in bidding for "wildcat" leases (i.e., leases for exploration on risky acreage).

Nevertheless, the academic debate continued: Dworin and Deakin (1983) suggested that previous studies had been swayed excessively by the pre-1970 data and that later evidence suggested the purchasers of the leases had made substantially higher returns. Moody (1993) showed that the alternative bidding systems have a significant impact on the degree of participation by smaller companies. However, in subsequent studies, Mead (1994) has re-iterated what is now the accepted position in the US, that pure bonus bidding approximates an optimal system.

In addition to the overall conclusion the studies have highlighted some other relevant issues. Gilley et al (1985) found evidence of financial constraints but that the operation of joint ventures is consistent with competitive behaviour in licensing auctions. Smith (1982) found that risk aversion is an important factor explaining the data, and that larger companies approach risk neutrality. In addition a number of studies have emphasised the importance of the release of pre-sale information. Kretzer (1994), and Porter (1995) noted the apparent delay of exploration decisions until the end of the lease term.

5 Summary

In the US context auctions have been shown to work efficiently. However, analysis of the much smaller sample of auctions in the UK has highlighted two potential shortcomings.

Firstly, the number of bidders in the UK auctions has been low, two to three on average, and this questions the competitiveness of the system in the UK context. Clearly, this raises another important question - why were fewer consortia on average bidding in the UK than in the US despite the international nature of the business? It may be that the industry perceived the UK risks/reward balance as worse because of the high investment costs in the North Sea. Alternatively it may be a function of the timing of the auctions. In the UK the auctions took place in conjunction with rounds of discretionary allocations, and it may be that the capacity of the bidders to undertake the work necessary to make applications and bids was fully utilised. These questions remain unanswered.

---

[2] Watkins (1975) and (1981) has confirmed a similar result in Canada: the ratio of bid to anticipated rent is sufficiently strong to allow efficient collection of ex ante revenue.

[3] A theme also taken up by Hendricks et al (1994).

Secondly, the UK experience has shown that greater activity (more second wells) takes place on discretionary blocks. One of the government's stated objectives relates to "expeditious" and "thorough" exploration and this aim is clearly better fulfilled by the discretionary allocations than by auctions. However, judging what is the economically efficient level of exploration effort - i.e., is the discretionary system leading to excessive activity, or auctions to sub-optimal activity? - is beyond the scope of this chapter.

The US literature has shown empirically that bonus bid auctions work well in the US. However, it has also served a useful function of drawing attention to those circumstances where it would not work so well:

- too few bidders;
- risk-aversion leading to low bids or entry barriers for small players;
- asymmetry of information, arising from operation of adjacent acreage for example;
- financial constraints.

Economic efficiency is clearly an important consideration for the choice of regulatory system but there are also others. Licence allocation is itself only part of an oil and gas tax regime, and a co-ordinated approach is necessary to ensure the appropriate level of overall government take and investment incentives given the attractiveness of the acreage. Finally, governments may wish to use the regulatory system to achieve other policy goals such as the rate of activity in the industry, or environmental protection. Some of these concerns are more directly addressed by discretionary than market-based systems.

References

Dworin and Deakin (1983), 'The Profitability of Outer Continental Shelf Drilling Ventures: an Alternative Approach', *National Tax Journal.*

Gilley et al (1985), 'Joint Ventures and Offshore Oil Lease Sales', *Economic Enquiry.*

Gilley, Karels and Lyons (1985), 'The Competitive Effect in Bonus Bidding: New Evidence', *The Bell Journal.*

Hendricks, Porter and Wilson (1994), 'Auctions for Oil and Gas Leases with an Informed Bidder and a Random Reservation Price', *Econometrica.*

Hughart (1975), 'Informational Asymmetry, Bidding Strategies, and the Marketing of Offshore Petroleum Leases', *Journal of Political Economy.*

Kretzer (1994), 'Exploration Prior to Oil Lease Allocation', *Resources Policy.*

Mead (1994), 'Toward an Optimal Oil and Gas Leasing System', *The Energy Journal.*

Mead et al (1984), 'Alternative Bid Variables as Instruments of OCS Leasing Policy', *Contemporary Policy Issues*.
Mead et al (1984), Competitive Bidding Under Asymmetrical Information: Behaviour and Performance in Gulf of Mexico Drainage Lease Sales 1959-1969.
Mead et al (1986), 'Competition in Outer Shelf Oil and Gas Lease Auctions: A Statistical Analysis of Winning Bids', *Natural Resources Journal*.
Moody (1993), 'Alternative Bidding Systems for Leasing Offshore Oil: Experimental Evidence', *Economica*.
Porter (1995), 'The Role of Information in the US Offshore Oil and Gas Lease Auctions', *Econometrica*.
Ramsey (1980), *Bidding and Oil Leases*, JAI Press.
Reece (1978), 'Competitive Bidding for Offshore Petroleum Leases', *The Bell Journal of Economics*.
Smith (1982), 'Risk Aversion and Bidding Behaviour for Offshore Petroleum Leases', *The Journal of Industrial Economics*.
Watkins (1975), 'Competitive Bidding and Alberta Petroleum Rents', *The Journal of Industrial Economics*.
Watkins and Kirkby (1981), 'Bidding for Petroleum Leases: Recent Canadian Experience', *Energy Economics*.

Appendix

Appendix Table 12.1: Elapsed time to drilling

	DISCRETIONARY			AUCTIONED		
	MONTHS ELAPSED TO			MONTHS ELAPSED TO		
	1ST WELL	2ND WELL	%2ND WELLS	1ST WELL	2ND WELL	%2ND WELLS
TOTAL 4TH ROUND	41	68	98%	79	81	100%
TOTAL 8TH ROUND	20	35	86%	34	37	60%
TOTAL 9TH ROUND	32	28	63%	22	17	33%

Source: Arthur Andersen Petroview

Appendix Table 12.2: Discoveries on auction vs. discretionary blocks

		4TH ROUND	8TH ROUND	9TH ROUND
BLOCKS OFFERED	DISCRETIONARY	421	169	180
	AUCTIONED	15	15	15
BLOCKS ALLOCATED	DISCRETIONARY	267	63	80
	AUCTIONED	15	7	13
NUMBER OF DISCOVERIES	DISCRETIONARY	118	16	12
	AUCTIONED	8	2	1
DISCOVERIES MMBOE	DISCRETIONARY	13,626	392	660
	AUCTIONED	1,772	44	198
AVERAGE DISCOVERY SIZE	DISCRETIONARY	115	25	55
	AUCTIONED	222	22	198
DISCOVERIES/BLOCK	DISCRETIONARY	0.44	0.25	0.15
	AUCTIONED	0.53	0.29	0.08

Source: Brown Book, Amerada Hess estimates

Appendix Table 12.3: Activity on auction vs. discretionary blocks

	DISCRETIONARY					AUCTIONED				
	BLOCKS	WELLS		WELL/BLOCK		BLOCKS	WELLS		WELL/BLOCK	
		EXPLORATION	E&A	EXPLORATION	E&A		EXPLORATION	E&A	EXPLORATION	E&A
4TH ROUND	270	611	896	2.3	3.3	15	58	114	3.9	7.6
8TH ROUND	63	77	100	1.2	1.6	7	10	11	1.4	1.6
9TH ROUND	80	103	119	1.3	1.5	13	16	32	1.2	2.5

Source: Arthur Andersen Petroview

Appendix Table 12.4: Fallow acreage

	SHARE OF BLOCKS CURRENTLY FALLOW	
	DISCRETIONARY	AUCTION
TOTAL 4TH ROUND	66%	58%
TOTAL 8TH ROUND	76%	80%
TOTAL 9TH ROUND	67%	67%
GRAND TOTAL	67%	63%

Source: Arthur Andersen Petroview

CHAPTER 13

UK NORTH SEA OIL PRODUCTION 1980-1996: THE ROLE OF NEW TECHNOLOGY AND FISCAL REFORM

STEVE MARTIN[1]

Allen Consulting Group Pty Ltd, 4th Floor, 128 Exhibition Street
Melbourne, Victoria 3000, Australia
Email: smartin@allenconsult.com.au

Keywords: North Sea oil; prices; production; taxation; technology.

1 Profile of UK North Sea oil production

In the 1970s and the beginning of the following decade, it was widely forecast that UK North Sea oil production would reach an early peak before declining rapidly. Certainly, as Figure 13.1 reveals, the evolution of output in the 1980s adhered to this pattern. In the first half of the 1980s, UK offshore oil production rose quickly, rising from 1.6 million barrels per day (b/d) in 1980 to a peak of around 2.6 million b/d in 1985. The latter part of the decade and the start of the 1990s witnessed falling oil supplies, which had slumped to about 1.8 million b/d by 1991. (A significant decline would have still occurred even without the Piper Alpha episode.)

In the remainder of the 1990s, there has been a revival of oil production, rather than a continuation of the downward trend. A second peak of around 2.5 million b/d was reached in 1995, and forecasts suggest that supplies may even reach a new high of close to 3.0 million b/d by the turn of the century. *So why were the pundits of the 1970s and early 1980s wrong in their predictions of output?*

To gain a better understanding of the dynamics of UK North Sea oil production, it is useful to make a distinction between two categories of oilfield:

- *The 1985 Group*, defined as those fields that were already on stream by 1985 (the year of the first peak in supplies); and
- *New Fields*, which are those fields that began producing oil after 1985.

[1] Steve Martin is currently an Associate at the Allen Consulting Group, Melbourne, Australia. This chapter was written whilst he was a Research Fellow at the Oxford Institute for Energy Studies (OIES). It is based on his study entitled *Tax or Technology? The Revival of UK North Sea Oil Production*, which was published by the OIES in October 1997. Details of the study can be obtained by faxing the OIES on (+44) (0)1865 310527.

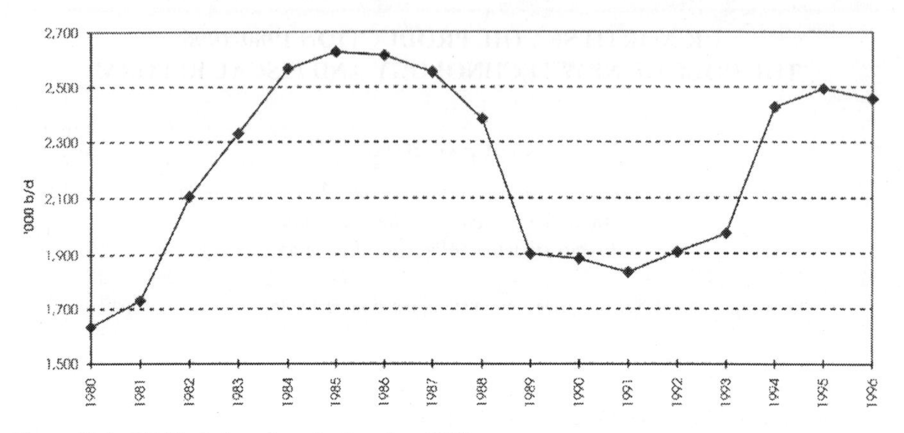

Figure 13.1: UK North Sea oil production since 1980

Figure 13.2 highlights the distribution of UK offshore oil production between these different groups of fields. Since 1985, combined output from *The 1985 Group* has steadily fallen, reflecting the mature nature of these developments. By 1985, many of the fields in this category had already passed their production peaks, and had entered the downward slopes of their production profiles. Meanwhile, the trend in output from the *New Fields* has moved in the opposite direction since 1985. Between 1986 and 1991, however, output from the *New Fields* was only able to offset *partially* the decline in oil production from *The 1985 Group*. As a result, overall UK oil output fell over this period. But the situation has changed since 1991, with increasing supplies from the rising number of *New Fields* being more than sufficient to compensate for dwindling production from *The 1985 Group*, thereby underpinning the revival in total UK oil output.

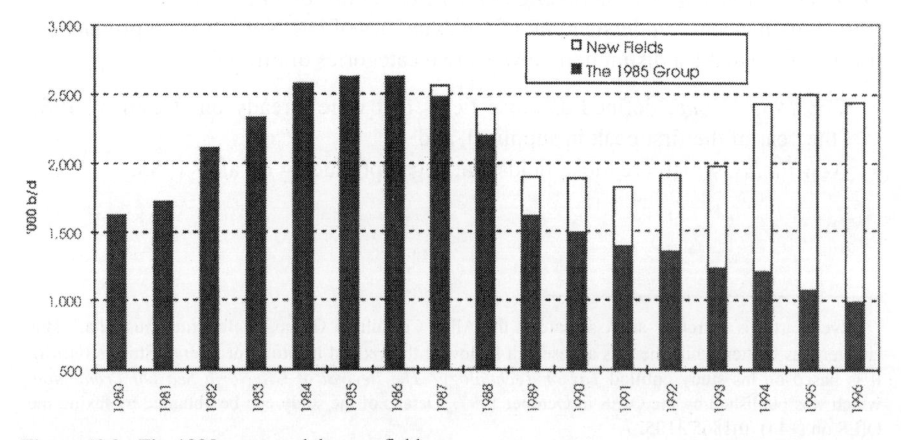

Figure 13.2: The 1985 group and the new fields

There is a marked contrast between the size characteristics of the members of the different categories of oilfields. *The 1985 Group* is dominated by much larger fields (average recoverable reserves of fields in this category are 477 million barrels of oil), whereas the *New Fields* group is characterised by small oilfields (an average of just 85 million barrels, with a large number of fields having reserves of less than 50 million barrels), although there are some important exceptions. This contrast in size structure should not be surprising. After all, given a discovered portfolio of different sized fields, oil companies will tend to exploit the larger, more productive fields before they turn their attention towards the smaller, more marginal ventures.

The emergence of a new generation of oilfields in the North Sea, coupled with the downturn in output from the more mature (and generally larger) oil provinces, mean that smaller fields are having an increasing powerful influence over total UK North Sea oil supply. The composition of annual production since 1985 by size of field is illustrated in Figure 13.3. It is evident from this figure that the influence of oilfields with recoverable reserves of less than 200 million barrels has increased significantly over the period under examination. In 1985, such fields accounted for little more than 10 per cent of annual output.

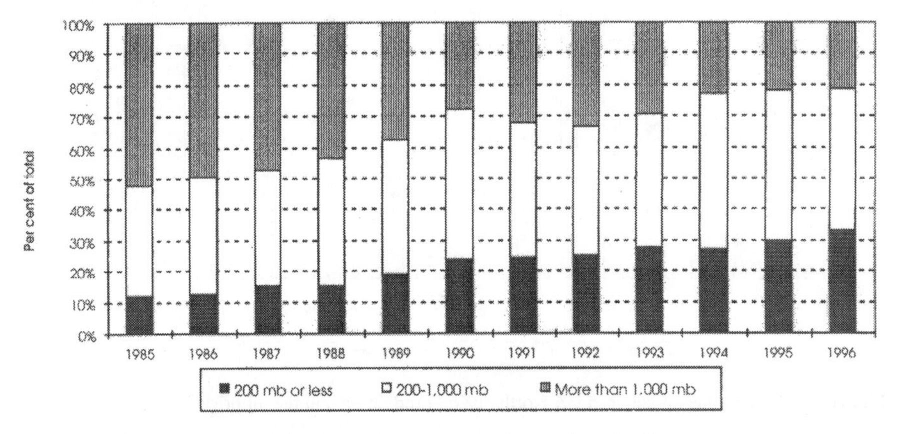

Figure 13.3: Trend in size distribution of annual UK North Sea oil production

However, over the course of the past decade, the sharp increase in the number of new fields that have come on stream (many of which fall within the 200 million barrel or less category) has lifted this share to current levels of over 30 per cent of annual production. This increase has come largely at the expense of very large fields — Beryl, Brent, Forties, Ninian and Piper — which have recoverable reserves in excess of 1 billion barrels. The role played by these very large fields, which dominated overall North Sea oil production at the beginning of the period under examination, is now much diminished. Indeed, their share of total production has slumped from 52 per cent in 1985 to little more than 20 per cent in 1995 and 1996.

Fields with recoverable reserves of between 200 and 1,000 million barrels, on the other hand, have successfully maintained their influence on overall oil supply from the North Sea: their share of total production has fluctuated within a range of 36 and 50 per cent over the past decade, with the underlying trend being upward. Their share has been underpinned in recent years by the Nelson and Scott fields coming on stream.

2 Factors influencing changes in North Sea oil production

So what factors have been responsible for the changes in UK North Sea oil production over the past decade or so? It is clear from the analysis presented above that output has been underpinned by the increasing number of new fields coming on stream. Many of these represent small developments which, in the past, would have been deemed uneconomic. However, for a number of reasons, the economic viability of such ventures has improved in recent years.

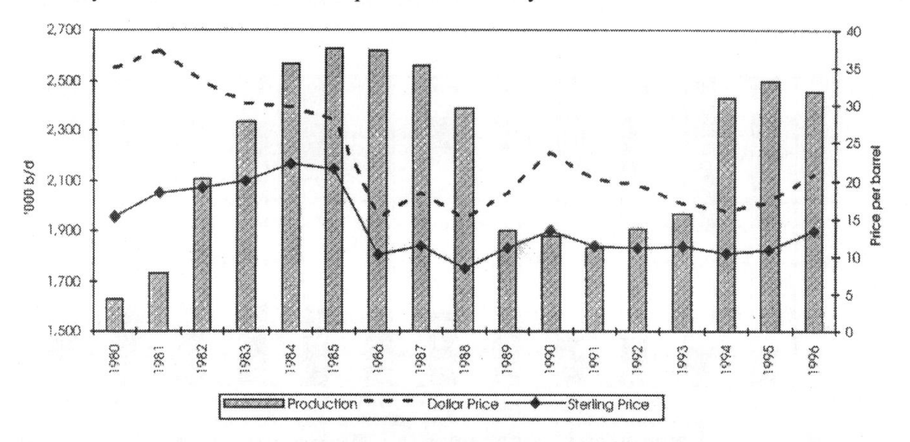

Figure 13.4: Relationship between UK North Sea oil production and the oil price

Economists tend to focus on *prices* to explain changes in output, but there is little evidence to suggest that the resurgence in oil production has been sparked by strong oil prices. Figure 13.4 charts the evolution of UK North Sea oil production against (nominal) oil prices (expressed in both sterling and dollar terms). There is no obvious direct relationship between the variables. Certainly, the rise in output in the 1990s has *not* coincided with particularly strong oil prices. One explanation for this apparent insensitivity is that, in the UK oil sector, capacity tends to be fully utilised, and capacity is developed under *long-run* price and cost considerations. Price increases may stimulate *exploration activity*, which can be expected to yield a larger number of discoveries, but *development* of these discoveries will depend on the oil companies' expectations of future prices and costs over the entire lifespan of the project.

Another reason for the insensitivity of oil production to the oil price is the existence of the *fiscal regime*, which can divorce the international price of oil from the price that is ultimately received by the producer. Past experience has shown that when oil prices have been high, the UK government has responded by *raising* the tax burden on companies. For example, during the high oil prices of 1979-82, there was a tightening of the UK fiscal regime:

- a new tax, the supplementary petroleum duty (later replaced by advanced petroleum revenue tax) was introduced in 1981;
- uplift on capital expenditure was reduced from 75% to 35% at the beginning of 1979;
- the oil allowance (which is a gross production relief that reduces a field's liability to petroleum revenue tax) was curtailed sharply in 1979;
- the rate of petroleum revenue tax (PRT) was increased (from 45 per cent in 1975, it was raised to 60 per cent in 1979, then to 70 per cent in 1980 before reaching 75 per cent in 1983); moreover, an element of pre-payment of PRT was introduced.

In the early 1980s, as oil prices eased back, there were calls for a relaxation in the UK oil taxation system, with many oil industry participants arguing the regime in place was deterring a continuing exploration effort and the development of new fields. Indeed, there was a dearth of Annex B approvals between 1979 and 1982, with only five field development applications being granted over this period. (To put this into some perspective, the same number was granted in 1978 *alone* — the year before the fiscal tightening.)

The UK government responded in the 1983 Finance Act with the following measures, which relaxed the tax burden and improved the economics of new field ventures:

- oilfields with Annex B after 1 April 1982 were exempt from paying royalty;
- the oil allowance of fields approved after 1 April 1982 was fixed at double the level for fields before that date — i.e., 500,000 tonnes for each six-month period, up to a threshold of 10 million tonnes; and
- expenditure associated with exploration and appraisal drilling was made available for PRT relief.

The PRT relief on exploration and appraisal drilling was, however, ended in the 1993 Finance Act. Nevertheless, in the same Budget, the economics of developing new fields were enhanced by the decision to abolish PRT for fields granted Annex B approval after 15 March 1993. Meanwhile, the rate of PRT on existing fields was lowered from 75 to 50 per cent (its lowest level since 1978).

In addition to the relaxation of the fiscal regime, the other important factor that is often cited as underpinning the resurgence in North Sea oil production in the UK is the impact of *new technology*. The oil industry has been to the fore in embracing new technology and methods which have helped to reduce costs and boost the profitability of current and future oilfield developments. As highlighted earlier, many of the fields that have come on stream since 1985 represent small developments, and are often marginal from an economic point of view. The volume of oil from such fields can be quite low (in some cases, estimated recoverable reserves may be as little as 5 million barrels or less), resulting in a relatively short lifespan of the field. Conventional means of developing an oilfield — i.e., by installing a fixed production platform — are rarely cost-effective under such circumstances. Instead, the development of small, marginal fields can often hinge on the ability to find alternative, low cost methods of exploitation. Fortunately, technological progress is providing such methods.

Technology is having an impact at every stage of the upstream oil industry — from discovery to development through to production. A comprehensive account of the various ways in which North Sea oil production has been affected by technological progress would span many pages, but the key developments can be summarised as follows:

- At the exploration stage, 3D (and more recently, 4D) seismic technology has helped in the discovery of new fields and has improved the knowledge of the structure of existing reservoirs. It has made an important contribution to the large number of 'satellite' field developments that have come on-stream in the North Sea in recent years.
- As far as production methods are concerned, the most popular method of exploiting the new fields that have come on stream since 1985 is by means of subsea systems. (This method accounts for around 40% of the total fields developed since 1985). These are wells that are drilled from a mobile rig and completed on the sea bed. Typically, the subsea equipment is connected to a series of 'umbilicals', which carry control signals and derive power supplies from a nearby platform. Production from the subsea well is carried through a flowline or pipeline to the platform. The major advantage of a subsea installation is that it is much cheaper than the construction and operation of a conventional platform.
- An increasing number of new fields are being exploited by means of floating and semi-submersible production systems. These are vessels that replace most, if not all, of the activities that are normally performed by a fixed production platform. They are particularly useful for fields that are remote from existing infrastructure. Their capital costs are often much lower than a fixed production platform, and they can be moved from field to field without the need for potentially expensive abandonment and removal of a fixed installation.

- Advances in technology have also enabled the drilling of extended reach and deviated wells. These are directionally drilled wells, often from an existing platform, which are designed to tap oil from outlying structures at a much lower cost than other options. The wells are often drilled horizontally, thereby providing greater contact with the producing zone, and resulting in a more efficient drainage of the reservoir.

As well as these new technologies, cost-savings are being realised through changes to the *internal organisation of the oil industry*. A joint, industry-wide, approach to the introduction of cost saving measures has been adopted in the form of the Cost Reduction Initiative for the New Era (CRINE), which has resulted in the simplification and standardisation of equipment and working practices.

3 The impact of fiscal reform

To determine the extent to which the increase in UK North Sea oil production can be attributed to lower taxes or to cost-saving technology, a cash flow model has been used. The purpose of the exercise is to assess how the economics of individual field developments after 1985 would look in the absence of the fiscal and technology changes that have actually taken place.

Because there exists no systematic information about the revenue and cost assumptions used by operators at the time the decision was taken to develop individual oilfields, the IRRs calculated in the cash flow model are based on Edinburgh-based consultant Wood Mackenzie's cost and revenue estimates and forecasts (as published in June 1996) — although it should be recognised that these may differ markedly from those actually used by operators during the development process of some fields. The tax burden of fields is estimated using the author's own simplified taxation model.

In the analysis, it is assumed that fields are developed if their calculated *internal rate of return (IRR) exceeds a certain target level*. In reality, of course, there may be other criteria by which projects may be judged. For example, oil companies may look at the absolute size of the net present value of a project, or more sophisticated profitability indices may be considered. Corporate considerations may also be taken into account, such as whether the proposed developments fits into the company's overall investment strategy, or whether it helps them achieve certain corporate aims, such as increasing market share. However, the great advantage of using the IRR method is that it enables a *systematic* analysis of a large number of fields.

For much of the analysis, a target IRR of 15 per cent is assumed. Of course, in reality, this will vary between operators and between projects. (In the study, the sensitivity of the results to a range of different IRRs is examined.) In some cases, fields have actually been developed in spite of IRRs that are apparently *lower* than 15 per cent, which may reflect a number of factors. For example, reserve/output projections may have been downgraded after a field has come on stream, which has the result of lowering revenue forecasts (and therefore the IRR of the field).

Another reason why fields with apparently unattractive IRRs reach the development stage is that the cash flows are calculated on a *stand-alone* basis, which sometimes does not fully reflect the economic attractiveness of a field at the *corporate* level. For example, transportation costs are deducted from the cash flow even though the same company operating the field may also have a financial interest in the pipeline, in which case the payment for the use of the pipeline represents an intra-company transfer cost.

By changing key parameters in the taxation model, it is possible to isolate the impact of changes to the fiscal regime on each of the *New Fields'* IRR. For example, to gauge the influence of the key changes of the 1983 and 1993 Finance Acts on the decision to develop the *New Fields*, we can compare the 'base case' IRRs (i.e, those calculated under *existing* fiscal and cost arrangements) with those IRRs calculated by the cash flow model in the absence of the key fiscal relaxation measures (but assuming that oil companies still benefit from new technology) — in other words:

- royalty is *not* abolished for fields with Annex B approval after 1 April 1982;
- the oil allowance is *not* doubled in 1983 and therefore remains at 250,000 tonnes for each six-month period, with a cumulative maximum of 5 million tonnes; and
- the rate of PRT remains at 75 per cent after 1983 and is *not* reduced to 50 per cent from mid-1993. Moreover, it is assumed that fields with Annex B approval after April 1993 are also subject to PRT.

Table 13.1: Impact of 1983 and 1993 fiscal changes on IRR of selected fields

	Base case	Without 1983 and 1993 fiscal changes
Alba	16.6%	13.8%
Brae East	15.8%	12.9%
Bruce	16.0%	12.8%
Dunbar	15.9%	12.4%
Eider	18.9%	15.0%
Miller	16.5%	12.6%
Strathspey	16.0%	11.9%

Table 13.1 highlights those fields where the fiscal changes have apparently influenced the development decision — i.e., *without* the fiscal changes (but given actual technology), the IRRs are below the target of 15 per cent, but *with* the fiscal changes (the base case), the target is exceeded.

4 The impact of cost-saving new technology

A similar approach can be used to consider the technology side of the equation. The cash flows of those fields that are exploited by unconventional means — i.e., without a fixed production platform — are adapted to reflect the situation that would exist in the absence of new technology. Clearly, this is a difficult modelling task since data on the costs of the alternative technologies are not available for the actual fields. To avoid this problem, the approach that has been chosen is to assume that, in the absence of the new technology, the only means of exploitation of an oil reservoir is through the installation of a conventional production platform. Thus, for all those fields that are actually exploited by other means, costs are re-estimated on the basis that a conventional platform was used instead. (The basis of this re-estimation is the costs of those 39 UK North Sea oilfields that *have* been exploited with production platforms.)

As before, by comparing IRRs both with and without the new cost-saving techniques, it is possible to identify those fields where new technology appears to have been influential in their development. These are identified in Table 13.2.

Table 13.2: Fields affected by cost-saving production technology

	Base case	Without new technology
New fields (oil)		
Angus	135.9%	negative
Arbroath	34.7%	14.5%
Birch	21.0%	-2.8%
Blenheim	236.8%	negative
Chanter	24.9%	negative
Columba D	75.6%	negative
Donan	63.0%	negative
Fife	206.1%	negative
Glamis	102.7%	negative
Leven	134.6%	negative
Medwin	38.6%	negative
Moira	33.2%	negative
Ness	910.0%	negative
Osprey	25.2%	11.3%
Pelican	27.2%	5.7%
Petronella	87.8%	2.2%
Strathspey	16.0%	5.2%
The 1985 group		
Deveron	207.6%	negative
Argyll	55.0%	14.3%
Scapa	46.1%	7.9%
Duncan	21.3%	negative
Innes	16.0%	negative
New fields (gas)		
Beinn	148.0%	negative
Ellon	47.6%	negative
Amethyst	19.9%	2.4%

5 Comparing the fiscal and technology effects

By pulling this analysis together, it is also possible to identify those fields whose IRRs, according to the model, would still exceed the 15 per cent target in the absence of *both* the fiscal relaxation and cost-saving new technology. Such fields include Scott and Nelson, which have had a great impact on UK North Sea oil production in recent years, jointly contributing over 300,000 b/d since 1994. What does all this mean for production? Figure 13.5 has four lines, showing:

- the actual evolution of production;
- the production profile if we exclude those fields whose development has been determined to have been triggered by the main fiscal changes of 1983 and 1993;
- the production profile if we exclude those fields which owe their development to cost-saving new technology; and
- the production profile if neither new technology nor the 1983 and 1993 fiscal changes had occurred.

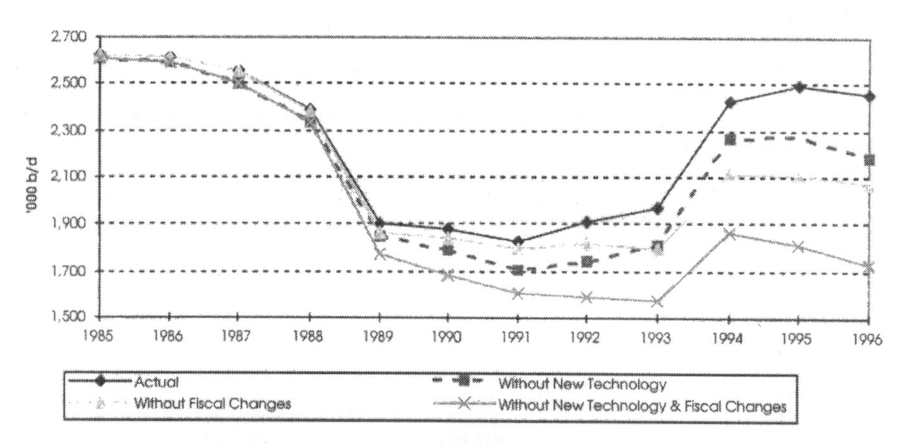

Figure 13.5: UK North Sea oil production under different scenarios

This figure demonstrates that in the absence of *both* new technology and the key fiscal changes, UK oil output from the North Sea would have been 720,000 b/d lower in 1995 (the year of the second peak) than was actually the case.

Although, according to the model, the impact of new technology has triggered the greater *number* of fields, many of these represent small developments, and their contribution to overall production levels has been relatively modest. Without the *New Fields* developed as a result of new technology, output in 1995 would have been approximately 220,000 b/d lower than actual levels. (It should be noted, however, that this does *not* take into account the impact of new technology on improved oil recovery from established fields.)

The fiscal changes appear to have had a stronger impact on output as the 1990s have progressed. Even though the development of a relatively small number of fields can be ascribed to the fiscal relaxation, some of the fields in question are relatively productive and, by 1995, they accounted for around 400,000 b/d of total UK North Sea oil production.

Another way to express the impact on production is to consider the composition of the *increase* in output between 1991 (the trough year) and 1995 (the recent peak year). The net increase in production was 655,000 b/d, with a decline of 312,000 b/d from the old fields (i.e., those in *The 1985 Group*) being more than offset by a contribution of 977,000 b/d from the *New Fields*. This 977,000 b/d increase can be attributed in the proportions shown in Figure 13.6.

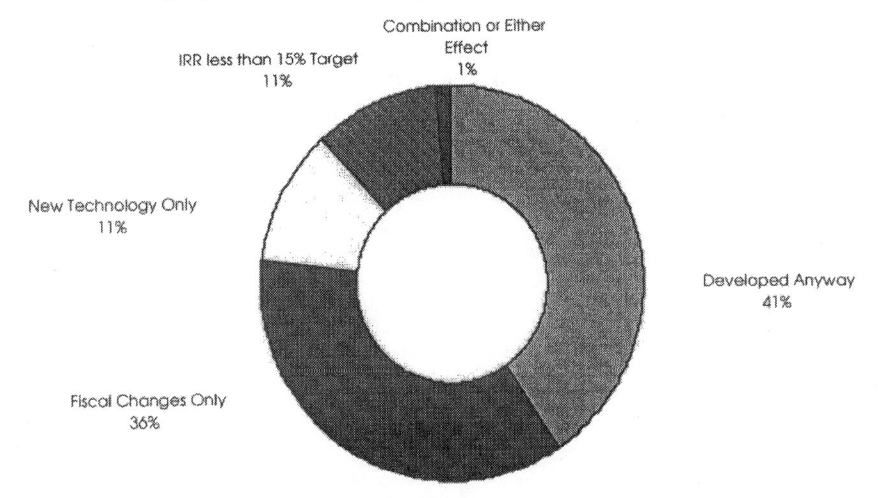

Figure 13.6: Breakdown of increase in new fields production (between 1991 and 1995)

As the figure illustrates, the major contribution to growth (41 per cent) came from fields that the model suggests would have been commercial anyway. Neither fiscal changes nor technological progress were necessary to move them above the 15 per cent threshold. Both factors, however, contributed to the profits of the companies involved. Of these fields, Nelson and Scott accounted for the lion's share. Fiscal relaxation comes next in order of importance (36 per cent). The contribution of new technology to the increase in production between the production trough of 1991 and the peak of 1995 is modest and is about as large as the contribution from those fields that have been developed even though their IRR *ex post* turned out to be less than 15 per cent.

6 Impact of new technology on established fields

There are a number of reasons why the above analysis underestimates the role played by new technology. For instance, the model does not take into account those fields that may only been developed because advances in exploration technology have enabled the discovery or appraisal of that field. Meanwhile, new technology has helped to boost output from older, established fields as a result of improved oil recovery techniques. Over time, there has been an increase in the estimates of total recoverable reserves from *The 1985 Group* of fields, which has helped to maintain oil production from these fields at higher-than-expected levels. An assessment of the impact of new technology on *The 1985 Group* of fields (which is measured in terms of the increase in estimates of recoverable reserves due to the reclassification of some of the original oil-in-place from uneconomic to economically recoverable as a result of technology-induced cost reductions) suggests that, since 1985, recoverable reserves have increased by around 2.7 billion barrels of oil. Once this is taken into account, the role of new technology in the UK North Sea oil sector is greatly enhanced.

CHAPTER 14

EXPLORATION AND DEVELOPMENT INVESTMENT AND TAXABLE CAPACITY IN THE UKCS UNDER DIFFERENT OIL AND GAS PRICES

PROFESSOR ALEXANDER G KEMP
University of Aberdeen, Department of Economics,
Edward Wright Building, Dunbar Street, Old Aberdeen AB24 3QY, Scotland
Email: A.G.Kemp@abdn.ac.uk

LINDA STEPHEN
University of Aberdeen, Department of Economics,
Edward Wright Building, Dunbar Street, Old Aberdeen AB24 3QY, Scotland
Email: pec078@abdn.ac.uk

Keywords: gas; North Sea oil; prices; profitability; taxation.

1 Introduction

In 1997/1998 the North Sea oil and gas industry has been the subject of a Government fiscal review. This period coincided with a major fall in the oil price which led to the announcement in September, 1998 that the Government was not proceeding with the proposed changes outlined in the March budget. The tax review produced a lively debate on the question of the taxable capacity of the North Sea industry. This is the subject of the present chapter.

2 Economic rents at development and exploration stages

The taxable capacity in the petroleum industry is usefully measured at the key decision points of (a) field development, and (b) exploration. Taxable capacity is directly linked to the economic rents expected at these decision points. At the field development stage these rents may be measured by the net present value (NPV) at the investor's discount rate or cost of capital.

At the exploration stage the economic rents are measured by the expected monetary value (EMV), again at the investor's discount rate. Reflecting the current realities of the North Sea, a distinction is drawn between the risks at both the exploration and appraisal stages. The size and costs of fields which may be discovered are also uncertain at this stage and this has to be reflected in the modelling. The investment situation facing the explorationist is shown schematically in Figure 14.1.

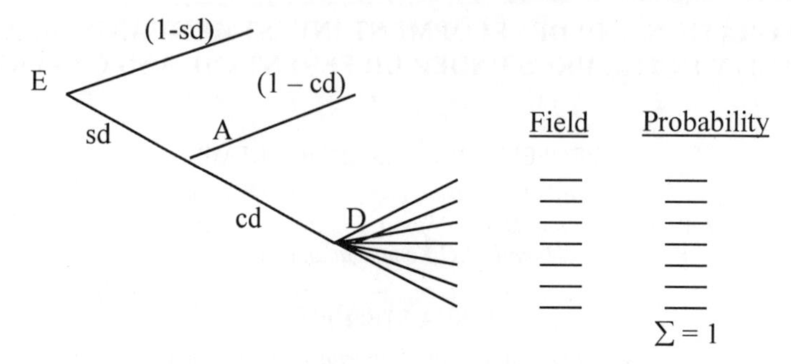

Figure 14.1: Schematic representation of investment situation facing explorationist

In the circumstances described above the $EMV = p_{cd} \left(\sum_{i=1}^{i=n} prob.NPV_{fi} \right) - E - psdA$

Where p = probability, sd = significant discovery, cd = chance of potentially commercial development, E = exploration costs, and A = appraisal costs.

3 Methodology and assumptions for analytical work

3.1 Data

The analysis of development economics was conducted on a set of 40 fields which have been identified by the licensees for serious consideration for development. The key field data on production, investment and operating costs for those and all existing producing fields and those under development were validated by the operators concerned.

For the analysis of exploration economics the period 1984–1994 was chosen for detailed investigation. The expenditures on exploration and appraisal and the numbers and dates of significant discoveries were taken from the official DTI *Energy Report*, Vol. 2, 1997. From the field database noted above the fields discovered in the period 1984–1994 inclusive, and those which were either developed or being seriously contemplated for development by September, 1997 were identified. This enables both the significant exploration and appraisal success rates to be ascertained. The exercise was conducted separately for the Southern North Sea and the rest of the UKCS. The results are shown in Table 14.1.

Table 14.1: Exploration and appraisal risks in the UKCS from 1984 -1994 experience

1. **SNS:**

Exploration success (significant discovery):	30.8%
Appraisal success:	56.8%
Potential commercial success:	**17.5%**

2. **Rest of UKCS:**

Exploration success (significant discovery):	19.3%
Appraisal success:	44.77%
Potential commercial success:	**8.64%**

3. **All of UKCS:**

Exploration success (significant discovery):	22.3%
Appraisal success:	49.3%
Potential commercial success:	**11%**

Notes: Exploration success based on 1984-1994 data
Potential commercial development by September 1997

From the field data probability distributions of potentially commercial fields according to size were determined. They are shown in Table 14.2.

Table 14.2: Probability distributions of field sizes in UKCS

Reserves: Southern North Sea	
BCF	**Probability**
0 – 75	0.429
75 – 125	0.071
125 – 275	0.357
275 – 475	0.0476
475 – 775	0.0476
775 +	0.0476
	$\Sigma = 1$

Reserves: Rest of UKCS	
MMBBLS	**Probability**
0 – 10	0.125
10 – 20	0.09375
20 – 50	0.3281
50 – 100	0.21875
100 – 180	0.1094
180 – 320	0.0625
320+	0.0625
	$\Sigma = 1$

Note: BCF equals billion cubic feet, and MMBBLS equals millions of barrels

From the field data a representative field was chosen for each band. As far as possible fields with reserves around the mean value in each band were chosen. The chosen fields were also representative in terms of their expected returns in relation to the portfolio of developments since 1984.

The prospective pre-tax and post-tax returns on the projects with the associated tax takes at the field development stage were calculated with a deterministic financial simulation model. The share of the rents taken in taxation was measured by the present value of the field lifetime tax payments as a proportion of the pre-tax field NPV. At the exploration stage the financial modelling calculated the pre-tax and post-tax EMVs and the associated tax takes to measure the expected economic rents and Government share. A range of discount rates was employed. For the Southern North Sea a cycle time from first exploration to first development of 4 years was assumed. For the rest of the UKCS the period was 5 years. The ratio of E : A expenditure was taken as 1 : 2. The average cost of an exploration well was assumed to be $8 million in the Southern North Sea and $16 million in the rest of the UKCS.

The investor was assumed to be in a tax-paying position at the time of all the investment expenditures. Thus he is able immediately to utilise the exploration and development tax allowances. The analysis was undertaken for oil prices of $12, $15 and $18 per barrel for oil and 10 pence, 13 pence and 15 pence per therm for gas. All are in constant real terms.

4 Results

4.1 Development stage

The pre-tax economic rents for the 40 fields under the 3 price scenarios at 10% cost of capital are shown in Figure 14.2 (Charts 1–3). The NPVs are shown in relation to field development costs. There is seen to be a very wide range of returns. Under the $18/15 pence price scenario the NPVs for 38 of the projects are positive. Under the $15/13 pence scenario the number is 37, but under the $12/10 pence case it falls dramatically to 22.

It is also seen from the charts that many of the projects are quite marginal. In situations of capital rationing, projects with a prospective positive but small NPV may not be acceptable. In such circumstances a higher project screening rate could be employed. If such a rate were 15% in real terms it was found that the number of projects with positive returns became 34 under the $15/13 pence scenario and 20 under the $12/10 pence one. Many of the projects offered only marginally positive returns under all these scenarios.

The post-tax returns under the current tax system are shown in Figure 14.2 (Charts 4–6). Although the returns are obviously reduced, for an investor already in a tax-paying position under the $18/15 pence scenario the number of fields having positive NPVs remains the same as pre-tax. Under the $15/13 pence and $12/10 pence scenarios the tax reduced the numbers of fields with positive NPVs to 36 and 20 respectively. With the 15% hurdle rate the numbers of fields with positive post-tax NPVs remain at 38 under the $18/15 pence scenario, falling to 32 under the $15/13 pence case, and 20 under the $12/10 pence case. Because the tax increases the number of fields coming under the marginal category (see Figure 14.2, Charts 4

–6) the adverse effect on investment incentives could be greater for investors faced with capital rationing.

The shares of the economic rents collected by the tax system are shown in Figure 14.2 (Charts 7–9). Despite the flat nominal rate there is a discernible tendency for the tax system to be regressive with respect to both oil/gas price and development cost variations. This results from a combination of the cost of capital and the 25% declining balance allowance for investment costs. In some cases where pre-tax profitability is low the tax take exceeds 100% of the economic rents. The regressive relationship with development costs is not highlighted distinctly in the charts because the capital intensity of different North Sea projects now varies considerably. The relative importance of project operating costs varies substantially depending on whether third party tariffs or lease costs are involved.

Under the $12/10 pence price scenario the effective tax takes in a substantial number of fields are negative. The present value of the investment reliefs (available to an existing taxpayer) exceeds the present value of any tax payments. Such projects are not viable before tax.

In the substantial number of cases where the pre-tax profitability is quite acceptable on the criteria used it is seen that the share of the economic rents collected by the state is moderate, frequently being in the 30%-40% range. While findings of this order have been used to support the view that the tax take in the UK is comparatively low by international standards, the present study also shows that the taxable capacity on many fields is low, especially under 1998 price conditions. In such circumstances the present system can produce effective tax takes well in excess of the nominal rate.

4.2 Exploration stage

The pre-tax and post-tax EMVs facing an explorationist in the Southern North Sea are shown in Figure 14.3. Under the 15 pence price the expected returns are positive with discount rates of up to just over 10%. With the 13 pence scenario the EMVs are positive with discount rates up to 7.5% and in the 10 pence case with rates up to 3%.

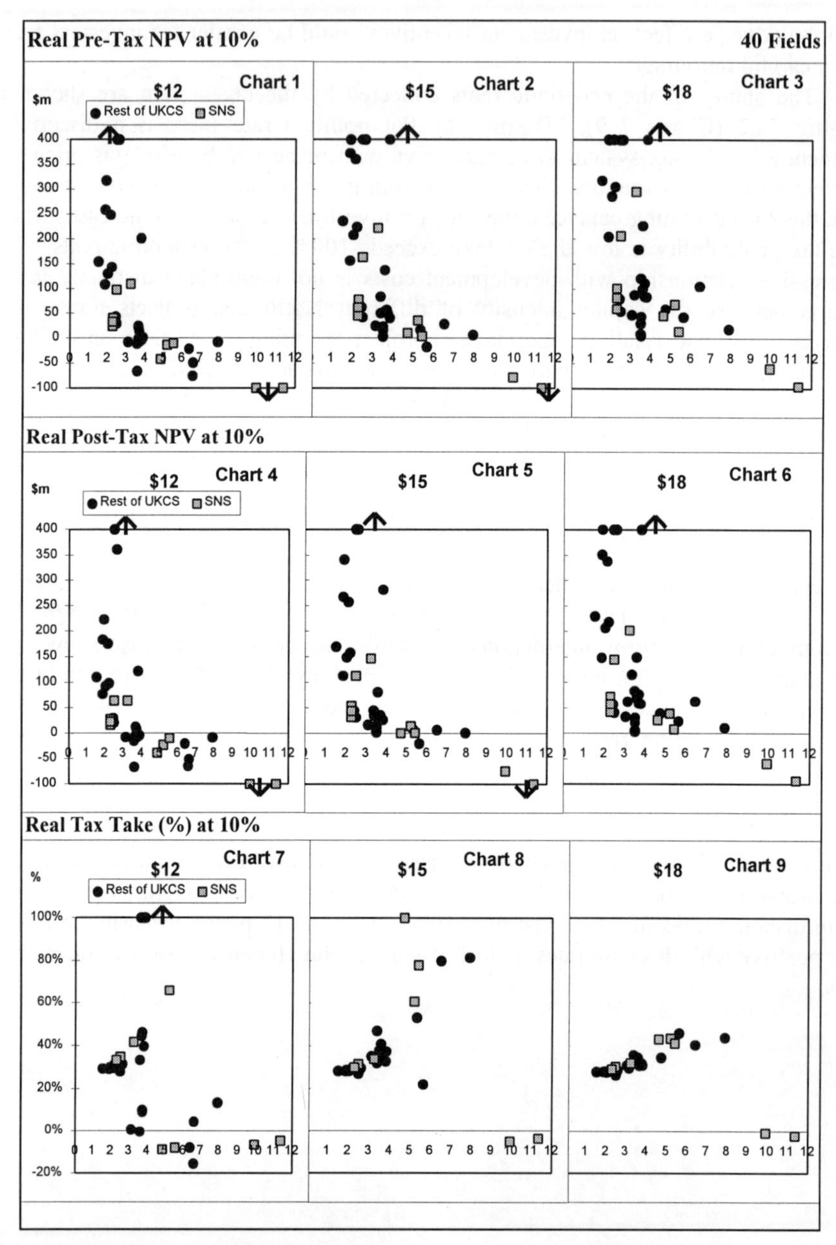

Figure 14.2: Development costs per barrel($)

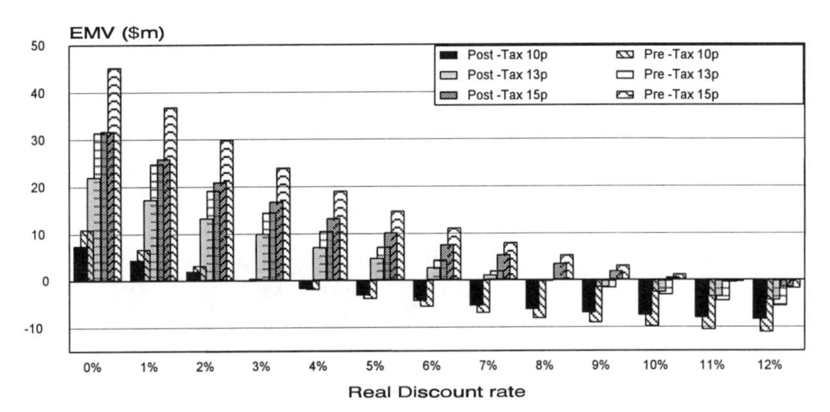

Figure 14.3: Expected monetary values Southern North Sea

The effective expected tax takes are shown in Figure 14.4. When the EMVs are positive they increase with the discount rate from around 30% when that rate is zero. The take is also regressive with respect to variations in gas prices, and indeed to other variables determining the profitability of the activity. When the pre-tax EMVs are negative the tax take also generally becomes negative. This reflects the sharing of the various costs at the exploration, appraisal and development stages.

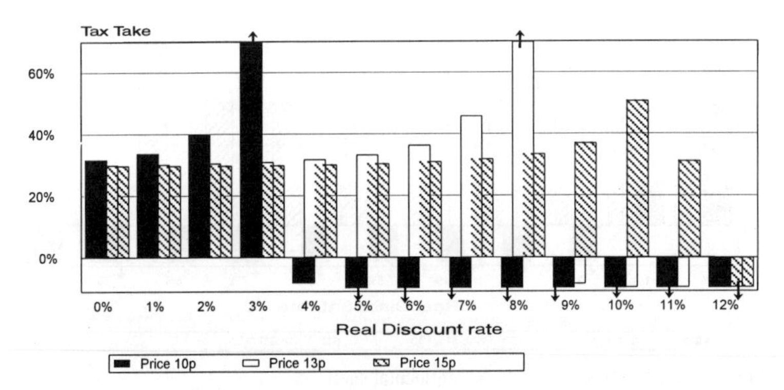

Figure 14.4: Expected tax takes Southern North Sea

The pre-tax and post-tax EMVs for the rest of the UKCS are shown in Figure 14.5. Under the $18/15 pence price case the returns are positive for discount rates just exceeding 8%. With the $15/13 pence case the corresponding rate is around 5%, and for the $12/10 pence scenario less than 1%.

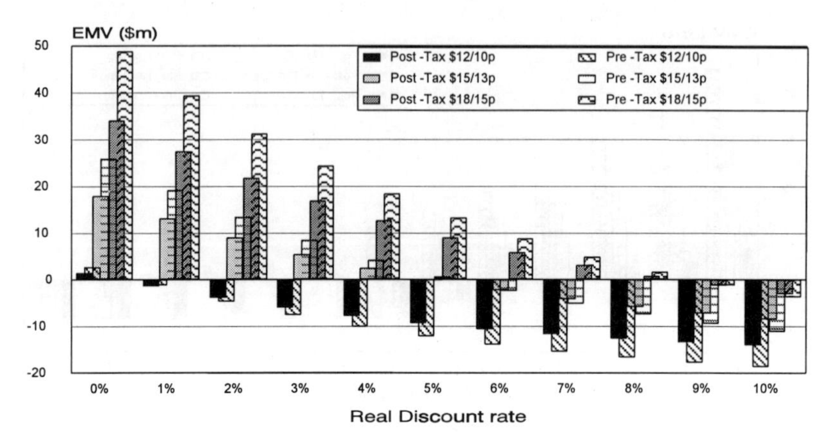

Figure 14.5: Expected monetary values rest of UK continental shelf

The corresponding effective expected tax takes are shown in Figure 14.6. The pattern is the same as for the Southern basin. As the expected rates of return are lower the effective takes tend to be higher in the rest of the UKCS for the same discount rate.

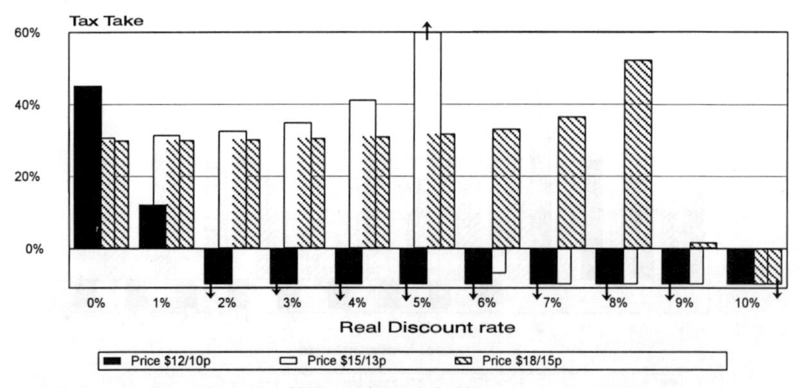

Figure 14.6: Expected tax takes rest of UK continental shelf

5 Returns to exploration with optimistic success rates

Both the exploration and appraisal success rates can fluctuate substantially. While it is felt that the rates employed above give a reasonable view of the prospects currently facing an explorationist it was decided to incorporate another scenario reflecting (a) more recent information on exploration success and (b) more optimistic assumptions regarding commercial success rates. The latter point can accommodate the notion that even finds which are not deemed to be commercially exploitable have some positive value. Over the last few years the trend in exploration success rates has been downwards. In the Southern basin, although the average success rate in the period 1984 – 94 was 30.8%, this figure has only been exceeded twice in the period 1988 – 1997. In the last few years it has fallen significantly, the values for the period 1995 – 1997 being 16.7%, 16.7% and 18.75%. For an optimistic scenario a potentially commercial success rate of 20% was chosen.

For the rest of the UKCS the recent trend in success rates has also been downwards. In the period 1994 – 1997 the exploration success rates were 14.6%, 6.25%, 15% and 13.3%. For the optimistic scenario a potentially commercial success rate of 15% was chosen. This is well above recent experience.

The pre-tax and post-tax EMVs for the Southern basin are shown in Figure 14.7. Under the 15 pence scenario the returns are positive for discount rates of up to 11%. With the 13 pence case the rate of return exceeds 8%, and with the 10 pence case it is under 4%. The corresponding tax takes are shown in Figure 14.8. The pattern is as before.

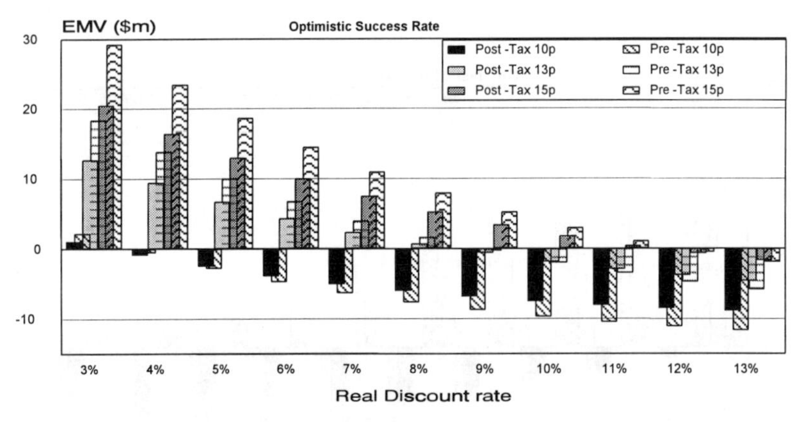

Figure 14.7: Expected monetary values southern North Sea

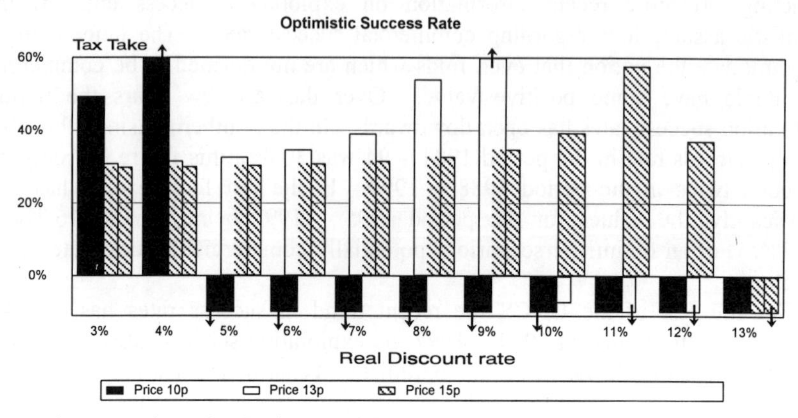

Figure 14.8: Expected tax takes Southern North Sea

The prospective returns for the rest of the UKCS are shown in Figure 14.9. Under the $18 price the expected rate of return exceeds 12%. Under the $15 price it exceeds 8%, and in the $12 case it exceeds 3%.

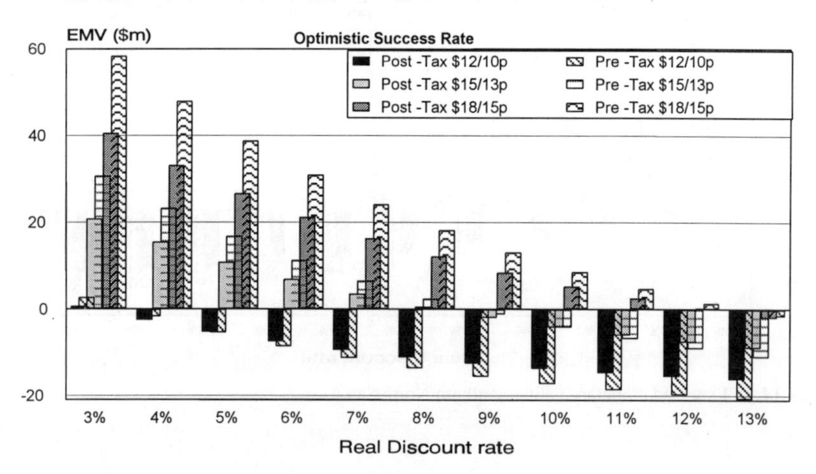

Figure 14.9: Expected monetary values rest of UK continental shelf

The corresponding expected tax takes are shown in Figure 14.10. The pattern is as before.

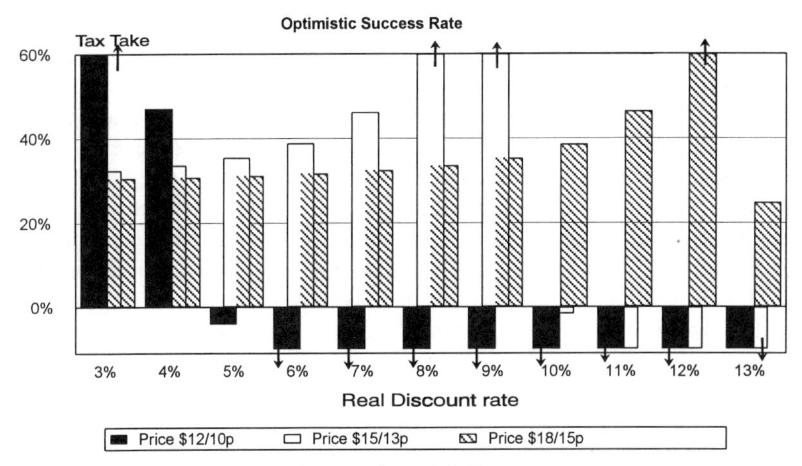

Figure 14.10: Expected tax takes rest of UK continental shelf

6 Summary and conclusions

In this chapter the expected returns from investments in the UKCS in (a) new field developments and (b) exploration have been examined. The expected economic rents and tax takes have been calculated. At the development phase under an $18/15 pence price scenario pre-tax returns on a substantial number of fields being considered for development are satisfactory from an investor's viewpoint. This remains the case after tax. On a significant number of fields the size of the expected (positive) NPVs is quite modest and investors faced with capital rationing might not find the prospective returns adequate. On such fields the share of the economic rents taken in taxation is considerably higher than the 30% - 40% range typically found on the more profitable fields. The regressive character of the tax system emanates from the 25% declining balance relief for investment and the cost of capital employed.

Under the $15/13 pence scenario the great majority of the fields being considered for development remain viable though more come into the marginal category. The regressive feature of the tax system is rather more pronounced, though generally for an existing tax paying investor the projects have positive post-tax returns when these were in prospect before tax. Under the $12/10 pence scenario the investment situation is transformed. Nearly 50% of the projects produce negative pre-tax returns. Many offer only very modest positive NPVs. The tax system is even more regressive, taking over 100% of the economic rents on some marginal fields. In very unprofitable fields the take is negative for an existing taxpayer.

At the exploration stage an investor in the Southern North Sea with success rates based on recent trends could expect a return of under 11% with a 15 pence per therm price. With a 13 pence price this falls to 7.5%, and 6 to 3% under a 10 pence price. Higher returns would depend upon a greater than average overall geological success and/or discovering fields above the average expected size.

In the rest of the UKCS on the basis of recent historical experience an explorationist could expect a return of around 8.5% under an $18/15 pence scenario, 5% under a $15/13 pence case, and less than 1% under a $12/10 pence case. Of course, if the explorationist had greater than average success in making a discovery and/or the size of a find was significantly greater than average, he could make a large return. For an existing taxpaying investor the current tax system generally does not significantly hinder the activity. For an investor without tax shelter the tax system bites more strongly.

When a case of exploration and appraisal success considerably in excess of these obtained in recent years was examined it was found that in the Southern North Sea under a 15 pence price the expected return was just over 11%. With a 13 pence price it became around 8% and with a 10 pence case under 4%. In the rest of the UKCS the optimistic success rate case produced an expected return of around 12% under an $18/15 price scenario, 8% under the $15/13 pence case and around 3% under the $12/10 pence case.

The conclusions from these finding are that the expected returns facing explorationists at 1998 oil and gas prices are often insufficient to cover the cost of capital involved. An investor needs to be optimistic about future oil and gas prices or confident that he will have more than average exploration success.

SECTION 5

RENEWABLE ENERGY

CHAPTER 15

RENEWABLES IN THE UK – HOW ARE WE DOING?

DR. CATHERINE MITCHELL
SPRU, University of Sussex, Falmer, E. Sussex BN1 9RF, UK
Email: cm@launde.u-net.com

Keywords: incentives; prices; renewable energy; subsidies; technology; UK.

1 Introduction

Renewable energy is supported in the UK by a market mechanism known as the Non-Fossil Fuel Obligation (NFFO). The previous Government's renewable energy policy was to award five NFFO Orders, the last of which was recently announced. This represents the end of the first phase of promoting renewables in the UK. A new (DTI Press Release, 1998) policy has to be put into place, although as yet it is unclear what that policy will be since it will depend on the outcome of a number of other, as yet unfinalised, policy decisions. With these uncertainties in mind, the Labour Government is undertaking a Renewable Energy Review (DTI, 1999) to examine the feasibility of formally adopting the Labour Party pre-1997 election policy of supplying 10% of electricity from renewable resources by 2010 (Labour Party, 1996).

2 Renewable energy in the UK - the promotional mechanism[1]

The NFFO obliges the Public Electricity Suppliers to buy a certain amount of renewable electricity at a premium price. The Non-Fossil Purchasing Agency (NFPA) reimburses the difference between the premium price and the pool selling price to the RECs[2] (see Figure 15.1) and this difference is paid for by a Fossil Fuel Levy on electricity, paid for by electricity consumers. Renewable energy projects received around £137 million in 1997-8 from the fossil fuel levy (FFL), with £116 going to the NFFO in England and Wales (see Table 15.1).

NFFO contracts are awarded as a result of competitive bidding within a technology band on a pre-arranged date. This means that wind projects compete against other wind projects but not against, for example, waste to energy projects. The cheapest bids per kWh within each technology band are awarded contracts and these are announced as an Order by the Secretary of State (for example, NFFO1).

---

[1] For more details, please see Mitchell (1998, 1997, 1996, 1995a, 1995b).

[2] Provided the electricity is non-pooled.

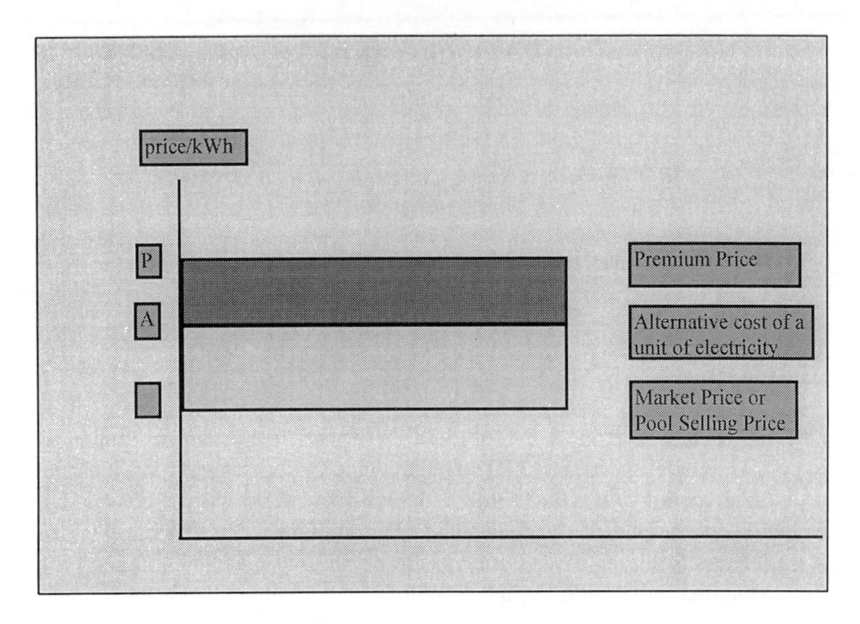

Figure 15.1: Value and subsidies

Table 15.1: The fossil fuel levy (£m)

Year	Total raised	Proportion for nuclear generation	Proportion for renewables	(%)
1990 - 91	1,175	1,175	0	0
1991 - 92	1,324	1,311	13	1
1992 - 93	1,348	1,322	26	2
1993 - 94	1,234	1,166	68	5.5
1994 - 95	1,205	1,109	96	8
1995 - 96	1,105	1010	95	8.6
1996 - 97	844	732.5	111.5	13.2
April 1996 - October 1996	633	570	63	10
Nov 1996 - March 1997	211	162.5	48.5	23
1997 – 1998	279	142.3	136.7	49

Source: OFFER Press Releases (Annual)

The previous Government's policy was to commission 1500 MW DNC[3] of new capacity (roughly 3% of electricity supply) by 2000 from five NFFO Orders

[3] MW DNC = Mega Watt Declared Net Capacity where DNC is the equivalent of base load plant that would produce the same average annual energy output

between 1990 and 1998. The five Orders have now taken place with over 3 GWs of contracts awarded (see Table 15.2). NFFO1 and NFFO2 contracts are until the end of 1998 while NFFO3 to NFFO5 contracts are for 15 years, following a maximum 5 year development period.

Table 15.2: Status of NFFO1-5

	Projects contracted		Projects generating		Projects terminated		Projects to commission		Completion rates (%)	
	Number	MW	Number	MW	Number	MW	Number	MW	Number	MW
NFFO1	75	152.12	61	144.53	14	7.58	0	0	81	93
NFFO2	122	472.23	82	173.73	40	298.49	0	0	67	37
NFFO3	141	626.91	58	191.4	2	1.9	83	460.99	40	26
NFFO4	195	842.72	10	18.46	0	0	187	828.96	4	2
NFFO5	261	1177	0	0	0	0	0	0	0	0
TOTAL	794	3270.98	211	528.14	56	307.97	270	1289.95	192	158

3 Successes of the NFFO

The NFFO has a number of very successful characteristics. It has flexible legislation which has meant that problems which occur in one NFFO Order can be rectified in the next and that the technologies eligible for each Order can be altered (i.e., sewage gas was excluded after NFFO2 and energy crop gasification included in NFFO4) (see Table 15.3). The NFFO has proved to be a relatively efficient way of giving out a large capacity of contracts, particularly given the more or less non-existent renewable energy activity in the UK prior to 1990. Competition, while being the cause of a number of problems, which are discussed below, is also the reason for its key success of rapidly falling prices (Table 15.4). The average price paid for the 1177 MW DNC of NFFO5 contracted capacity in 1998 was 2.71p/kWh (see Table 15.5). This was significantly lower than the average purchase cost of electricity (3.52p/kWh) by the RECs in 1996-1997 (OFFER, 1997). Moreover, the tough NFFO world has ensured that the UK renewable industry is the most market-orientated within Europe.

Another important characteristic of the NFFO has been its acceptance by the electricity companies. As shown in Figure 15.1, the RECs buy the NFFO electricity at the pool selling price which is generally cheaper than the cost of an alternative unit. Thus, the NFFO provides an incentive for the RECs to take the NFFO electricity. It is clear that the low-cost aspects of the NFFO are extremely attractive to Governments and electricity companies in countries, such as Germany and Italy, which are moving towards Phase 2 of their renewable promotion policies. At a recent seminar held in Italy to discuss the future Italian renewables policy (ENEA Seminar, 1998), it became clear that wind electricity in the UK was being paid no more than half the price per unit of electricity than in Germany, Denmark and Italy (i.e., 0.04 ECU in the UK compared to 0.08-0.1 ECU in the other countries).

Table 15.3: Eligible technologies by NFFO order

Technology	**NFFO1**	**NFFO2**	**NFFO3**	**NFFO4**	**NFFO5**
Wind	*	*	*	*	*
Wind - sub bands	-	*	*	*	*
Hydro	*	*	*	*	*
Landfill gas	*	*	*	*	*
Sewage gas	*	*	-	-	-
M&IW (mass burn)	*	*	*	-	-
M&IW (fluidised bed)	-	-	-	*	*
M&IW/CHP	-	-	-	*	*
Biomass (steam generation)	*	*	-	-	-
Biomass (gasification)	-	-	*	*	-
Wet Farm Wastes (anaerobic digestion)	-	-	-	*	-

* = eligible, - = ineligible
M&IW = Municipal and Industrial Waste
M&IW/CHP = Municipal and Industrial Waste with Combined Heat and Power

Table 15.4: NFFO prices

Technology band	**NFFO1** Cost-justification	**NFFO2** Strike price (p/kWh)	**NFFO3** Average price (p/kWh)	**NFFO4** Average price (p/kWh)	**NFFO5** Average price (p/kWh)
Wind	10.0	11.0	4.43	3.56	2.88
Wind sub-band	-	-	5.29	4.57	4.18
Hydro	7.5	6.0	4.46	4.25	4.08
Landfill gas	6.4	5.7	3.76	3.01	2.73
M&IW (mass burn)	6.0	6.55	3.89	-	-
M&IW (fluidised bed)	-	-	-	2.75	2.43
Sewage gas	6.0	5.9	-	-	-
EC&A&FW (gasification)	-	-	8.65	5.51	-
EC&A&FW (residual)	-	5.9	5.07	-	-
EC&A&FW (AD)	6.0	-	-	-	-
M&I W with CHP	-	-	-	3.23	2.63
TOTAL	7.0	7.2	4.35	3.46	2.71

M&IW = Municipal and Industrial Waste
EC&A&FW = Energy Crops and Agricultural and Forestry WasteAD = Anaerobic Digestion
Source: Adapted by Author from OFFER

Table 15.5: NFFO5

Technology	No. of projects contracted	Capacity of projects contracted MW	Lowest contract price p/kWh	Average contract price p/kWh	Highest contract price p/kWh
Landfill gas	141	314	2.59	2.73	2.90
Energy from waste	22	416	2.39	2.43	2.49
Energy from waste using CHP	7	70	2.34	2.63	2.9
Small scale hydro	22	9	3.85	4.08	4.35
Wind energy exceeding 0.995	33	340	2.43	2.88	3.1
Wind energy not exceeding 0.995 MW	36	28	3.4	4.18	4.6
Total	261	1177	-	2.71	-

4 The problems of the NFFO process

Success tends to bring parallel problems, and the NFFO is no exception. The simple NFFO goal of price reduction through competition has meant that other goals, such as industrial policy, have been neglected. For example, the NFFO has not only not led to an improved manufacturing industry (although there have been one or two exceptions) but it has also undermined it. This is because in order to win an NFFO contract, the cheapest technology must be used and this has tended to be from non-British sources.

The competitive base of the NFFO has also led to large developments (to maximise economies of scale) by developers which are subsidiaries of major companies (and which have access to cheap finance). Thus, the NFFO has not supported a diverse set of developments or types of developers. The bureaucratic application process, which lasts about a year and which provides no certainty of a contract, exacerbates this trend.

Partly because of this, the NFFO has had little success in developing new entrant 'mentors' for renewable electricity. The majority of NFFO power plants are owned by subsidiaries of the major generators, the RECs, or subsidiaries of privatised water companies. These companies, while promoting renewables themselves, do not lobby for mechanisms to support renewables which would be detrimental to their parents 'core' businesses. Moreover, there is a very limited number of individuals who feel part of the renewable community and who lobby for renewables. This should be compared with other countries. For example, 100,000 households have invested in wind turbines in Denmark out of a population of 5 million (Helby, 1998). Those 100,000 are an effective, unconstrained lobbying association for renewable energy. The UK renewable industry would benefit from mentors: either a powerful lobby of individuals or a company (or companies) which has/have market power which could lobby for alterations to electricity system regulation which would benefit distributed generation and energy services (and by extension would benefit renewable energy).

The NFFO has been successful in deploying projects of near-market technologies but less successful in deploying projects of emerging technologies, such as energy crops. The competitive nature of the NFFO does not allow the creation of a pool of companies which can build up expertise. Thus, the NFFO has so far only been successful with technologies which had already developed the basis of an industry or which were able to benefit from the high prices of NFFO1 and NFFO2 (i.e., wind and landfill gas).

Moreover, the awarding of contracts as a result of price alone does not complement the planning process. In addition, there are no incentives within the NFFO system to site power plants where it is beneficial to the electricity system. While this is not a problem at the moment, it is likely to become more important as more power plants are deployed to make up the 10% target.

Finally, and possibly most importantly, a question mark hangs over the extent to which the competitive NFFO system is able to deliver deployment. Of the 3271 MW DNC given out in contracts, 530 MW DNC has been commissioned in England and Wales (see Table 15.6). This poor commissioning rate is partly explained because NFFO3 to NFFO5 projects do not have to be commissioned for 5 years, from the awarding of the contract. Thus, NFFO3, NFFO4 and NFFO5 contracts do not have to be commissioned until the end of 1999, 2002 and 2003 respectively. Moreover, because the premium payments are index-linked, there is an incentive to starting later rather than earlier in those five years. As competition became stronger, NFFO bidders will have been tempted to 'bid in' prices based on expectations of technology cost reduction over the next four or five years. This means that there may be a risk that the low NFFO contract prices are below those than are economically viable.

Table 15.6: NFFO 1-5 status as at 30 June 1998

	Live projects	
	Number	Capacity (MW DNC)
Hydro	38	28.621
Landfill Gas	90	171.19
M&IW	11	147.449
Other	5	57.98
Sewage Gas	24	25.039
Large Wind	38	93.766
Small Wind	5	4.098
TOTAL	211	528.143

Thus, the commissioning rate for NFFO3 is beginning to look rather worrying. Of the 141 contracts given out, 56 are generating and 2 have terminated (OFFER, 1998). This means that 83 projects of 461 MW DNC are still to be commissioned within the next year. This is not much less than the 530 MW DNC which has been commissioned in England and Wales through the NFFO process since 1990. The poor deployment rate of NFFO3 does not seem to be linked to projects being turned down by the planning process. It may be that next year will see an increased commissioning rate as the NFFO3 projects come on line to maximise their index-linked payments. If a good proportion of the NFFO3 projects are commissioned next year, then poor deployment may prove not to be a problem of the NFFO process. However, if the commissioning rate of NFFO3 is poor, then it will bring into question the ability of a competitive system to deploy renewables.

5 Meeting a 10% by 2010 target

The net electricity supplied in 1997 was 326 TWh (Dukes, 1998). One would hope this would be reduced by 2010, although predictions are generally that it will rise. Of this, large hydro provided 4.1 TWhs while renewables provided 3.2 GWhs or about 1% of supply. Thus, a 10% supply target means that renewables must provide 32 TWhs +/- 10% , depending on whether large hydro is counted towards the target and whether demand rises or falls.

Translating this figure into installed capacity is complex, depending as it does on technologies and load factors. Originally, it was said that the 1500 MW DNC of NFFO1-5 would produce around 9 TWh annually or about 3% of supply (DTI, 1994). The NFFO, Scottish Renewable Order and Northern Ireland NFFO has so far led to the commissioning of around 580 MW DNC[4] which generates the current 1% supply. Overall, therefore it can be argued that 10% of electricity supply will be the equivalent of about 5000-6000 MW DNC. If large hydro is counted towards that, and given the 580 MW DNC which is already commissioned, it means that around another 350-450 MW DNC needs to be commissioned annually between 1999 and 2009 throughout Britain. This is a challenging target, although it ought to be possible given that Germany deployed 570 MW of wind energy alone in 1996-7 and that about 2000 MW DNC has already been given out as contracts through NFFO4 and NFFO5. Understanding the percentage rate of deployment of NFFO3 contracts will be vital in understanding how likely it is that the UK is able to meet that target.

---

[4] NFFO Status Summary as at 30 June 1998, available from ETSU Commercialisation Programme, Harwell, UK.

6 UK energy policy uncertainties

In addition to the difficulties of accurately assessing the success of the NFFO, a number of energy policy uncertainties exist which, when resolved, will have implications for the final characteristics of the renewable energy policy to be put in place to follow-on from NFFO5. Effective support could be provided for renewables through changes in the electricity trading arrangements; the forthcoming utility regulation bill; the climate change review; and from recommendations of the Marshall Inquiry on economic instruments and the business use of energy. On the other hand, these factors could all undermine the renewable policy as well.

The proposals for electricity trading arrangements have major implications for renewable policy. Currently, the legislation sets the subsidy as the difference between the pool selling price and the NFFO premium price (see Figure 15.1). Under the new trading arrangements, a pool selling price would not exist and, it is hoped, a transparent market price would become apparent. A new reference price may have to be decided upon in order to assess the cost of the renewable subsidy. If this reference price reflects an average purchase cost of electricity to RECs or suppliers, the total subsidy to renewables would be low and only necessary for the more expensive technologies. On the other hand, a lower reference price, which reflected off-peak electricity purchase costs, would increase the cost of subsidy. Numerous other decisions effecting renewables will be taken as a result of the changes in the electricity trading arrangements. It is vital that renewables are thought of as those arrangements are put in place; not so that they have special treatment but so that mechanisms which discriminate against renewables as a result of thoughtlessness are excluded.

In addition, the utility regulation bill will also effect renewable energy policy and its cost. It has been widely argued that distributed generation (generation which feeds directly into the distribution system) allows a number of savings to distributors or suppliers[5]. Under the current regulation system, those savings do not accrue to the generator. If these savings did accrue to the generator, then the subsidy could be removed from certain technologies and would considerably cost less for others. Moreover, distributed power plants are able to act as reinforcers of the distribution system. This means that distribution companies could save or defer operation and maintenance costs through appropriate siting of renewable power plants. Currently, the level of information required for appropriate siting in the distribution system is not in the public domain, despite distribution companies being regulated monopolies, so that prospective distributed power plant operators are unable to pinpoint the places where the should go. Thus, the utility regulation bill could go a long way towards helping renewables compete, and be incorporated, within the

[5] REVALUE Final Report, DG12 Joule 3 Project 1996-1998, available from c.mitchell@sussex.ac.uk; IEE, 1998, Colloquium on Economics of Distributed Generation, Savoy Place, London, 29 October.

electricity system. The regulator could (1) require a 7 Year Distribution Statement, analogous to the NGC's 7 year Statement, from all the distribution companies so that renewables can site themselves appropriately; (2) require that the savings of distributed power return to the generator; (3) provide incentives for distribution companies to move from their 'passive' role of takers of electricity from the national transmission grid towards a more 'active' role where they start to develop the possibilities within their region.

Thus, both the changes in electricity trading arrangements and the new utility regulation will have fundamental effects on renewables and their competitiveness. Moreover, the policies espoused within the Government's climate change review and the recommendations of the Marshall Inquiry with respect to environmental taxation will also have implications for renewable policy.

However, other Government measures have made the problems of promoting renewables more complex. The recent Comprehensive Spending Review (CSR) transferred the cost of the NFFO as a new budget line to the Department of Trade and Industry. This means that the Treasury has allotted the DTI an acceptable annual cost of the NFFO (even though the NFFO is paid for by electricity consumers). Any extra spend on the NFFO over the allotted amount will have to be paid for out of another DTI budget area. The receipts of the fossil fuel levy, however, have not been transferred to the DTI. This means therefore that the NFFO is no longer a closed budgetary circle from the perspective of the DTI or the Treasury and has concentrated minds on the cost of the renewables policy in a way that did not occur before. Furthermore, the agreed maximum annual £150 million cost of the NFFO will have to be re-negotiated in the next CSR (Energy Report, 1997).

Finally, the electricity industry in the UK has significantly reduced its CO_2 emissions from 1990 levels because of the switch from coal to gas. Although the wider energy market will require deployment of renewables to meet the domestic CO_2 reduction target, the electricity sector does not. This will not increase the electricity industry's interest in renewables.

7 The EU influence

The European Union (EU) may also influence the development of renewable energy policy in the UK as well as other European countries. The EU has signed a legally binding agreement to an 8% reduction in a basket of six emissions in the commitment period 2008-2012 as a result of the Framework on Climate Change Convention (FCCC). The deployment of renewables is a central tool for the EU in meeting this target.

The EU White Paper for a Community Strategy and Action Plan (EC, 1997) proposed a target of 12% of energy supply from renewable resources by 2010. This is a doubling of energy supplied from renewables in the EU. Much of this current European supply is derived from large hydro and traditional biomass. Some European countries, for example Austria, considerably exceed the 12% of energy supply target yet only supply a small percentage of it from 'new'[6] renewable resources. The UK, on the other hand, only provides around 0.5% of energy supply from large hydro and therefore has a very long way to go to reach the 12% target, most of which will have to be derived from 'new' renewables. Clearly, much negotiation will take place as each country agrees its own commitment to this proposed target.

Furthermore, a Directive on Fair Access for Renewables to the European electricity market was published for consultation in 1998. It argued for common rules for supporting renewables within the internal market for electricity, including a standard minimum payment for renewables across Europe. The Directive was shelved following protests, particularly from Germany. A new Working Paper (DG17, 1999) has recently been published which puts forward a number of options for meeting the EU target and which allows more freedom for States to promote renewables in the way that they wish. However, crucially, it also argues for minimum standards for the payment to, and connection of, renewable energy power plants.

The extent to which the EU is able to promote their policies within Member States is therefore crucial.

8 The 2nd phase

Renewable energy policy in the UK is therefore at an important stage of development. It can be argued that Phase 1 of renewable energy promotion in the UK occurred not because of a burning desire by the then Conservative Government to support renewables but more as a bi-product of the necessity of obtaining a subsidy for nuclear power. Thus, the renewable NFFO came into being with little thought and renewables benefited from the political determination of the then Government to push through electricity privatisation. It is unlikely that the new policy drivers of Phase 2 have as much importance to the present Government. Furthermore, those drivers are more disparate.

[6] 'New' renewables is a recognised term for renewables which are deployed as a result of promotional mechanisms and which exclude large hydro and conventional biomass.

9 Future policy: lessons to be learned

It can be argued that there are four main lessons to be learned from the NFFO process so far:

1. The UK must alter the primary goal of the promotional mechanism from price reduction to deployment. Experience from the Continent shows that a pre-known standard payment is successful in attracting all types of developers and in producing a high rate of deployment. Thus, a future UK promotional mechanism should incorporate some sort of price stability and it is important to note that this does not necessarily mean a high price. In the short term, until regulatory change takes place, one possibility for the UK would be an obligation on suppliers to buy renewable electricity at the equivalent of a market price plus a payment to reflect the benefits of distributed power. Once regulatory change takes place, these benefits should accrue to the generator and so the obligation would only be an obligation to purchase.
2. Any future policy must include some means of ensuring that renewables are sited appropriately for the electricity system. Siting based on price is acceptable for a small percentage of electricity supply. It is unacceptable if renewables are to reach 10% of supply by 2010. In this case, they must be sited to complement planning and electricity system requirements. A key way forward is for reform through the utility regulation process, as discussed above. This policy requirement therefore requires Government guidance to the regulator.
3. Similarly, it is important that the means of incentivising distribution companies to develop active distribution networks and to act as market facilitators is enhanced. Again, this requires Government guidance to the regulator.
4. Finally, the promotional mechanism should provide incentives for developing 'mentors' for renewables so that the requirements of renewables can be lobbied for. This will occur through a network of measures. On the one hand, incentives for 'active' distribution systems may support the development of larger companies with market power which may also act as mentors for the renewable industry. In addition, a range of other factors would promote individuals becoming involved in renewables. For example, a transparent and easily understandable verification procedure for green electricity should promote the purchase of green electricity. Local authority guarantees for purchasing a percentage of their demand from green electricity schemes would involve the local community and would permit suppliers to provide a long-term contract to generators, which in turn would allow the financing of the schemes. The price of green electricity need not be at a premium to the local authorities current tariff. Experience from the Continent shows that tax exemptions are the

most successful way of attracting individuals to investment in green schemes (FIRE, 1998).

10 Conclusion — how well has the UK done?

The UK is coming to the end of the first phase of promoting renewables. The NFFO has been extremely successful in reducing prices and creating a market-orientated industry. However, the UK has yet to prove that the competitive system is able to deliver deployment and until that is the case, the NFFO cannot be said to be a resounding success. If the 10% of electricity from renewables by 2010 target is agreed by the Government, the mechanism must ensure that renewables are sited to complement planning requirements and to suit the electricity system. Without this, public acceptance is unlikely to be forthcoming. Thus, the UK has done well so far but it is entering a more complex and difficult policy era.

References

DG17 (1999), *Electricity From Renewable Energy Sources and the Internal Electricity Market*, Working Paper of the European Commission, March.
DTI (1994), New and Renewable Energy: Future Potentials in the UK, EP62, March.
DTI Press Release (1998), *John Battle Makes Greatest Commitment Ever to Renewable Sources of Energy*, 24 September (Announcement of NFFO5).
DTI (1999), *New and Renewable Energy - prospects for the 21st century*, March.
DUKES (1998), Annual, HMSO.
EC (1997), *Energy for the Future: Renewable Sources of Energy*, White Paper for a Community Strategy and Action Plan, COM (97) 599,26.11.97.
ENEA Seminar (1998), Conferenze Nazionale Energia and Ambiente, IEFE, Universita Comerciale Liugi Bocconi, Milan, Italy, 7 October.
Energy Report (1997), Annual, HMSO.
FIRE (1998), Financing Renewable Energy - Opportunities and Challenges, The White House, London, 1 October.
Helby P. (1998), *Denmark - A Case Study*, ENEA Seminar, 1998, Conferenze Nazionale Energia and Ambiente, IEFE, Universita Comerciale Liugi Bocconi, Milan, Italy, 7 October.
Labour Party (1996), *In Trust for Tomorrow*.
Mitchell C. (1995a), *Support for Renewable Energy in the UK - Options for the Future*, available from CPRE; Mitchell C. (1995b), *The Renewable NFFO - A Review*, Energy Policy, December; Mitchell C. (1997), *Future Support of Renewable Energy in the UK - Options and Merits*, Energy and Environment,

Volume 7, No. 3, 1996; Mitchell C. (1998), *Renewable Energy in the UK - policies for successful deployment*, available from CPRE.
OFFER (1997), *The Competitive Electricity Market from 1998: Price Restraints*, Fifth Consultation, August.
OFFER (1998), *Fifth Renewables Orders for England and Wales*, September.

CHAPTER 16

FLUCTUATING RENEWABLE ENERGY ON THE POWER EXCHANGE

KLAUS SKYTTE

Risø National Laboratory, Systems Analysis Department
P.O. Box 49, DK–4000 Roskilde, Denmark,
Email: klaus.skytte@risoe.dk

Keywords: Denmark; electricity markets; Nord Pool; power exchange; renewable energy; wind power.

Many countries around the world liberalise their electricity markets by introducing power exchanges. At the same time the share of renewable energy supplies is steadily increasing.
Renewable energy supplies, such as are derived from sunlight or wind, have more fluctuating production patterns than those derived from more conventional power generation.
Power exchange is one way to balance out the fluctuations in energy production. Generators of fluctuating power can incorporate their production offers on the daily spot market on the power exchange at the same prices as other generators. The only extra expenses for fluctuating power arises if the generators are unable to fulfil the commitments made on the spot market when the actual deliveries take place. This expense comes about from the regulation expense the system operator encounters by maintaining the total balance between supply and demand on the spot market.
The goal of this chapter is to examine this type of expense and how the use of a power exchange tends to balance out the effect of fluctuations in wind power production. To illustrate this, the Nordic power market is used as the main example.
The chapter briefly describes the different markets of Nord Pool, the Nordic power exchange, and their potential use for enabling wind power to be utilised fully. By use of research results on the prices on Nord Pool and a Danish case study, different scenarios are set up for the wind power producer's use of the power exchange. It is found that not only does the accuracy of the prediction influence the use of the power exchange, but the structure of the power exchange itself may also play an important role.

1 Introduction

Renewable energy is playing an increasingly important role in energy planning in most countries around the world, especially in consideration of the need to stabilise and eventually reduce the emission of greenhouse gases. As this role increases many countries are liberalising their electricity markets and introducing power exchanges.

Electricity generation from some technologies may be more predictable than others in an electricity market which combines several generation technologies. Wind power is one of the technologies where the electricity supply is difficult to predict on either a short- or long-term basis. The results of research in this area (see e.g., Landberg 1997) have indicated that wind power predictions made from meteorological forecasts at best can have an accuracy of approximately 90% (up to 36 hours following the prediction).

On the one hand, should technologies with fluctuating power production account for a significant market share, market balance could well be disturbed. Both the system operator and power generators have therefore a common wish that fluctuation in electricity supply is small. On the other hand, many technologies (for example, hydropower from high dams) can adjust their power generation in a very short time in order to compensate for the unpredictable fluctuations of, for example wind turbine output, and thereby re-establish market balance. This regulation possibility especially matches power plants with rapid regulation properties, e.g., not only hydropower plants, but gas turbines and combined heat and power (CHP) plants, where in the latter case heat storage facilities can be used as short-term buffers for regulating electricity generation. This is especially true in the case of extraction CHP plants where the proportion between heat and power production can be varied (Grohnheit 1993).

Power exchange is one way to balance out the fluctuations in energy production. A power exchange is an organised marketplace for wholesale purchasers and sellers of electricity. Power exchanges often consist of a spot market for balancing physical supply and demand, and a futures market for hedging against price fluctuations. At most power exchanges the spot market closes some hours before the actual physical delivery takes place in order to clear the supply against the demand and thereby create a market balance.

The primary function of the power exchange is to mediate electricity trades and prices. The prices on the power exchange reflect the marginal electricity prices on the market if all the actors in the market have free access to the power exchange. The concentration of power dealers on the power exchange enables power to be offered in small amounts. This means that generators of fluctuating power can incorporate their production offers on the daily spot market of the power exchange at the same prices as other generators.

The only extra expense for fluctuating power occurs if the generators are unable to fulfil the commitments made on the spot market when the actual deliveries take place. This type of expense comes from the regulation cost the system operator incurs in keeping the total balance between supply and demand on the spot market.

It is reasonable to assume that fluctuating renewable energy suppliers, such as wind power suppliers, are price-takers on the spot market. This is partly because, at least at present, they are relatively small on the total market. Also, it is more economical to sell the wind power at a low price rather than damp down production. The wind power producers therefore make offers on the spot market at a low price in order to ensure that their wind power is going to be sold. When the spot market closes for offers and clears supply and demand, the price of wind power is settled with the total balance price on the spot market (spot price).

Besides the spot price, wind power can be supported directly via subsidies or indirectly via the sale of green certificates, etc. Due to administrative costs these supports are often independent of the regulation of short-term fluctuations, and are therefore not discussed in this chapter.

2 The Nord Pool power exchange

Norway, together with England and Wales, are among the first countries in the world that have liberalised the electricity market and introduced power exchanges (Newbery and Pollitt 1996, Skytte and Grohnheit 1997). Even though the power exchange in Norway (Nord Pool) and the one in England and Wales (The Pool) were launched almost simultaneously, they were built up independently of each other and therefore have different structures (see Knivsflå and Rud 1995, Grohnheit and Olsen 1995).

From being a national Norwegian power exchange, the Nordic power exchange Nord Pool was extended in 1996 to cover both the Swedish and Norwegian electricity markets. The Danish and Finnish utilities are active buyers and sellers at Nord Pool as well. Nord Pool is composed of a common Norwegian and Swedish spot market for physical trade, as well as a common financial futures market. The regulation of deviations from the spot market balance is made individually in each of the participating countries. Sweden uses a regulation system almost similar to that of Britain, whereas Norway has kept the original regulation system derived from its national power exchange with a regulating power market.

The Nordic spot market closes at noon every day. At closing time the supply and demand bids are cleared against each other (balanced) and commitments are made for delivery the following day on an hourly basis. The interval between the time the bids are made and actual trading takes place is at least twelve hours. A certain amount of fluctuation in the actual supply and demand is therefore unavoidable compared with the commitments made on the spot market.

Unlike the British structure where the balance is centrally controlled, Norway has a **regulating power market,** where supply and demand bids determine the price for regulation. This means that the supply of up- and down-regulation services are cleared against the net need for these services in order to maintain the market balance found at the spot market (see Nord Pool ASA's homepage). There is only one price for regulation services per hour, since the regulating power market is a balance market, i.e., prices are determined on the basis of the net needs.

If a power supplier delivers less or a buyer uses more than the amount agreed upon on the spot market (excess demand), then the supplier has to pay for **up-regulating power** in order to be able to fulfil his agreement on the spot market. Other suppliers get paid to deliver the supply deficit, or some buyers get paid to decrease their demand for power.

If an amount of electricity is supplied more, or used less, than that agreed upon on the spot market (excess supply), then **down-regulating power** is implemented to maintain the power balance in the market. The excess supply is sold to buyers who then increase their purchases. Alternatively, suppliers buy the excess supply in order to reduce their own supply.

The regulating power market closes two hours before the actual trades take place, but the clearing does not take place until fifteen minutes before the trades takes place. The suppliers of regulating services on the regulating power market therefore have to be able to fulfil their bids within fifteen minutes of notice.

Payments on the spot and regulating power markets are made separately, i.e., a payment for a commitment on the spot market is made with no attention given to the actual trade. Any deviations are then paid on the regulating power market via the balance price between supply and demand for regulating power.

The price for up-regulating power is often larger than the spot price, i.e., a producer must pay more for up-regulating power than he gets on the spot market, and thereby he suffers a loss. The price for down-regulating power is often less than the spot price, which means that a producer must sell his excess supply at the lower down-regulation price instead of the spot price. He therefore suffers a potential loss. In this way, how a producer uses the spot and regulating power markets becomes important.

The regulation of deviations from the spot market balance in Sweden is different from the regulating power market in Norway. The Swedish bids of regulation services are arranged in merit order and are centrally dispatched via the system operator. The Swedish regulation system is not a balance market as it is in Norway. This means that it is possible to have different prices for up- and down-regulation services at the same time. The last section of this chapter discusses the different ways of handling balance payments and their importance for fluctuating power.

This chapter mainly focuses on the regulating power market in Norway. The regulating power prices analysed below are therefore valid only for the Norwegian system.

3 Spot and regulating power prices

It is necessary to analyse the spot and regulating power prices in order to find the expenses for fluctuating power if the generators are unable to fulfil the commitments made on the spot market at the time when the actual deliveries take place.

The price level on the Nordic spot market strongly reflects the total energy demand (consumption) in winter, where the inter-median power plants are price setters. On the other hand, in early summer, when the demand is low and there is usually plenty of water in the reservoirs of the high dams, the spot price reflects the total demand weakly. However, other physical and economic variables may also influence the spot price (see Johnsen 1996).

Analyses have been made of the spot markets but almost none have looked at the regulation of the market balance, i.e., the regulating power market.

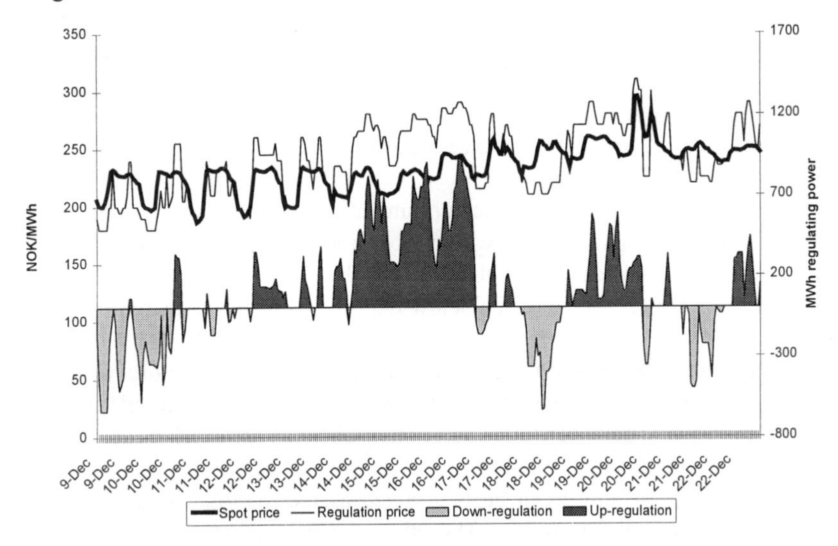

Figure 16.1: Regulating power in December 1996 (1 NOK ≈ 0.12 EURO)

The regulating power price follows the spot price and thereby indirectly reflects the price setters though the spot price. From Figure 16.1 it is seen that the difference between the spot and regulating power prices depends on the amount of regulation. It cannot be stated whether the connections between the spot price and the up- and down-regulating power prices are the same or not. Since there might be different buyers and sellers who bid for up- and down-regulation, the regulating power prices may be more sensitive to the amount of either up- or down-regulation. In addition, the dependence of the spot price may also be different for up- and down-regulation.

It seems therefore reasonable to set up a hypothetical relation as follows:

$$\begin{aligned} \mathrm{PR}(\mathrm{P}_t,\mathrm{S}_t,\mathrm{D}_t) &= \eta\cdot \mathrm{P}_t \\ &+ 1_{S_t<D_t}\cdot(\lambda\cdot \mathrm{P}_t + \mu\cdot(S_t-D_t)+\eta) \qquad (1)\\ &+ 1_{S_t>D_t}\cdot(\alpha\cdot \mathrm{P}_t + \gamma\cdot(S_t-D_t)+\beta). \end{aligned}$$

where PR_t is the price of regulating power, P_t the spot price, S_t the amount announced at the spot market and D_t the actual delivery. $(S_t - D_t)$ is the amount of regulation. The values of P_t and S_t are known when the regulating power price is determined, since the spot market will close before the regulating power market starts. The only unknown variable is therefore the actual delivery, D_t.

There is an excess demand for power when $S_t > D_t$. This is, e.g., the case when some producer has delivered less than promised on the spot market. He therefore has to buy up-regulating power in order to fulfil his promise. Likewise, there is an excess supply of power when $S_t < D_t$, which means that the producer buys down-regulating power, i.e., he sells the excess power at the price for down-regulation, which is lower than the spot price.

The **1** in relation (1) is an indicator function, i.e., equal to 1 when the sub-statement is true, and equal to 0 elsewhere. Relation (1) therefore states: when there is neither up- nor down-regulation, then the regulating power price equals the spot price scaled by a factor. We will see below that this factor is estimated to be equal to 1.

The indicator functions are included in order to accentuate more voluminous oscillations in regulating power prices for either up- or down-regulation. The indicator function will be superfluous if the coefficients in the brackets are estimated to be statistically identical.

The coefficients μ and γ can be interpreted as the marginal regulating power prices per unit of regulated power. The other coefficients are independent of the amount of regulation. These coefficients can be interpreted as determining a **premium of readiness** paid to the suppliers of regulation services. This may be an important factor, since the suppliers have to be able to regulate within fifteen minutes of notice, compared to the spot market where the time period between the acceptance of the bids and the time of the physical trades is at least twelve hours.

The premiums of readiness for, respectively, up- and down-regulation services consist of a constant term and one connected to the spot price. This means that part of the premium is common to all suppliers of regulation services and another part depends on the price level on the spot market.

Skytte (1998)[1] used data series from Nord Pool (the Oslo area, week 50, 1996 till week 21, 1997) to estimate the hypothetical price relation (1) for regulating power. Relation (1) was estimated to explain more than 99% ($R^2 = 0.998$) of the fluctuation in the regulating power prices. The estimated relation is shown in relation (2).

[1] Some of the estimation results are also reported in Nielsen et al. (1997).

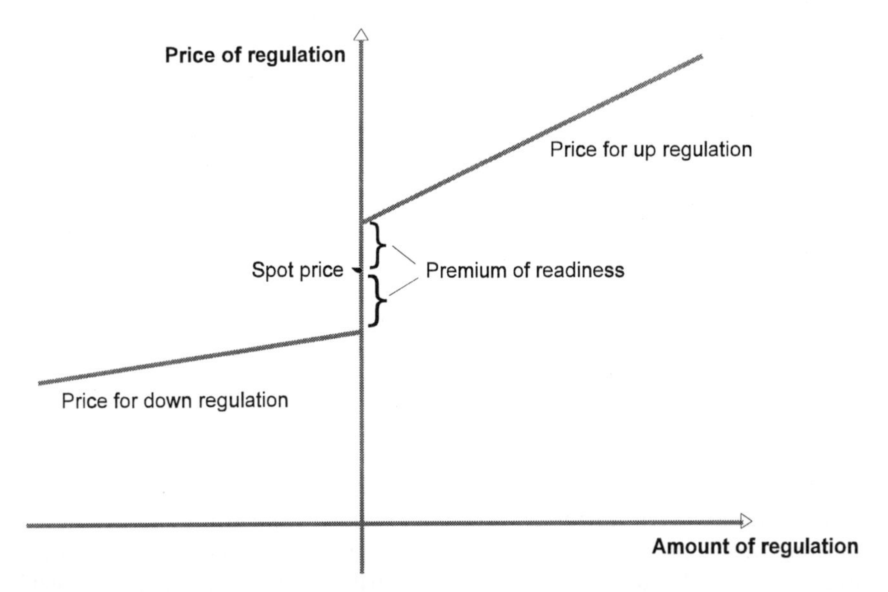

Figure 16.2: Price of regulating power

$$
\begin{aligned}
\mathrm{PR}(\mathrm{P}_t, S_t, D_t) = \mathrm{P}_t & \\
& + 1_{(S_t < D_t)} \cdot (-0.069 \cdot \mathrm{P}_t + 0.023 \cdot (S_t - D_t) - 4.3) \\
& + 1_{(S_t > D_t)} \cdot (0.028 \cdot \mathrm{P}_t + 0.042 \cdot (S_t - D_t) + 13.07).
\end{aligned}
\tag{2}
$$

First of all, it is seen that η in relation (1) was estimated to be equal to 1 (t-value = 1000). This means that the regulating power price equals the spot price when the amount of regulation is zero.

Secondly, it is seen that the use of indicator functions is justified, since the coefficients in the brackets are significantly different. Note that down-regulation is represented by a negative amount of regulation, which means that the down-regulating power price is always less than or equal to the spot price, which is less than or equal to the up-regulating power price. This is illustrated in Figure 16.2. The regulating power price is seen to be twice as sensitive to the amount of up-regulation relative to the amount of down-regulation.

Thirdly, the premiums of readiness are seen to be different for up- and down-regulation. The premiums were estimated (in NOK/MWh) to be

$$
\begin{aligned}
\mathrm{Premium}_{\mathrm{Down}} &= 0.069,\mathrm{P}_t + 4.3 \\
\mathrm{Premium}_{\mathrm{Up}} &= 0.028,\mathrm{P}_t + 13.07
\end{aligned}
\tag{3}
$$

4 Danish case study

Denmark is one of the countries in which energy planning relies strongly on wind power. The total installed wind power capacity in Denmark is (mid-1997) approaching 1000 MW, and according to the latest Danish energy plan, Energy21 (1996), this is expected to increase to approximately 1700 MW by 2005. In the long term, wind turbines are expected to play an even more important role with a projected capacity of 5400 MW by 2030. This implies that wind-generated electricity will supply more than 50% of the total electricity demand in Denmark by that year.

A research study was conducted in 1997 (see Nielsen et al., 1998), which among other things looked into the introduction of large-scale renewable energy on the Danish electricity market and the use of the power exchange Nord Pool as one means of balancing out fluctuations in energy production.

Parameters used in the study were the progress scenarios for wind power from the Danish energy plan Energy21. The price relation (2) was used to describe the regulating power prices, in order to calculate the costs and benefits of using the power exchange Nord Pool as one way to balance out fluctuations in energy production.

A portfolio of energy models was used in order to describe the total electricity market as realistically as possible. Two operating models, the Samkjøring and Sivael models, were used to simulate trade and electricity prices on the Nordic market until year 2005. The Samkjøring model was used to link the Scandinavian countries. It optimises the total power and heat supply on the basis of a superior description of the supply system. In addition, the model determines the marginal weekly production price and the amount of electricity exchanged between the countries.

The Sivael model is a detailed operating model for the Danish power and heat systems. With the prices and exchange-amounts found from the Samkjøring model, the Sivael model was used to optimise the Danish power and heat supply on an hourly basis.

The disclosed price relation (2) for regulating power was implemented in a third model ES^3, together with the calculated spot prices and trade pattern found in the first two models. In addition, a time series from wind parks in Denmark (1996) was scaled according to the progress scenario and implemented in the model. The ES^3 was used to calculate the consequences, on an hourly basis, of introducing large-scale wind power as well as the economics involved.

It was assumed that the actual amount of wind energy produced could be predicted to within 90% accuracy (with reference to research studies by Landberg et al., 1997). It was therefore incorporated into the study that 10% of the offered renewable energy did not succeed in providing sufficient electricity and another

10% yielded more than had been offered on the spot market. In other words, an amount corresponding to 20% of the wind power was settled on the regulating power market.

The three models were run three times for the year 2005, since the price level in the Nordic electricity market depends on the yearly precipitation (due to the large share of hydropower). The runs were made for a normal rainfall year, and wet and dry years.

Table 16.1 shows the average total revenue per MWh wind power in the year 2005. It is seen that the expenses incurred by fluctuations in all the Danish wind power generators are less than 4% of the revenue that would be produced in the absence of fluctuations. These are the expenses that result from using the regulating power market instead of the spot market to fulfil a commitment.

Table 16.1: Average total revenue per MWh wind power in the year 2005 (NOK/MWh)

	Normal	Wet	Dry
Without any fluctuations	176	113	226
With 20% fluctuations	170	109	219
Expenses of fluctuations	6	4	7
In percent of sales price	3.4%	3.5%	3.1%

If no prediction of wind power is given, and the offers are made with respect to the previous day's production, then almost 50% of the wind power offers on the spot market cannot be fulfilled. In this case the expenses of the fluctuations are between 15 and 18 NOK/MWh, i.e., between 8 and 13% of the revenue in the absence of any fluctuations.

The numbers found in this case study support the results found by Skytte (1997), where price relation (2) was implemented in a spreadsheet together with real Nord Pool data.

5 Different ways of handling balance payments

The above-described case study has shown that the power exchange is a relatively cheap means of balancing out the fluctuations in energy production. The results depend not only on the accuracy of the prediction, but also on both the structure of the power exchange and handling of the balance payments.

At the time a wind power producer announces his production on the spot market he does not know his actual delivery, if he has a fluctuating production. A producer may have revenues from his sale on the spot market and costs from regulating his delivery in order to fulfil his commitments (bids) on the spot market.

For a wind producer the use of the different markets on the power exchange depends on the forecast of his actual delivery, i.e., on the wind forecast. The prediction of wind power generation depends on the time horizon for the prediction. A longer time horizon gives poorer predictions. The power exchange Nord Pool has a time horizon between 12 and 36 hours between bidding on the spot market and the actual delivery. Unpredictable fluctuations are therefore generally inevitable for wind power on Nord Pool.

Apart from the problems involved in the uncertainty of the forecast, the extra expense created by experiencing fluctuations will also depend on the way the balance market handles the regulation costs. The regulating power market at Nord Pool is a kind of balance market where the price of regulation is determined by the total need for regulation (market balance between supply and demand).

The fluctuation from the individual wind power producers is therefore settled in accordance with the total net need for regulation. If, for example, there is a total need for up-regulation and a wind power producer generates more energy than he promised on the spot market, he can then sell his excess power at the higher up-regulation price instead of a lower down-regulation price.

As mentioned in the beginning of this chapter, Sweden and several other countries have different ways of handling balance payments compared with Norway.

Figure 16.3 illustrates three different ways of handling balance payments compared with the spot price. The dotted line illustrates the spot price, which is known at the time of balance regulation. The full-drawn lines illustrate the total revenue for the individual wind power producer from the power exchange. This total revenue is determined as the revenue from the spot market minus regulation costs.

The figure illustrates a case with total excess demand compared with the spot market balance, i.e., total need for up-regulation. The wind power producer increases the total need if he generates less than the amount agreed upon (offered) on the spot market. He therefore has to pay the price for up-regulation for the amount of energy that is lacking. The wind producer suffers a loss, since this price is higher than the spot price.

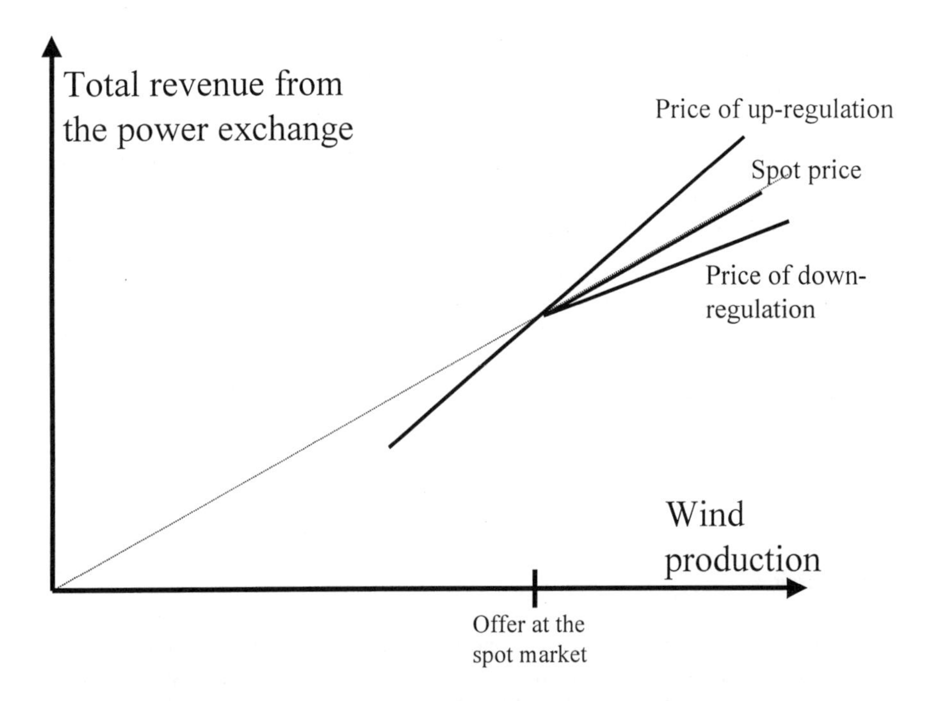

Figure 16.3. Different types of handling balance payments. There is a net need for up-regulation.

The wind power producer reduces the total need for up-regulation if he generates more energy than the amount agreed upon (offered) on the spot market. The figure indicates three different ways in which the wind power producer can be paid for his extra production:

- He gets the up-regulation price (the method of the regulating power market in Norway).
- He gets the spot price.
- He gets a price for down-regulation (the method of the balance market in Sweden).

Similar considerations can be made if there is a total need for down-regulation. In this situation the individual wind power producer increases the total need for down-regulation if he generates more than is offered on the spot market. He decreases the total need if he produces less than offered.

The three ways of handling balance payments can briefly be described below:

1. If the producer should deviate from his offer in the same direction as the total market, he would be punished by being obliged to pay the regulation price.

2. If the producer should deviate from his offer in the opposite direction as the total market, then he would be either:
   - rewarded by using the regulation price based on the total net need; or
   - punished by using a regulation price based on the total gross need in his direction; or
   - indemnified by using the spot price.

If the total need for regulation is made up from many technologies and consumer groups, then there is no reason to assume that the fluctuations from wind power generators are correlated with the total need. It can be assumed that the fluctuations from the individual wind power producer are half the time in the same direction as the total market and half the time in the opposite direction.

The structure of the power exchange and the handling of the balance payments can therefore play an important role in the economics of balancing out the fluctuations in energy production.

6 Discussion

This chapter examines how the use of a power exchange can tend to balance out the effect of fluctuations in wind power production. One question still remains, however: is wind power "competitive" in a liberalised market?

Since wind power encounters very low marginal costs it is competitive in the short term where prices are made with respect to the price setter's marginal cost. In the long term, however, wind power will also have to cover its investments and other costs. As a consequence, the competitiveness of wind power in the long run will depend on the total cost of conventional power plants.

Besides the spot price, wind power can be supported directly via subsidies or indirectly via the sale of green certificates, etc. The long-term competitiveness can thereby be ensured.

In a liberalised market, the success of renewable energy supplies like wind power will depend on the revenue produced. Since the revenue of wind power generators depends on the fluctuations and predictions of wind turbine generation, this form of electricity generation may set up claims on the structure of the power exchange. An obvious claim is a reduction of the time horizon between the close of the spot market and the actual energy delivery.

Another claim on the power exchange structure is the way balance payments are handled. The economics of wind power is more dependent on regulation cost than other technologies which do not encounter many fluctuations. If balance payments are handled incorrectly the competitiveness of wind power can be severely compromised.

In the case study described in this chapter the conclusion was reached that in the Norwegian way of handling balance payments the regulation costs of the fluctuations were less than 3.5% of the spot price. This cost can be assumed to be higher if the handling of balance payments were structured differently.

References

A Northern European Power Exchange, http://www.risoe.dk/sys-esy/elbr.htm.

Grohnheit P.E. and O.J. Olsen (1995), 'Electricity liberalization and export of hydro power from the Nordic countries', *Pacific and Asian Journal of Energy*, Vol. 5(2), Special Issue December, pp. 285 – 300.

Grohnheit P.E. (1993), 'Modelling CHP within a national power system', *Energy Policy*, Vol. 21, No. 4, April, pp. 418-429.

Johnsen T.A. (1996), 'Demand, generation and spot price in the Norwegian market for electric power', Working paper, Statistics Norway.

Knivsflå K.H. and L. Rud (1995), 'Markets for electricity: structure, pricing, and performance', *Pacific and Asian Journal of Energy*, Vol. 5(2), Special Issue December, pp. 261-284.

Landberg L. and S.J. Watson (1994), 'Short-term prediction of local wind conditions', *Boundary-Layer Meteorology*, Vol. 70. pp. 171–195, Kluwer Academic publishers.

Landberg L. et al. (1997), 'Implementing Wind Forecasting at an Utility', Risø report R-929(EN).

Newbery D.M. and M.G. Pollitt (1996), 'The restructing and privatisation of the CEGB: Was it worth it?', *Journal of Industrial Economics*, Vol. XLV No. 3, pp. 269-303.

Nielsen L.H. and P.E. Morthorst (eds.), *Fluktuerende vedvarende energi i el- og varmeforsyningen - det mellemlange sigt.*, Risø Report 1055 (1998), 152 p. (in Danish).

Nord Pool ASA – Nordic Power Exchange, http://www.nordpool.no/eng.htm.

Skytte K. and P.E. Grohnheit (1997), På vej mod et frit elmarked. *Electra* Vol. 1, January.

Skytte K. (1997), 'Fluctuating renewable energy on the power exchange. Conference proceedings', Energy Economics Conference, The International Energy Experience: Markets, Regulation and Environment. University of Warwick, December.

Skytte K. (1999a), 'Market imperfections on the power market in northern Europe. A survey paper', *Energy Policy*, Vol. 27 No. 1, March, pp. 25-32.

Skytte K. (1999b), 'The regulating market on the Nordic Power Exchange. An econometric analysis', Forthcoming in *Energy Economics*.

The Danish Ministry of Environment and Energy, *Energy21*. Denmark, 1996.

CHAPTER 17

LESSONS FOR THE UNITED KINGDOM FROM PREVIOUS EXPERIENCES IN THE DEMAND FOR RENEWABLE ELECTRICITY

ROGER FOUQUET

Centre for Environmental Technology, Imperial College of Science, Technology & Medicine, 48 Prince's Gardens, London SW7 2PE, UK
Email: r.fouquet@ic.ac.uk

Keywords: green marketing; markets; price premium; renewable electricity; UK; USA.

Liberalisation of the electricity market in 1998 offers an opportunity for a major diffusion of renewable technology. With inelastic supply, demand may be vital for determining the future development of renewable electricity market. This chapter examines previous experiences of the demand for renewable electricity, mainly in the U.S. and in the Netherlands, to draw lessons for the British market and recommendations for the electricity regulator, if it seeks to promote these alternative technologies. This chapter finds that, although small, a niche demand is likely to actually (as well as hypothetically) exist for 'green' electricity in the United Kingdom at the usual premium prices charged. Raising awareness and generating confidence amongst customers, as well as keeping price differentials low, are likely to be vital to increasing demand for renewable electricity.

1 Introduction

The United Kingdom Government's plans for renewable sources to provide 10% of the country's electricity by 2010 will require a major development of renewable energy technology and markets (DTI 1999). The majority of the research into renewable electricity has tended to focus on the supply-side of the market. This may be an accident or due to a belief that demand is unimportant. With relatively inelastic supply, however, demand could play an important part in influencing the future development of the United Kingdom renewable electricity market especially after liberalisation in 1998 (Fouquet 1998). The current chapter presents a simple model of the private contribution to environmental services (Sections 2-3), examines previous experiences of residential demand for renewable electricity, mainly in the U.S. and the Netherlands, and tries to use this information to understand crucial factors related to demand in the United Kingdom in the post-1998 liberalised market (Section 4). This understanding will provide a basis for recommending policies related to demand that could become vital in determining the future development of renewable electricity in the United Kingdom (Section 5).

2 Household production of services

Based on the household production function approach, economic theory considers that customers try to maximise their utility U with services (Zi) (Lancaster 1966, Becker 1976); i.e.,

$$U = u(Z1, \ldots, Zz) \tag{1}$$

Customers produce these services through the combination of goods (Xj) used and time (Tt) spent with human capital (HKi) and an environment (Ei), which determines the productivity of the consumer for service Zi; so,

$$Zi = f(Xj, Ti, HKi, Ei) \tag{2}$$

Subject to time constraints, they have to split their time between earning income in the labour market (Tw) and producing services (Ti),

$$T = Tw + Ti \tag{3}$$

And to income constraints, they earn money in the labour market (w.Tw) and from other sources (V),

$$Y = w.Tw + V \tag{4}$$

Utility is maximised where the ratio of the marginal utility of service Zi divided by the marginal utility of service Zk is equal to the marginal cost of service Zi divided by the marginal cost of service Zk; which can be represented by

$$MU(Zi)/MU(Zk) = MC(Zi)/MC(Zk) \tag{5}$$

The marginal costs are, therefore, the shadow prices for the services Zi, which are determined by the price of goods, the time spent producing them and the productivity of each customer at producing the services.

3 Value of environmental services and free riding

One of the services individuals desire is environmental quality (Xe). Many studies have shown that people value the environment and are willing to pay to improve its quality (Carson and Mitchell 1989, Worcester 1996). This is because environmental services provide valuable inputs into other services being produced by individuals, such as health and well-being (e.g., Zh).

An individual's combination of goods, time and know-how can only provide the smallest improvement in environmental quality since environmental services (such as air, land, water) are the result of processes within the natural environment. These processes are affected when large numbers of individuals and firms produce certain services and goods; the effects are dependent on the types of activities, the level of activity and of associated pollution, and the assimilating capacity of these natural processes (i.e., the ability to get rid of any pollution) (Arrow et al 1995). Thus, in most cases, individuals do not try to improve environmental quality because they are (usually) aware that the level of environmental services (Ze) will change only slightly and so, utility from Ze (i.e., U(Ze)) will be small. Thus, customers tend to 'free ride' from improving the environment.

Consumers' decisions to fail to free ride must be, therefore, because: they perceive the costs (i.e., MC(Ze) in equation (5)) of acting in an environmentally friendly manner to be very low; they (falsely) perceive their action will improve the environmental service and their utility (MU(Ze) in equation (5)); they perceive other benefit associated with environmentally-friendly behaviour (which we could call MU(Zre)). Thus, despite a large majority of the population not choosing to buy environmentally-friendly goods, some customers will, if they are in either of the three cases[1].

4 Previous experiences of the demand for renewable electricity

Residential demand for electricity is dependent on the efficiency of appliances and the demand for heat, power and light, which is influenced by market structure (including number of choices available), disposable income, the real price of electricity, the real price of other sources of heat, weather conditions (mainly temperature and daylight), and beliefs about features of electricity generation, distribution and use (including social and environmental impact). For renewable electricity, much of the demand will simply alter the share of traditional (i.e., from fossil fuels and nuclear power) electricity consumed rather than overall demand. Thus, the key determinants of renewable electricity demand will be customer beliefs about its attributes (such as its environmental benefits), its price relative to traditional electricity, and the value of free-riding (i.e., not buying an environmentally-friendly product and hoping others will buy it because the environmental benefit from one person's choice is negligible). These features will be examined with reference to previous experiences.

4.1 Information and perception of renewable electricity attributes

The key to creating a demand for 'green' electricity is raising awareness and persuading customers of the value of its benefits and of actually buying it. In most countries, it seems that there is growing media coverage related to the impact of

---

[1] There are numerous explanations in economics for the private provision of public goods, which this chapter does not have the space to explore. The main explanations are: the warmglow effect (Andreoni 1990), signalling (Glazer and Konrad 1996), morality or reciprocity (Sugden 1984), confusion (Andreoni 1995) and docility (Simon 1993). These can be considered explanations for the three cases mentioned above.

fossil fuel use on the environment, especially focussing on air pollution, acid rain and climate change.

Not all information has the same impact on customer perceptions. The factors influencing beliefs are likely to include: whether the information has been acquired actively or passively, the clarity and dramatic content of the message, the frequency with which it is received and the credibility of the source. In the United Kingdom, it seems, for example, that customers are acquiring information passively (e.g., from television), rather than actively seeking sources and are more likely to believe information from non-governmental organisations, rather than from companies or government (Worcester 1996). Indeed, surveys suggest that concise and understandable information is paramount, and this includes the need to reassure customers that 'green' electricity does lead to environmental improvements and that it is sold at as low a premium as possible.

It would seem that the public does believe in the importance of the environmental impact of energy use. Both in the U.S. (77%) and the United Kingdom (67%), a majority of the population considered that renewable sources were the least damaging energy source for the environment (Farhar and Houston 1996, PRASEG 1996).

There appears to be considerable concern, however, that locals are not willing to have new plants 'in their back yard' principally because of the lack of compensation for externalities. The ecological effects of large scale hydropower projects are now well versed. Other renewable energy installations are starting to come under attack, such as wind power, leading to negative externalities, e.g., associated noise and landscapes. In the Netherlands, of the 1225TJ of renewable electricity generated, 70% is from wind energy; and concerns about associated externalities are slowing the expansion of and increasing the installation costs of renewable plants (Slingerland 1997).

The perceived value of renewable electricity could also be influenced by marketing strategies. In the US, companies such as those in Wisconsin (WEPCO) and in New England, have focussed on marketing existing renewable electricity as green. This marketing strategy is an attempt to increase their customer share by providing customers with renewable electricity at no extra cost. Surveys indicate that customers want to "feel confident they are getting a high quality green energy product, and that the company selling the product is reliable and ethical" (Rabago et al 1998, p.39). Customers may become cynical of companies trying to dupe them into explicitly buying what was sold before. Electricity suppliers in the United Kingdom that currently have a large supply of renewable electricity, such as Scottish Power, may seek such a strategy.

In California, concerns about product quality and marketing strategies have led to a renewable-based green power certification programme, which labels appropriate products as 'Green-e'. Within a month of its introduction, the six main renewable electricity suppliers in California sought to be recognised by the Eco-label scheme. Because the programme provided information on the nature of the energy sources used, it raised the quality of products supplied. This programme has also increased customer confidence in the market for renewable electricity, while reducing the costs to firms of supplying and to customers of gathering information about these products; the consequence of the 'Green-e' labelling is a raised interest in renewable electricity in California (Rabago et al 1998).

Despite concerns about 'unethical' marketing strategies, it is probable that even without such schemes many customers will prefer to buy renewable electricity if it can be provided at no extra cost. If a sufficiently large proportion of the electricity can be provided in this way and if customers become aware of this no-premium renewable electricity, an information cascade effect may develop and all customers will demand renewable electricity at no extra cost. While this might drive up costs, it will put considerable pressure on suppliers to rapidly install new renewable plants; the implications of a rapid growth in demand could be both positive, as it creates incentives, and negative, due to upward pressure on prices and a tendency to develop short term strategies towards these plants (as was the case in the dash-for-gas).

4.2 Willingness to pay and the price of renewable electricity

Most plans to sell 'green' electricity are preceded by an assessment of consumers' willingness to pay, usually using the contingent valuation method. Despite doubts about their ability to capture actual behaviour, contingent valuation studies provide an early indication of the scale of demand. For example, this method was used to estimate customer participation in a 'green-pricing' programme in Colorado (Bough, Byres, and Jones 1994). Responses indicated that 82% of customers would contribute \$1-\$4 per month (approximately 10%) more than standard prices to support the development of renewable energy installations. In Sacramento, residents were surveyed; 26% of the general population (57% of the 'green' population) suggested they would be willing to pay an additional 15% for photovoltaic electricity installed on their roofs (Farhar and Houston 1996). Reviews of these estimation exercises suggest that 'green' consumers and older respondents tended to be more willing to pay for renewables (Farhar and Houston 1996). And, although income and years of education appear to be unrelated to willingness to pay levels, studies have also concluded that willingness to pay increases with awareness of environmental damage (Bergstrom et al 1989, Ajzen and Brown 1996), which supports the above argument that the provision of information is likely to be influential in determining future demand. In addition, it seems that the majority of the demand will come from residential rather than commercial consumers (Wiser and Pickle 1998).

In the United Kingdom, the Parliamentary Renewable And Sustainable Energy Group (PRASEG 1996) commissioned a contingent valuation study indicating that a third of the sample population believe that the government should maintain the 10% levy on electricity to support the use of environmentally-friendly electricity. One-fifth said they would be willing to pay a premium for environmentally friendly electricity. Five percent of these domestic customers - equivalent to 4.5 TWh per year, if extrapolated to the whole UK electricity market - would be willing to pay more than a 20% premium. Figure 17.1, constructed from the responses, shows the curve representing the UK public's apparent willingness to pay a premium for green electricity above the average domestic price paid for standard electricity (Fouquet 1998).

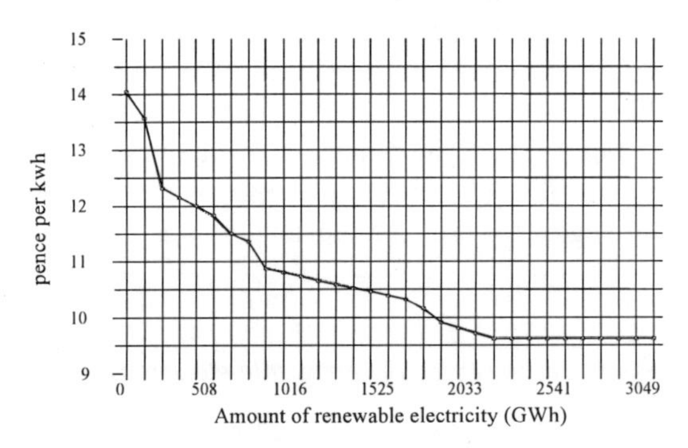

Figure 17.1: Estimates of the willingness to pay for renewable electricity by UK domestic customers

In the U.S. and in the Netherlands schemes have been set up and customers have been faced with prices to actually buy 'green' electricity. Table 17.1 shows the monthly premiums paid by customers in various U.S. utilities. The premiums represent either the average contribution (in the case of a donation program), the monthly premium for an average customer (e.g., based on typical energy use), or the actual monthly fee for participation (if a fixed fee) or for a unit of participation (where a unit might be 100 Watts of capacity).

This evidence indicates that customers paying a premium for specific electricity tend to pay more than those contributing to unspecified future projects. This is borne out by the likely schemes in the United Kingdom. For example, Eastern Electricity are offering customers the option of paying 5% or 10% on top of their bills as part of a scheme for contributions to a charitable trust that will support future projects and research. This is unlikely to be more than British customers will have to pay for specific 'green' electricity in a competitive market[2].

[2] It has been argued, however, that if utilities acquire renewable plants, for example, to build fuel diversity into their portfolio, then customers should not be charged a premium (Eber 1995).

Table 17.1: Premiums/contributions proposed by suppliers for 'green' electricity in the U.S.

	Type of project and source	**Premium/Contribution paid: dollars/month**	**% of bill***
Detroit Edison Company	Premium/Solar	6.59	16.5%
Gainesville Regional Utilities	Contribution/Solar PV	3.27	8.2%
Niagara Mohawk Power Corp.	NA/NA	6.00	15.0%
Public Service Co. of Colorado	Contribution/NA	1.73	4.3%
Sacramento Municipal Utilities District	Premium/Solar PV	6.00	15.0%
Traverse City Light & Power	Premium/Wind	7.58	19.0%
Wisconsin Public Service Corp.	NA/NA	1.85	4.6%

*means the average bill is assumed to be $40 per month.
NA means that the information is not available.
Source: Holt (1996), Farhar and Houston (1996)

4.3 *Actual demand for renewable electricity*

But will customers actually pay even a small premium for renewable electricity? Customers have an incentive to free-ride, that is, to buy the cheapest electricity and hope that others buy the environmentally-friendly type. Since most customers are likely to think this way, there are doubts about whether any customers will actually buy green electricity at a premium. The proportion of customers buying will be very small, certainly at first; in the long run, it is still very uncertain.

Reviews of the U.S. evidence conclude that the actual demand is considerably lower than the willingness to pay studies indicate (Farhar and Houston 1996, Holt 1996). These differences are due to differing incentives to free ride in hypothetical and real situations. In addition, there may be problems with the surveys or with the products (Wiser and Pickle 1997). Nevertheless, there does exist a niche demand for renewable electricity as a differentiated product and, if appropriately marketed, this could develop into a substantial proportion of the overall electricity market (Nakarado 1996).

Table 17.2: Participants, premiums and marketing strategies for 'green' electricity in the U.S.

	Marketing Strategy	**Participation**	**Premium***
Detroit Edison Company	Based on non-targeted mailings	0.3%	16.5%
Gainesville Regional Utilities	Of all customers	1.0%	8.2%
Niagara Mohawk Power Corp.	Based on one targeted mailing	0.6%	15.0%
Public Service Co. of Colorado	Of all resident customers	1.4%	4.3%
Sacramento Municipal Utilities District	Based on telemarketing effort	29.0%	15.0%
Traverse City Light & Power	Of all customers	3.1%	19.0%
Wisconsin Public Service Corp.	Based on market simulation	9.0%	4.6%

* See Table 17.1 for details.
Source: Holt (1996)

Table 17.2 presents projects which did generate an actual demand at higher prices than standard electricity. Wiser and Pickle conclude that: first, a non-negligible proportion of residential customers, if switching suppliers, will base their decision partly on environmental considerations; second, a majority of customers are likely to remain with their current suppliers, however, thus limiting renewable demand; third, there is no single 'green' electricity product, enabling suppliers to use various environmental claims and products.

For example, it seems that suppliers selling contribution-based projects have attracted greater demand. This may be because of the lower price of contribution based-projects. It may also be because product-specific projects can only supply a limited number of customers. In fact, the waiting lists to join such projects, as capacity expands, are lengthy (Farhar and Houston 1996). It is difficult to draw more detailed conclusions though. Many other factors can influence participation levels, including programme design, ease of participation, customer awareness and marketing effort, as well as other features bundled with the renewable energy supply that increase its value (Holt 1996).

Overall though, initial demand in the U.S. has been weak and few suppliers have been willing to risk entering the markets - and those that did were mostly able to re-market existing generation capacity. Future returns will need to be substantial to justify the development of new capacity necessary for the market to grow. The fragility of the market means that demand and supply will need to grow at similar rates, avoiding excess strains to either consumers or suppliers. So, despite the existence of a demand for renewable electricity, it is highly uncertain whether these markets will be successful, although the uncertainty can be reduced by the appropriate public policy that minimises transaction costs faced by firms and customers, controls market power and minimises market barriers to entry (Wiser et al 1998).

The creation of a market for 'green electricity' in the North of the Netherlands also shows that consumers are buying electricity from renewable sources. Over 10,000 customers are buying renewable electricity from wind power and incinerators from two suppliers, PNEM and EDON. In early 1996, they paid on average a premium of 15-20% above their normal bill. This differential has fallen after the introduction of an eco-tax on the consumption of non-renewable electricity (ENDS March 1996).

It would appear that electricity customers in certain countries are actually buying renewable electricity. The Renewable Energy Company Ltd and South Western Electricity Plc have already had consumers demanding renewable electricity. Eastern Electricity plc is only expecting around 10,000 customers (about 0.3% of their current total customers) a couple years after the beginning of its campaign for contributions to develop a trust for investing in 'green' electricity. It is clear as the overall electricity market opens in the United Kingdom (as well as in

Europe), actual demand will exist, even if not on the scale suggested by Figure 17.1 (and its underlying survey of hypothetical demand).

5 Lessons, conclusion and recommendations

Evidence shows that the market in the United Kingdom is not likely to be one-sided: consumers can influence the development path of renewable technologies and markets, and regulators should bear them in mind when formulating policies about energy and associated environmental effects. The purpose of this chapter is to examine the evidence related to the demand for renewable electricity, use it to draw lessons for the future development of the market in the United Kingdom and recommend appropriate policies for promoting successful development.

Most of the evidence comes from the U.S. or the Netherlands, and some preliminary information within the United Kingdom itself. In these countries, there exists a hypothetical willingness to pay a premium for renewable electricity; the related demand curve appears to be price-sensitive. The premium generally proposed by suppliers seems to be at about 10% for contributions to unspecified future projects and 10%-30% for specific renewable electricity. This higher value is within the range of the hypothetical willingness to pay in the United Kingdom. Studies also suggest that the actual demand is considerably lower, certainly at first, than the hypothetical estimates. Nevertheless, a niche demand is likely to exist in the United Kingdom as more companies develop their marketing strategies and as the whole of the electricity market becomes more competitive.

Each country's endowment of renewable resources is unique. In particular, the U.S. evidence is mainly based on solar power; and in the Netherlands, almost exclusively wind generated electricity. The United Kingdom has a considerable amount of hydro power, although there appears little potential for any large scale installations. The future sources of renewable electricity are likely to be mainly wind, wave and waste. Nevertheless, the specific resource endowments of a country are likely to be less relevant for understanding the demand side than for understanding the supply side. More important are the different budgets, different tastes and different perceptions in each country. But because there are still many common features between the countries observed, an understanding of these previous experiences does have, however, some value and provides sufficient grounds to anticipate an actual demand for renewable electricity in the United Kingdom.

In the United Kingdom, the Government has taken on the responsibility to ensure an adequate standard of environmental quality (Department of the Environment 1996). Since few individuals and firms take account of the environmental effects of their choices, there is a case for the Government to regulate those economic activities in order to reduce market failures and improve overall well-being. The Government has passed on the responsibility of managing pollution from electricity generation, distribution and use to the Office of Electricity Regulation (OFFER).

One approach that is expected to reduce electricity generation pollution would be through a shift in sources of fuel used. The dash for gas amongst generators in the early 1990s will be one of the main reasons the UK manages to meet its agreement to stabilise to the 1990 level of CO_2 emissions by 2000 (Department of the Environment 1997). The UK Government is planning that by 2010 renewable energy will provide 10% of the UK's electricity mix (DTI 1999). In 1998, it provided around two percent. Such a dash for renewables would have a substantial impact on emissions. Thus, if current institutional barriers and market failures can be minimised, the environmental and social benefits from ensuring the successful development of a market for renewable electricity would be considerable.

To harness the potential benefits, the Government, as well as companies marketing 'green' electricity, will need to take account of features on the demand-side of the market[3]. This review of previous experiences related to renewable demand has provided some insights. The chapter indicates there is likely to be a genuine demand for renewable electricity in the United Kingdom, for which a very small proportion of the customer base may be willing to pay a premium of approximately 20% more than standard electricity. To ensure that these customers actually buy 'green' electricity, they should be provided with clear information about the costs and environmental benefits of producing electricity from renewable sources. There is likely to be a period of experimentation to target the appropriate customers and encourage them to buy the product. This will also involve developing trust in customers that are likely to be cynical about information from companies. One way to ensure both issues are dealt with is for an outside body, such as a non-governmental organisation, to be encouraged to monitor suppliers' decisions and impact upon the environment. Such a scheme currently being developed is a form of eco-labelling accrediting electricity suppliers as 'green' by the Energy Savings Trust. To increase demand, it might be necessary to create financial incentives. These could include an environmental tax on non-renewable electricity; naturally, this means that electricity becomes more expensive than other competing fuels, which may herald the need for taxes on all energy sources. Also, supply companies may seek to reduce free-riding as much as possible. This could include: raising social pressures associated with environmental-friendly behaviour by appealing to customers' sense of community, and creating local subsidiaries more in contact with community; convincing customers that their actions have an effect; encouraging longer term contracts; bundling private goods (e.g., personal

[3] Fouquet (1998) examines related policy recommendations in more detail.

energy services) with the public good of environmental improvements (Wiser and Pickle 1997). In any case, information provision and financial incentives are likely to be vital in determining the future demand for renewable electricity in the United Kingdom, especially after the liberalisation of the electricity market.

References

Ajzen I. and T.C. Brown (1996), 'Information bias in contingent valuation: effects of personal relevance, quality of information and motivational orientation', *Journal of Environmental Economics and Management*, 30, pp. 43-57.

Andreoni J. (1990), 'Impure altruism and donations to public goods: a theory of warm-glow giving', *The Economic Journal*, 100, pp. 464-77.

Andreoni J. (1995), 'Co-operation in public good experiments: kindness or confusion?', *American Economic Review*, 85, pp. 891-904.

Arrow K., Bolin B., Costanza R. et al. (1995), 'Economic growth, carrying capacity, and the environment', *Science*, 268, pp. 520-1.

Baugh K.A., Byrnes B. and C.V. Jones (1994), 'Developing a Customer Driven Renewable Program: Public Service Company of Colorado's Use of Market Research Findings', *6th Biennial Marketing Research Symposium*, EPRI TR-104558. Palo Alto, Calif. Electric Power Research Institute. pp. 55-68.

Becker G.S. (1976), *The Economic Approach to Human Behaviour*, London: University of Chicago Press.

Bergstrom J., Stoll J. and A. Randall (1989), 'Information effects in contingent markets', *American Journal of Agricultural Economics*, 71, pp. 685-691.

Carson,T.A. and R.C. Mitchell (1989), *Using Surveys to Value Public Goods: The Contingent Valuation Method*, Resources for the Future, Washington D.C.

Department of the Environment (1996), *Indicators of Sustainable Development*, HMSO, London.

Department of the Environment (1997), 'UK On Course to Meet International Obligations on Climate Change', News Release, 18 February, London.

DTI (1999), *New and Renewable Energy. Prospects for the 21st Century.* Department of Trade and Industry, London.

Eber K. (1995), 'Green Pricing Opens New Markets for Renewables'. This chapter provided the basis for an article published in *Electrical World*, September 1995 (available on the Internet on http://www.eren.doe.gov/greenpower/library. html).

Farhar B. and A. Houston (1996), *Willingness to Pay for Electricity from Renewable Energy*, NREL/TP-460-21216, National Renewable Energy Laboratory, Golden, Co.

Fouquet R. (1998), 'The United Kingdom demand for renewable electricity in a liberalised market', *Energy Policy*, 26(4), pp. 281-93.

Glazer A. and K.A. Konrad (1996), 'A signalling explanation for charity', *American Economic Review*, 86(4), pp. 1019-28.

Holt E.A. (1996), 'Green Pricing Experience and Lessons Learned', 1996 ACEEE Summer Study, August (available on the Internet on http://www.eren.doe.gov/greenpower/library.html).

Lancaster K.J. (1966), 'A New Approach to Consumer Theory', *Journal of Political of Economy*, 74, pp. 132-57.

Nakarado G.L. (1996), 'A market orientation is the key to a sustainable energy future', *Energy Policy*, 24(2), pp. 187-93.

PRASEG (1996), 'Consumers are Willing to Pay for a Green Energy Future', Parliamentary Renewable and Sustainable Energy Group Press Release, 30 October, London.

Rabago K., Wiser R.H. and J. Hamrin (1998), 'The Green-e program: an opportunity for customers', *The Electricity Journal*, 11(1), pp. 37-45.

Simon H.A. (1993), 'The Economics of Altruism', *American Economic Review Proceedings*, 83, pp. 156-60.

Slingerland S. (1997), 'Energy conservation and organisation of electricity supply in the Netherlands', *Energy Policy*, 25(2), pp. 193-203.

Sugden R. (1984), 'Reciprocity: the supply of public goods through voluntary contributions', *The Economic Journal*, 94, pp. 772-87.

Wiser R.H. and S.J. Pickle (1997), *Green Marketing, Renewables and Free Riders: Increasing Customer Demand for Public Goods*, Lawrence Berkeley National Laboratory, LBNL-40632.

Wiser R.H. and S.J. Pickle (1998), *Selling Green Power in California: Product, Industry and Market Trends*, Lawrence Berkeley National Laboratory, LBNL-41807.

Wiser R.H., Pickle S.J. and J.H. Eto (1998), 'Detail, Details... The Impact of Market Rules on Emerging 'Green' Energy Markets', *Proceedings for ACEEE 1998 Summer Study on Efficiency in Buildings*, 23-28 August, Pacific Grove, California.

Wood L., Kenyon A., Desvousges W. and L. Morander (1995), 'How Much Are Customers Willing to Pay for Improvements in Health and Environmental Quality?', *Electricity Journal*, 8(4), pp. 70-77.

Worcester R. (1996), 'In the Aftermath of Brent Spar and BSE'. Paper given at the Business & the Environment Programme, 16 September, University of Cambridge, Cambridge.

245-59

CHAPTER 18

MODELLING THE PROSPECTS FOR RENEWABLE AND NEW NON-RENEWABLE ENERGY TECHNOLOGIES IN THE UK AND SOME OF THE CONSEQUENCES IMPLIED

Q42 L94 D12

REINHARD MADLENER[1]

Institute for Advanced Studies Carinthia, Domgasse 5, A-9010 Klagenfurt, Austria
Email: madlener@carinthia.ihs.ac.at

Keywords: cost-benefit analysis; emissions; employment; energy modelling; renewable energy technologies; SAFIRE.

1 Introduction

At present less than 1% of the primary energy requirements in the UK and less than 2% of the total electricity supply are met by renewable energy sources (RES), despite a large estimated theoretical potential in the range of 700-1,100 TWh per annum (cf. Table 18.2), and the fact that the UK wind, tidal and wave energy resources are among the finest in the world. Over the late 1980s and early 1990s, respectively, the contribution of energy generated from RES towards total inland energy consumption remained essentially unchanged at an astonishingly low level of around 0.5%.

In recent years, there has been an increasing commitment towards the use of RES, both from the European Community (ALTENER, JOULE and THERMIE Programme, Declaration of Madrid, 1995 White Paper on energy, 1997 White Paper on renewable energy sources, etc.) and the UK government, whose introduction of the Non-Fossil Fuel Obligation (NFFO) by means of the 1989 Electricity Act has turned out to be quite a successful approach in promoting renewable energy projects—although the original intention was the support of the nuclear power sector only.

Policy issues relatively high on the political agenda both in the UK and the EU are competitiveness, reduction of long-term structural unemployment, and—somewhat further down the list—protection of the environment. The increased use of renewable and new non-renewable energy technologies (RETs and new non-

[1] The author gratefully acknowledges financial support received from the European Commission (Human Capital and Mobility Grant No. 940613a) during his stay at the Macroeconomic Modelling Bureau, University of Warwick, Coventry, UK. The study was enabled through the provision of the SAFIRE software and documentation by ESD Ltd., Corsham, U.K., and continuous product support given by Mark Whiteley and Alastair Gill. The chapter is a considerably shortened version of the paper presented at the BIEE conference "The International Energy Experience: Markets, Regulation and Environment", University of Warwick, 8-9 December, 1997.

RETs) could help to achieve all three of these goals and hence yield a "triple dividend".

The aim of this study is to investigate the market potential as well as the expected market penetration of RETs and new non-RETs in the UK for the period 1993-2020, by using the energy substitution model SAFIRE[2]. Outcomes from scenario simulations are presented which were used for an assessment of the potential first-order impacts on the economy and environment that result from the market penetration of RETs and new non-RETs. An extensive study from ETSU (1994ab) was used to contrast certain assumptions and outcomes.

The chapter is organised as follows: Section 2 introduces the main features of SAFIRE. Section 3 contains a description of the three scenarios studied for this particular modelling exercise. Section 4 reports on the main results gained, and Section 5 concludes.

2 The SAFIRE approach

SAFIRE is an engineering-economic bottom-up model for the assessment of first-order impacts of "rational" (i.e., renewable and new non-renewable) energy technologies on a national, regional or local level against a background of different policy instruments and scenario assumptions. Due to a lack of space, the model description given here only covers the bare essentials needed to understand the main principles of SAFIRE (for more details see ESD, 1995ab, 1996ab; or Madlener, 1997a).

The time horizon of the model used is 28 years (1993–2020), and the various outputs are made available at five-year intervals. SAFIRE is data-driven and includes an extensive database for 22 RETs and 9 new non-RETs (cf. Table 18.1). Moreover, 10 conventional primary and secondary energy sources are included, all assumed to be available in unlimited amounts: peak and off-peak electricity, natural and derived gas, hard coal/coke, lignite/peat, light and heavy oil, petrol, and diesel. Input data in SAFIRE are grouped into base (or scenario-independent) data and scenario data. They are used jointly to calculate the substitution potential and impacts of the RETs and new non-RETs, first for *decentralised* heat and electricity generation and then—in order to account for the residual electricity demand—for *centralised* electricity generation (i.e., all heat energy is assumed to be provided decentrally).

Starting from the base year of a particular scenario, the main calculation sequence for *decentralised use of RETs and new non-RETs* is the following: (1) determine the energy demand for each end use/sector combination (activity indicator × specific energy consumption; cf. Section 4.2); (2) determine the Technical

[2] SAFIRE (*S*trategic *A*ssessment *F*ramework for the *I*mplementation of *R*ational *E*nergy) has been continuously modified in the past. The version employed in this study, v1.31, was released in September 1996.

Potential (TP) for each RET (demand- or resource-constrained), serving as an upper limit for the Market Potential (MP) calculation (steps 3–5); (3) calculate the payback period for each new technology and permissive end use/sector combination that would arise if the technology was used to replace a certain conventional technology; (4) determine the maximum long-term proportion of the achievable MP (the Market Fraction—MF), using the MP/payback period curve chosen for the scenario; (5) repeat stages 2–4 for each competing technology; (6) compute the actual Market Penetration (MPen) for the current year from the MPen in the previous year, a user-defined S-shaped market diffusion curve, and the long-term MP; (7) repeat stages 2–6 for all new technologies that can be used decentrally in order to get the contribution made by RETs and new non-RETs towards the various end-use energy needs (electricity, heating, cooling, transport, etc.).

Co-generation (CHP) is assumed to be all decentralised and restricted to the industrial and commercial/institutional subsectors (cf. Table 18.1 for the admissible technologies). In order to assess the CHP potential, SAFIRE calculates average unit sizes for each subsector (thermal and electricity requirements divided by the number of plants in the subsector)[3].

For *centralised electricity generation* and national level studies, the mix of plants is computed on a least-cost dispatching basis. Particularly, the calculation of the demand for centrally generated electricity is based on the generation level in the base year (historical data), the estimated electricity demand growth over the model horizon (user-entered by sector), the retirement of old plant stock, and the contribution made by RETs and new non-RETs for decentralised use.

Seven different *cost-benefit (C-B) indicators* are calculated by SAFIRE: pollutant emissions, employment, government revenues, energy import dependency, value added, capital expenditures, and external costs. These C-B indicators are calculated as *net* effects, taking into account the impact of the new technology on the one hand and the impact caused by the displaced conventional technology on the other hand (see ESD, 1996a, 1996b; or Madlener, 1997a, for details).

[3] In particular, the database contains capital cost figures for each CHP technology for three different plant sizes. SAFIRE then makes an approximation for the capital cost of the computed CHP plant size by ordinary least squares estimation. There are three options: (i) no buyback, no third-party access (TPA); (ii) buyback, no TPA; (iii) TPA.

Table 18.1: SAFIRE technology/sector matches

Technology / sector	dom.	comm./ inst.	ind-ustry	agric.	transp.	centr. el.
(a) renewable energy technologies						
wind power	•	•		•		•
large-scale hydro (≥5 MW)						•
small-scale hydro (< 5 MW), photovoltaics (PV)[†]	•	•	•	•		•
active solar thermal	•	•	•			•
passive solar design (PSD)[‡]	•	•				
forest residues[*]	•		•			
woody energy crops, ethanol, biodiesel[*]	•		•		•	•
solid agricultural wastes[*]		•		•		
liquid agricultural wastes				•		
solid/liquid industrial wastes[*]			•			
MSW[*]		•				•
MDW[*]		•				
landfill gas[*], geothermal electric, geothermal heat			•			•
wave power, tidal power						•
(b) new non-renewable energy technologies						
Fuel cells[*], heat pumps		•				
gas co-generation		•	•			
IGCC, PFBC, HOCC, OCGT, CCGT, PWR						•

Notes: [†] *For decentralised use assumed to be building-integrated systems.* [‡] *PSD only comprises passive heating for the domestic and commercial/institutional sectors and daylighting & cooling for the commercial/institutional sector. Technologies marked with an* [*] *can also penetrate as CHP technologies. IGCC=integrated gasifier combined cycle, PFBC=pressurised fluidised bed combustion, HOCC=heavy oil combined cycle, OCGT=open cycle gas turbine, CCGT=combined cycle gas turbine, PWR=pressurized water reactor (nuclear).*

3 The scenarios investigated

- *Base Case (BC):* For the BC scenario, which serves as a benchmark against the two alternative scenarios, representative data were used and parameter values we considered reasonable. In particular, many of the judgements made by the model proprietor with regard to the commercial and technological developments of the various technologies over the model horizon have been adopted and—in cases we deemed it necessary—adjusted accordingly. An example is the assumed development of the sectoral energy demands and conventional fuel prices, which has been brought closely in line with that used in Energy Paper 65 (DTI, 1995). This is in contrast to Madlener (1997b), where energy demand and price projections from a MIDAS study were taken (MIDAS is the main energy forecasting model used by the EC).
- *High-Fossil-Fuel-Prices (HFFP):* Under the HFFP scenario, we suppose that all fossil fuel prices covered by the model rise faster than under the Base Case, which makes the new technologies relatively more competitive (either because

they use no or hardly any fossil fuels, or because they have higher conversion efficiencies than the conventional technology replaced). Particularly, compared to the Base Case, the annual growth rates for fossil fuel prices are 1% higher (assumed to raise electricity prices by 0.7% p.a.), while the price for nuclear fuel is assumed to remain constant.

- *High-Environmental-Concern (HEC):* Under the HEC scenario, societal, political, and managerial decisions are assumed to be more strongly influenced by environmental concerns than under the Base Case. Consequently, we assume that the government undertakes political action in order to promote the use of RETs and to reduce the barriers for their increased implementation (e.g., by campaigns, payment of small subsidies, etc.). Relative to the Base Case, four particular assumptions have been made: (i) consumers are assumed to be willing to pay a 5% premium for energy produced from RES (calculated as a mark-up on the cost per usable kWh produced from conventional sources); (ii) the MF curve is chosen such that the MP for a given payback period is slightly larger; (iii) the MPen rate is assumed to be somewhat higher; (iv) the manufacturing, construction & installation (MC&I) and operating & maintenance (O&M) costs of the RETs and new non-RETs are assumed to be subsidized by 5% (except for the MC&I costs for biofuel production, which are assumed to be subsidized by 37 ECU/t, 0.15 ECU/l and 0.36 ECU/l, respectively, from 1993-2020, instead of until the year 2000 as in the Base Case).

Table 18.2: Accessible resource (MARKAL) vs. technical potential (SAFIRE), in 1,000 GWh p.a.

technology	TP	AR	technology	TP	AR
onshore wind	267.5–286.2	340.0	forestry wastes	6.7	5.0
offshore wind	-	380.0	agric. solid wastes	20.3	11.2
small-scale hydro	3.9	3.9	agric. liquid wastes	0.7	2.9
large-scale hydro	7.7	6.9	municipal and industrial wastes	120.8–139.5	36.2
PV	27.9–33.8	84.0	landfill gas	22.9	5.3
photoconversion	-	84.0	tidal	40.0	19.0
active solar thermal	68.1–78.1	30.8	shoreline wave	50.0	0.4
passive solar design (PSD)	0.0	10.0	offshore wave	-	0.03
energy crops	49.0–54.8	194.0	geothermal heat	1.3	1.3

Notes: Where a range of TP figures is given, the former refers to 1993 and the latter to 2020. AR figures provided are estimates at an assumed discount rate of 8%.

4 Results

4.1 Technical potential

The assessment of the TP for RETs undertaken in SAFIRE is based on an estimate for the energy a technology could supply if there were no economic constraints, market barriers, or competition from other energy sources. It can be *demand-*

constrained, as in the case of solar power, for example, or *supply-constrained*, as in the case of a more finite energy resource such as forest residues[4]. Running SAFIRE for the Base Case produced an estimated annual TP of 687 TWh for the base year 1993, expected to rise gradually to about 746 TWh by 2020[5]. Table 18.2 shows a comparison between ETSU's "Accessible Resource" (AR)[6] using the MARKAL[7] model (ETSU, 1994b; DTI, 1994) and the TP calculated with SAFIRE.

Although the definitions underlying the TP and the AR concept are slightly different and hence comparisons should be made with care (the AR assessment, for example, is based on generation costs < 10p/kWh, while that of the TP is not), it is interesting to contrast the outcomes in terms of the relative importance of the various technologies considered. For the AR (discount rate 8%), wind power has by far the greatest resource potential (59.2%), followed by energy crops (15.9%), photoconversion (6.9%), and PV (6.9%). All other technologies contribute less than 3% each (at the 15% discount rate, the picture remains very much the same, apart from an AR that is by a factor of twenty smaller for PV and zero for tidal power). By contrast, regarding the TP for 1993, wind energy again is the dominating power source (with a share of "only" 38.9%, mainly because the SAFIRE version used does not include off-shore wind), followed by municipal and industrial waste combustion (17.6%), AST (9.9%), wave power (7.3%), energy crops (7.1%), and tidal power (5.8%). All other RETs contribute 4% or less each. In sum, the biggest difference in total numbers stems from the non-availability of off-shore wind in SAFIRE and a seemingly more conservative view about the potential for energy crop production,[8] which is only partly offset by a more optimistic view regarding municipal and industrial waste combustion.

4.2 *Energy demand*

In SAFIRE, the energy demand for every subsector and end-use is computed as the product of an activity indicator (e.g., floor space, industrial output) times a corresponding specific energy consumption (SEC) factor. Due to the great difficulties encountered in obtaining activity indicator and SEC data for the

---

[4] In some cases, the TP may be further constrained by other factors, such as planning restrictions (e.g. National Parks) or physical restrictions (e.g. maximum feasible number of wind turbines on a given area); see ESD (1996b) for details.

[5] The TP of some RETs is allowed to vary over time, e.g., for wastes to reflect different waste policies, for wind to reflect technological progress, and for PSD as a function of demand for space heating, cooling and daylighting.

[6] The Accessible Resource represents the theoretical resource available for exploitation by mature RETs after only primary constraints are considered (e.g. excluding wind power use in National Parks or areas used for housing, roads and lakes). In most cases full exploitation of the AR is unlikely to be acceptable (cf. ETSU, 1994b, p.26).

[7] MARKAL ("Market Allocation") is the energy systems model of the International Energy Agency, used in many variants and countries throughout the world.

[8] The assumption used in SAFIRE was a land area available for crop production of 3 mio. hectares, while ETSU apparently used their long-run estimate of 5 mio. hectares.

43 industrial subsectors that are the SAFIRE defaults, we decided to use only 12 instead, in line with those published in the Digest of UK Energy Statistics (DTI, 1997, Table 9)[9]. For the Base Case energy price and demand trajectories have been calibrated against those in DTI (1995), taking into account the exclusion of air transport energy demand in SAFIRE due to the lacking substitutional potential for RES.

4.3 *Market potential*

Figure 18.1 depicts the estimated MP over time for each RET for decentralised use, both for heat and electricity generation, under the Base Case. Concerning *electricity generation* (left plot), it can be seen that industrial solid waste, woody energy crops, and forest residues have the largest MP. Both for the HEC and the HFFP scenario (not shown), LFG turns out to have a remarkably high potential towards the end of the modelling horizon. Furthermore, considerable and fairly steady potentials in the range between 400 and 1,000 GWh p.a. exist for municipal solid, industrial liquid and agricultural solid wastes. The MP for wind power rises considerably under the HEC scenario, where PV is also predicted to make a modest contribution. Regarding *heat generation* (right plot), agricultural and industrial solid wastes and wood energy crops turn out to be most important. As for electricity generation, the growth of the MP for LFG both under the HEC and the HFFP scenario are striking (not shown). The MPs for MSW, industrial liquid waste and forest residues are in the range of 2,000 to 3,000 GWh per annum. Active solar thermal makes a modest contribution under the HEC scenario only.

The total MPs for decentralised electricity generation from RETs turn out to be between 5,000-6,000 GWh p.a. in 1990, reaching about 8,000 (BC), 12,000 (HFFP), and 15,000 (HEC) GWh p.a. by the year 2020. For heat generation, the market potential for decentralised use of RETs is predicted to be between 27,500 (BC), 37,000 (HFFP), and 43,500 (HEC) GWh p.a., respectively.

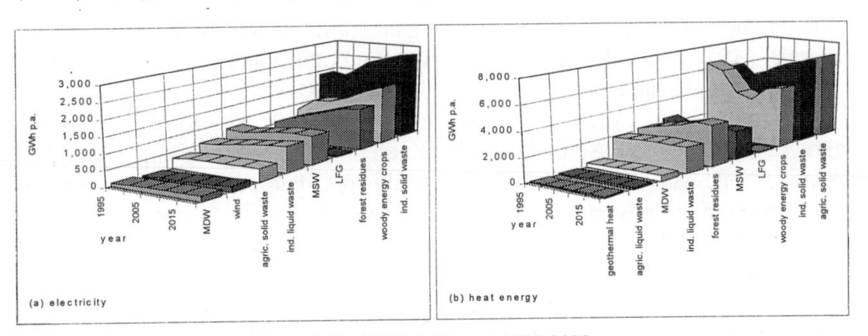

Figure 18.1: Estimated market potential by RET, base case, 1995-2020

[9] The commercial/institutional sector is divided into seven, the agricultural sector into six, and the transport sector into two subsectors (petrol and diesel).

4.4 Market penetration

The Market Penetration (MPen) reflects the speed at which a technology is assumed to take a place in the market. For the RETs the total MPen for electricity generation is predicted to grow rapidly from 5,700 GWh p.a. in 1993 (all three scenarios) to 24,700 GWh p.a. for the Base Case (HEC: 38,700 GWh p.a.; HFFP: 25,800 GWh p.a.) in 2020, while that for heat generation is about 3,700 GWh p.a. in 1993 (all three scenarios), and saturates at a level of some 23,600 GWh p.a. for the Base Case (HEC: 29,900 GWh p.a.; HFFP: 27,000 GWh p.a.) by 2020. Quite remarkable in the case of the RETs and for all scenarios considered is the fact that centralised electricity generation makes a significant contribution to the overall MPen towards the end of the modelling horizon, mainly due to the contributions made by large-scale hydro and wave power stations (see below).

With regard to the new non-RETs the differences between the three scenarios are marginal, with both electricity and heat generation shown to grow at a decreasing pace. Rather unsurprisingly, in contrast to the RETs, electricity generation is much more important from the outset for the new non-RETs, because the majority of these technologies are used for centralised electricity generation (and hence predominantly for large-scale applications).

Figure 18.2 presents the expected development of the MPen by RETs and by technology over time, while Figure 18.3 reports the same for the new non-RETs. As shown in Figure 18.2 (plots a/c/e), the *electricity generated from RETs* stems mainly from large-scale hydro power. Rather surprisingly, wave power is predicted to make a very substantial contribution to the total MPen figure by 2020. The use of wind power and LFG rises rapidly both for the HEC and HFFP scenarios. Industrial and municipal solid waste, woody energy crops, forest residues and MDW are shown to account for an appreciable share, while the predicted contribution of all other RETs is either very little or zero.

With regard to *heat generation from RETs* (Figure 18.2, plots b/d/f), one can see that agricultural and industrial solid wastes, woody energy crops and forest residues dominate the picture. LFG utilization gains importance rapidly for the HEC scenario. MSW and industrial liquid waste make contributions at a smaller scale, while geothermal heat, agricultural liquid waste and AST are almost negligible.

The plots for *electricity generation from new non-RETs*, reported in Figure 18.3 (a) for the Base Case, are dominated by the growing importance of the combined cycle gas turbine (CCGT), the steady contribution made by nuclear power stations and the initially rapidly growing and later on saturating contribution of gas co-generation. Both heavy oil combined cycle (HOCC) power stations and the use of fuel cells account for a very small share only.

For the two alternative scenarios HEC and HFFP, open cycle gas turbine (OCGT) power stations turn out to have a non-zero MPen. As Figure 18.3 (b) shows, the only new non-RET penetrating the market in terms of heat generation is (decentralised) gas CHP, and there is little variation among the three scenarios in terms of the MPen.

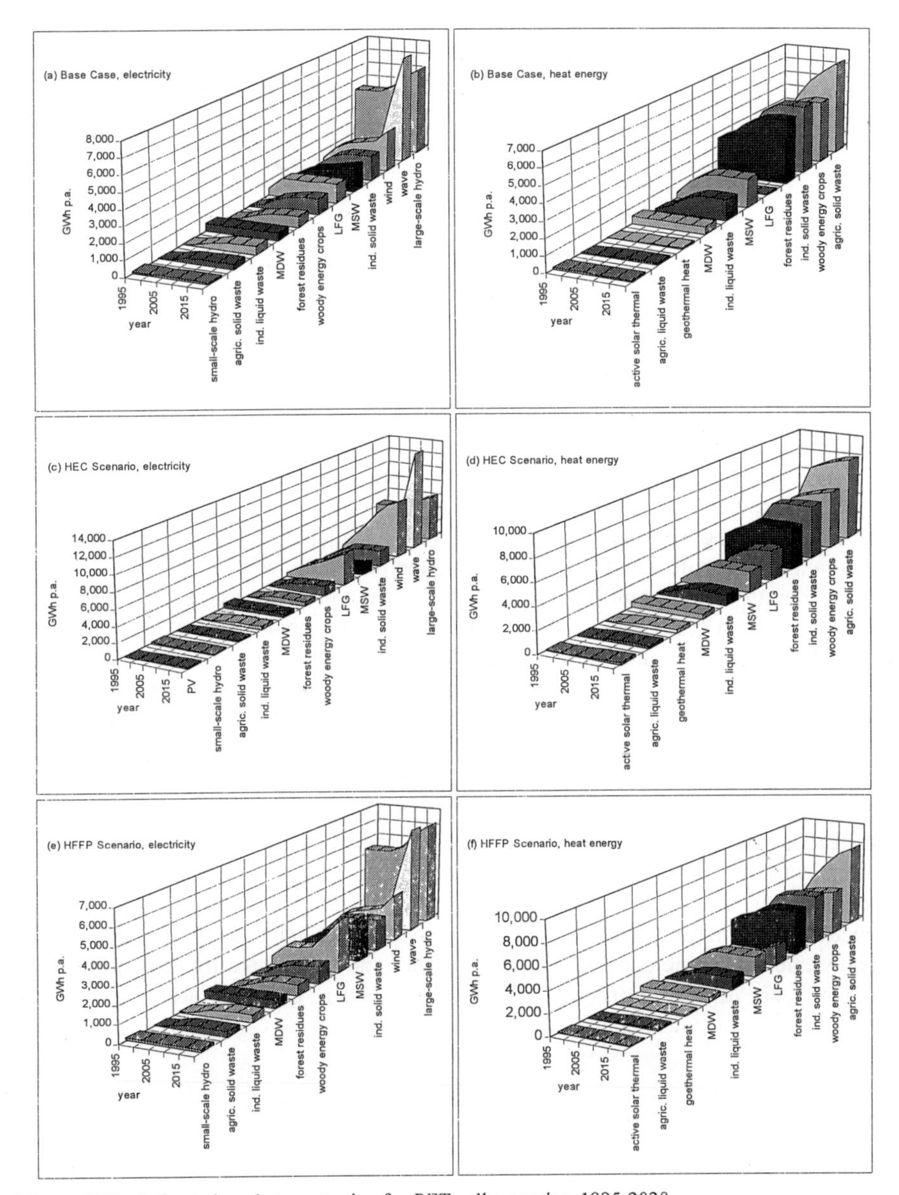

Figure 18.2: Estimated market penetration for *RETs*, all scenarios, 1995-2020

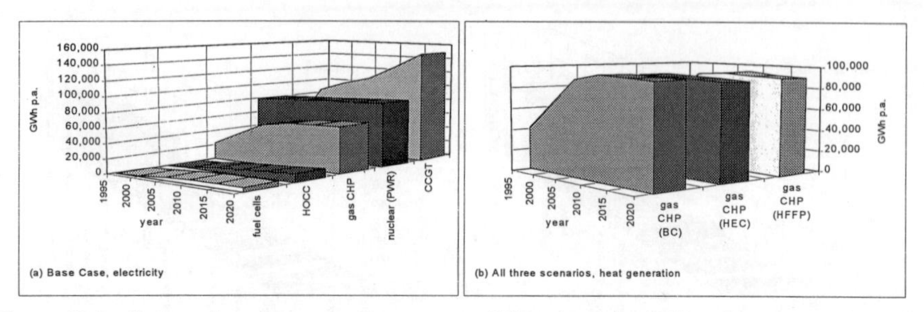

Figure 18.3: Estimated market penetration *new non-RETs,* electricity (BC) and heat generation (all), 1995-2020

4.5 *Cost-benefit assessment*

SAFIRE has been developed as a model for the analysis of the EU12 countries. The seven different cost-benefit (C-B) indicators produced (employment, emissions, value added, net government revenues, import dependency, capital expenditures and externalities) are calculated from the MPen and nationally adjusted C-B coefficients taken from an input-output (I/O) model of the German economy of 1987. The C-B indicators give a crude indication of the direct substitutional effects that arise from the MPen of RETs and new non-RETs[10]. For obvious reasons, even when employing adjustment factors, the C-B indicators can only represent very coarse signals as to how and by how much the economic and environmental situation might change as a result of this penetration[11].

SAFIRE only calculates the changes that might occur, as compared to a situation where no MPen by RETs and new non-RETs has occurred. In other words, no absolute levels are computed. The C-B impacts are calculated separately as effects caused by MC&I on the one hand and O&M on the other hand (sometimes including fuel combustion). Because every beneficial effect caused by the penetration of new technologies is linked with an adverse effect due to the displacement of some conventional fuel(s), the reported figures for the various C-B indicators are *net* effects (referenced to the base year). Due to a lack of space only employment and emission effects can be reported here.

[10] Average coefficients were used for each branch in the I/O table, so that the I/O model operates at constant returns to scale. The C-B coefficients used in SAFIRE have been calibrated for 1 bn ECU of final demand.

[11] What SAFIRE urgently needs, in our opinion, are either I/O-coefficients derived from country-specific I/O tables or at least an update with I/O-coefficients taken from a more recent I/O-table. First steps in this direction, for selected RETs, electricity generation, and UK employment impacts only, have been undertaken by ECOTEC (1995), see Section 4.5.1 below.

4.5.1 Employment

Net employment created, in person-years, is computed as the sum of jobs created by new MC&I activities and the number of jobs created by O&M, minus the jobs lost by the replacement of conventional technologies, plus the amount of jobs created by indirect effects in other industrial sectors producing inputs to the "new" energy sector and, where relevant, by biofuels.

Regarding the *RETs*, Figure 18.4(a) depicts the net employment effects caused by their MPen for the BC. Striking is the strong positive contribution from woody energy crops, the large (initially positive, later slightly negative) employment effect caused by the MPen of industrial solid waste (MSW has a similar impact at a smaller scale), and the large negative employment effect caused by the penetration of wave power. For the HFFP scenario, a much stronger positive employment effect caused by LFG towards the end of the model horizon occurs, which pushes the total net employment figures up by some 2,500 person-years in 2015 and 2020. For the HEC scenario, corresponding to the higher MPen found in Section 4.4, wind energy and LFG account for a stronger positive employment effect (e.g., 2010: +2,900 person-years for wind, relative to the BC, and +2,150 person-years for LFG), only partially offset by the increasingly adverse impact of wave power (-2,600 person-years in 2020, relative to the BC). Total net employment created ranges from 1,400-17,500 for the BC, 4,700-18,000 for the HFFP, and 3,200-25,900 for the HEC scenario.

Considering the development for the *new non-RETs* in Figure 18.4(b), it is important to note that the results only reflect the positive impact caused by the new non-RETs (and not the negative effects caused by the replacement of conventional technologies, which are reported separately in SAFIRE). The striking feature of the Base Case outcome is the large positive employment effect both of CCGT and gas co-generation, dwarfing the relatively small positive impact of the HOCC power plants and the negative impact of nuclear power stations of a similar magnitude. By technology, decentralised gas CHP is responsible for the largest increase in employment creation (e.g., more than 9,000 person-years in 2020, relative to the BC), while relative to the baseline scenario between 750 (1995) and 3,600 (2020) job-years are lost due to the lower penetration of HOCC power stations. For CCGT power stations, the difference in magnitude of the employment effect is most striking for the year 2020 (+8,500 person-years, relative to the BC). For the HEC scenario, we get the interesting result that several thousand job-years less are created from the year 2000 onwards, relative to the Base Case, mainly on account of the lower MPen of CCGT and HOCC power stations for heat generation. Over the total model horizon, total employment created by new non-RETs ranges from 55,900-88,200 (BC), 60,000-87,400 (HFFP), and 46,400-84,900 (HEC).

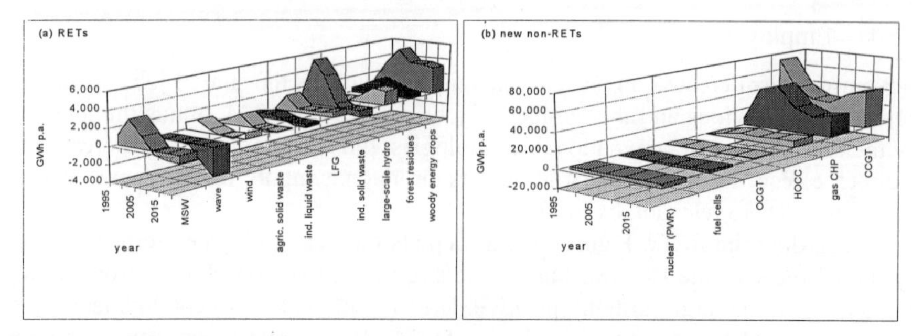

Figure 18.4: Estimated net employment creation by technology, base case, 1995-2020

For a comparison, the only other recent quantitative analysis on employment effects caused by the use of RETs in the UK we are currently aware of is a preliminary study by ECOTEC (1995). Although of limited scope (only a relatively small group of RETs is covered by the analysis which is confined to electricity generation), two features make the study attractive: (i) a simple and relatively up-to-date (spreadsheet) I/O model of the UK economy is used; and (ii) a spatial analysis is undertaken. The outcome, remarkably positive as well, is of a similar magnitude to ours: by the year 2005, 11,600 jobs are created (SAFIRE Base Case, heat and electricity generation: 15,193), of which some 3,500 jobs could be created in areas of greatest employment need.[12]

4.5.2 Emissions

Emissions may occur either from MC&I or O&M activities, or from fuel consumption. SAFIRE distinguishes between six different types of emissions: CO_2, CO, SO_2, NO_x, VOC and particulates. The emissions caused by O&M activities and by fuel consumption are combined to a single component (i.e., using a single C-B coefficient), as in cases where fuel consumption is involved, the proportion of emissions arising from fuel burn is much higher in comparison with the O&M related emissions. Hence total net emissions comprise those caused by MC&I and O&M (including fuel consumption) caused by the MPen of a new technology, minus those avoided by replacing a (presumably more polluting) conventional technology. Finally, it should be noted that SAFIRE does *not* calculate absolute emission figures. Hence, if energy demand growth is large enough to cause additional emissions that outweigh the emissions saved by the substitution of new for conventional technologies, then net overall emissions can be higher—despite the emission reductions achieved.

Figure 18.5 shows the Base Case outcome for (a) RETs and (b) new non-RETs in terms of the net total emission changes for the various pollutant emissions covered. Perhaps most striking is the predicted high amount of SO_2 emissions

[12] Apparently, this conclusion is based on a very simplistic back-of-an-envelope calculation (cf. ECOTEC, 1995, pp.30-1).

avoided. Comparing the outcomes from the alternative scenarios with the baseline scenario, we found that the differences between the HFFP scenario and the Base Case are minor for the RETs and between the HEC scenario and the Base Case for the new non-RETs. By contrast, the additional amount of emissions avoided by the MPen of RETs is remarkably high for the HEC scenario (e.g., the amount of total SO_2 emitted is lower by more than 36,000 t and that of CO_2 by more than 11,100 t), and also very high in the case of the MPen of new non-RETs for the HFFP scenario (e.g., SO_2 emissions are 60,000 t lower than for the BC).

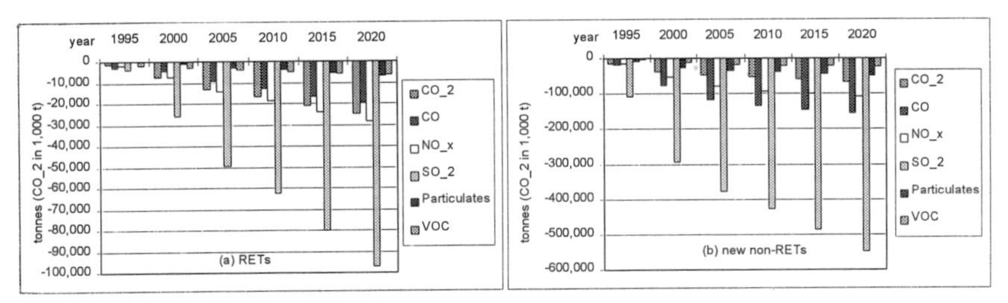

Figure 18.5: Estimated net total emission changes, base case, 1995-2020

5 Summary and conclusions

The aim of this study was to model the impacts of an increased penetration of renewable and new non-renewable energy technologies in the UK over the period 1993-2020 by using the model SAFIRE, v1.31. Because SAFIRE has originally been developed as a model for the EU12, and not just the UK, it is not particularly well suited to reflect the special features of the UK energy policy of Renewables Obligation Orders (e.g., bidding procedure, inclusion of certain RETs only, guaranteed prices for contracted projects, etc.). Consequently, what the results of the modelling exercise show is rather what the situation would be *without* any Renewables Obligation Orders in place. Moreover, we caution against reading too much into single numbers.

The total *Technical Potential* for RETs turned out to be around 700 TWh, i.e., somewhat lower than the "Accessible Resource" reported by ETSU (1994b). Two major reasons for the difference seem to be that SAFIRE neither includes off-shore wind power nor imposes any cost cap.

The total *Market Potential* for decentralised use of RETs for electricity generation was shown to rise from 4.4–8.5 TWh from 1993 to 2020 (BC), from 4.4–11.8 TWh (HFFP), and from 5.7–14.8 TWh (HEC). Concerning heat generation, the model predicts the MP to grow from 22.6–27.4 TWh from 1993 to 2020 (BC), from 22.7–37.3 (HFFP), and from 30.6–43.9 TWh (HEC). The RETs with the highest MP for electricity generation are industrial solid waste, woody energy crops, forest residues and, for the HEC and HFFP scenarios, also LFG. Regarding heat generation, agricultural and industrial solid waste and woody energy crops turn out to be the most promising technologies.

The *Market Penetration* speed was assumed to be at the lower end of the spectrum allowed for in SAFIRE. For the reference scenario, the total MPen turned out to be 9.4 TWh in 1993 and 48.3 in 2020 for RETs, and between 327.1 TWh (1993) and 615.9 TWh (2020) for new non-RETs. The main contributors are large-scale hydro, wave and wind power (RETs, electricity); agricultural and industrial solid waste, woody energy crops, and forest residues (RETs, heat); CCGT, PWR and gas CHP (new non-RETs, electricity); and gas CHP (new non-RETs, heat).

The *C-B indicators* reported show that despite a high degree of uncertainty in the analysis, caused especially by the neglect of multiplier effects and the sometimes weak underpinnings of the data, we found clear signs for benign (albeit often modest) total net effects on both the labour market and the environment. However, an inspection of the job-creation results by technology revealed that some of the new technologies exhibit a negative job impact over the whole modelling period (e.g., wave power, fuel cells, nuclear power), while others turned out to have an increasingly negative effect after about 2010 (e.g., municipal and industrial solid waste). Sometimes these adverse effects can considerably offset the positive impacts of other new technologies.

The study showed further that higher environmental concern and/or fossil fuel prices would greatly help in establishing a variety of RETs as viable alternatives to investors in the energy sector and to put their MPen on a more sustainable growth path. A clear commitment of the government, however, will be necessary for the detected beneficial effects to materialize, coupled with the provision of a true "level-playing-field" (e.g., by reflecting the true social costs in the energy prices and by removing non-market barriers). Further research will be needed in order to refine the results and their interpretation—this, however, was well beyond the scope of this exploratory study.

References

DTI (1994), *New and Renewable Energy: Future Prospects in the UK*. Department of Trade and Industry, Energy Paper 62, HMSO, London.

DTI (1995), *Energy Projections for the UK. Energy Use and Energy-Related Emissions of Carbon Dioxide in the UK, 1995-2020*, Department of Trade and Industry, Energy Paper 65, HMSO, London.

DTI (1997), *Digest of United Kingdom Energy Statistics 1997*, Department of Trade and Industry (DTI), HMSO, London.

ECOTEC (1995), The Potential Contribution of Renewable Energy Schemes to Employment Opportunities. Report No. ETSU K/PL/00190, ECOTEC Research and Consulting Ltd. for the Energy Technology Support Unit (ETSU) on behalf of the Department of Trade and Industry, ETSU, Harwell OX11 0RA, U.K.

ESD (1995a), *SAFIRE Final Report*, report prepared for the Commission of the European Community, Directorate-General for Research and Development (DG XII), by Energy for Sustainable Development (ESD) Ltd., Overmoor Farm, Neston, Corsham, Wiltshire SN13 9TZ, U.K.

ESD (1995b), *SAFIRE Users Manual (Draft)*, report prepared for the Commission of the European Community, Directorate-General for Research and Development (DG XII), by ESD Ltd., Corsham, U.K.

ESD (1996a), *SAFIRE Cost-Benefit Coefficients Report*, report prepared for the Commission of the European Community, Directorate-General for Research and Development (DG XII), by ESD Ltd., Corsham, U.K.

ESD (1996b), *SAFIRE Methodology Report*, report prepared for the Commission of the European Community, Directorate-General for Research and Development (DG XII), by ESD Ltd., Corsham, U.K.

ETSU (1994a), *An Appraisal of UK Energy Research, Development, Demonstration and Dissemination*, Energy Technology Support Unit, Report R83, HMSO, London.

ETSU (1994b), *An Assessment of Renewable Energy for the UK*, Energy Technology Support Unit, Report R82, HMSO, London.

European Commission (1995), An Energy Policy for the European Union, COM(95)682: 13.12.95, Office for Official Publications of the European Communities, Luxembourg.

European Commission (1997), Energy for the Future: Renewable Sources of Energy, White Paper for a Community Strategy and Action Plan. COM(97)599: 26.11.1997, Office for Official Publications of the European Communities, Luxembourg.

Madlener, R. (1997a), SAFIRE—A Review of the Energy Substitution Model. ESRC Macroeconomic Modelling Bureau, University of Warwick, Coventry, CV4 7AL (unpublished mimeo).

Madlener, R. (1997b), Job creation effects, in the UK through a more sustainable energy system, ESRC Macroeconomic Modelling Bureau, University of Warwick, Coventry CV4 7AL (unpublished mimeo).

SECTION 6

ENVIRONMENT AND ENERGY EFFICIENCY

CHAPTER 19

CHINA'S ENERGY SECTOR AND ITS ENVIRONMENTAL IMPACT[1]

RALPH W. BAILEY and ROSEMARY CLARKE

Department of Economics, University of Birmingham, Birmingham B15 2TT
Email: R.W.Bailey@bham.ac.uk – ClarkeR@bham.ac.uk

Keywords: abatement costs; carbon dioxide emissions abatement; China; coal; environment; pollution.

1 Introduction

After the introduction of economic reform in 1978, China's GDP growth averaged 9.4% up to 1995 (World Bank 1996). This rapid growth has been fuelled by coal which is China's main energy resource: in 1995, approximately 77% of its commercial energy came from coal, 18% from oil, 2% from gas and 2% from hydro (IEA 1997). China is second only to the USA in its commercial energy consumption but per capita consumption at 28GJ is very low compared with the world average of 61GJ in 1993 (World Resources Institute 1998).

This heavy reliance on coal means that China is now the world's second largest emitter of carbon dioxide: in 1995 it generated 14% of total emissions while USA emitted 24%. Apart from this global pollution, coal also generates other pollutants, including sulphur dioxide, nitrogen oxide and total suspended particulates (TSP), which have local and regional impacts. Many major cities suffer ambient concentrations of particulates and sulphur dioxide which often exceed WHO recommended limits by considerable margins. Other damage includes acid rain with critical loads (i.e., the highest deposition of acidic compounds that can occur without causing harmful effects to an eco-system) probably exceeded in some regions, including Korea and Japan.

Our chapter commences with a brief discussion of the energy sector and of the problems of meeting the energy demand of the rapidly expanding economy. This is followed by a review of energy-related environmental impacts. The fourth section reports simulation results for carbon dioxide abatement scenarios with different reduction targets and we end by drawing a few brief conclusions.

[1] We are grateful to the Economic and Social Research Council (contract L 320 253 199) for financing this research, to Peter Pearson for comments, Alexander Smith for computing assistance, the Oxford Institute for Energy Studies for allowing us the use of its library and Lavinia Brandon for help in obtaining material.

2 The energy sector

Under state planning, extensive subsidies to industry and consumers for fuels, especially coal, encouraged inefficient energy use and distorted the wider economy (Clarke and Winters 1995). Fears of the adverse political impact of fuel price rises and the recurrence of periods of inflation have meant that price reform and market liberalisation have come more slowly to energy than to other sectors of the economy. The central government's stated intention is to leave coal and oil prices to be determined in the market but only coal prices were freed in 1995 and the oil market is still subject to regulation.

2.1 Coal

Coal production doubled in the period from 1979 to 1995 but an increasing share of output has been produced by the rapidly growing collective and individually owned mines which were not subject to tight state control. Whereas in 1979 state mines produced 83% of output, by 1993 this proportion had fallen to just under 58% (Thomson 1996). The 1985 reforms allowed state mines more flexibility: low prices continued for state quotas but output above quota, yet within target, attracted higher "negotiated" prices. Production above target could be sold on the market; as a result productivity in state mines has risen over time along with investment and, by the end of 1997, they were making profits. However, because of the Asian financial crisis, demand at home and for exports slackened, stock piles increased and for the first time production actually fell, 44 million fewer tonnes being produced in 1997 than in 1996 (Han Zhenjun 1997).

About three-quarters of main coal reserves are in the north, north-east and north-west, far from most consumers who are located mainly in coastal provinces. Coal is transported across China, mainly by rail or by coastal shipment and the average rail distance is around 550 km. The rail system is still subject to price regulation and suffers from insufficient investment so that rail freight capacity is inadequate to meet demand. Freight capacity increased by 35% in the ten years to 1990 while coal carried rose by 53% (Todd and Jin 1997, Table 19.3). Only about 18% of coal is washed and much of this freight includes large amounts of waste. As over 40% of rail freight capacity is occupied by coal, coal washing would probably pay for itself and bring environmental benefits, as various studies have indicated (see, for example, Xie and Kuby 1997). One result of the distances between supplier and consumer is that coal prices can vary considerably: in 1995 the price of steam coal in Guangzhou, in the south, was more than double that charged by Datong, the largest mine in the north, some 2,400 km distant.

2.2 *Oil*

After experimenting briefly with price liberalisation, controls on imports and prices of both crude oil and oil products were re-introduced in 1994, imports being managed by state companies. While China's oil production increased from 40 million toe (tons of oil equivalent) in 1980 to 150 million toe in 1995, existing oil fields, located mainly in the north-east, have passed peak production and, as yet, reserves in the remote Tarim Basin, in Xinjiang province, remain undeveloped for lack of capital (Wang 1995). Demand, which increased at an average of 6.2% p.a. from 1985-95 (Horsnell 1997, p. 23), exceeds supply. Those provinces where economic growth is most rapid are in the south, far from the refineries which are mainly sited in the north, near the oil fields. Transport problems and costs, similar to those experienced for coal, mean that for these coastal provinces it is cheaper to import oil. Since 1990 imports of both crude oil and oil products have grown fast and China has had an overall oil product deficit since 1991 despite the re-introduction of restrictions following rising inflation in 1993. In 1995, it exported 5.3 million toe of oil products and imported 18.2 million toe (IEA 1989, 1997). The restrictions on imports have brought quantity rationing of oil products and, with low world oil prices, the result has been extensive black markets and smuggling, especially in the southern Fujian and Guangdong provinces where demand is strongest; in most regions internal prices are currently much higher than world prices.

2.3 *Other energy sources*

Currently only 2% of energy supplies come from natural gas. Known reserves of natural gas within China are few and the government has signed agreements with other countries for both oil and gas. However, natural gas, piped from Shanxi and Gansu provinces, is to replace coal in central Beijing from 2000, with coal burning banned (SWB 17 June 1998). Approximately one-third of coal is used in power generation and 73% of electricity is produced by coal fired plants. Hydro provides some 19% and has excellent potential in the south and south-west, an area distant from the main consumers. The first nuclear power plant commenced operation in 1994 and others will come on line in the next few years. Installed capacity has increased by nearly 250% since 1979 but China's rapid economic growth has placed great strains on the industry and electricity power cuts are experienced in all parts of the country. Estimates suggest that production falls short of demand by 10% and peak power capacity by 20-30% (Yang and Yu 1996, Fesharaki et al. 1994). The removal of coal subsidies has increased fuel prices and reduced profitability so that new power plants are unable to cover their costs unlike older plants which continue to benefit from state subsidies (Fesharaki et al., op. cit.).

A large proportion of the rural population rely on biomass or low grade coal for both heating and cooking and it has been calculated that rural families consume some 250 million tons of fuelwood a year, amounting to 25% of their total energy needs (SWB 9 July 1997). The remote inland high plain and mountain areas have high insolation levels and strong winds and a start has been made on introducing various forms of renewable energy (SWB 25 March and 22 July 1998). Apart from the small hydroelectric plants which have been built in areas with water resources, solar and wind energy are the only practical alternatives to biomass for these sparsely populated areas.

Overall, since reforms commenced in 1978/9 there has been a move towards greater market forces but the continuing dominance of coal means that if China is to reduce carbon dioxide emissions and urban pollutants, it will have to switch from this dirtier fuel to the cleaner fossil fuels, oil and gas. However, the switch into oil is currently impeded by the government's policy of tightening oil market regulation as a means of restricting general demand in periods of inflation. In the longer run, it will be difficult to contain the rising demand for oil products, especially given the conflict with other objectives such as the government's decision to encourage car production and ownership, and the continued lack of significant investment in the railway system.[2]

3 Environmental impacts

Fossil fuels generate many different pollutants, both atmospheric and local. China has signed and ratified the UN Framework Convention on Climate Change and its declared intention is to rely on improved energy efficiency to reduce carbon dioxide emissions (China's Agenda 21, 1994). While various studies show that already there have been marked improvements in energy efficiency (Lin 1992, Sinton and Levine 1994), due in no small part to the declining share of subsidised coal, it will also be necessary in the longer run to switch from coal to oil and natural gas which emit less carbon dioxide.

Fossil fuels also emit other local and regional pollutants, such as sulphur dioxide, nitrogen oxide and particulates. Sulphur dioxide emissions have risen over time as few power stations have abatement technology but particulate emissions have remained nearly constant. These emissions can damage health, vegetation, ecological systems and materials. As Table 19.1 indicates, urban ambient concentrations are extremely high, even at danger levels in many cities and often exceeding Chinese standards by significant margins. Indoor pollution levels can

[2] Demand for gasoline and diesel grew at an average rate of approximately 7.5% during the period 1985-94 (Horsnell 1997, p. 24).

also be very high in both urban and rural households which use inefficient wood and coal burning stoves (Mumford 1987, Smith 1988). The most significant costs are those affecting health. Wells et al. (1994) estimate that the value of reducing mortality and morbidity by abating TSP and SO_2 emissions in Beijing and Shenyang could amount to 530 and 165 yuan per ton at 1990 prices (US$435/ton and US$113/ton). Acid rain, from SO_2 and NO_x, affects some 29% of China's land area including some 5.3 million hectares of farmland (Yang and Yu 1996) and for 49% of 73 cities surveyed, pH was less than 5.6 (the definition of acid). Estimates of the damage to farming and forestry amounted to $4.36 billion in 1995 (World Bank 1997, p. 27).

Table 19.1: Ambient concentrations of SO_2 and TSP

	SO_2 ($\mu g/m^3$)	TSP ($\mu g/m^3$)
1994 - 77 cities		
Average daily concentration	8-451	108-815
Northern	100	407
Southern	96	251
1991		
Beijing	110	400
Shenyang	150	500
Xian	50	400
Shanghai	90	230
Guangzhou	50	200
1988		
Chongqing - winter/summer	260/660	600/870
Guiyang - winter/summer	189/571	358/365
WHO recommended limits (1 year)	40-60	60-90
Chinese standards	20-100	60-150

Other environmental impacts include coal waste of 117 million tons of ash, 79 million of slag and 118 of gangues. The disposal of this waste causes many problems. China has relatively little cultivable land from which to feed its large population but arable land is being used for waste disposal, especially around power stations and near urban areas. Water is another very scarce resource in the north and mines are a source of pollution as only about 7% of polluted waste water is treated before being discharged into rivers, lakes, reservoirs and the sea (China Environment Yearbook 1996). If China were to introduce a carbon tax to abate carbon dioxide emissions, this would generate secondary benefits from the linked reductions in other pollutants and, in some studies, such benefits have been shown to be significant (Clarke and Edwards 1997, Ekins 1997).

4 The simulations

China is now amongst the top ten exporting countries in the world, its export growth averaging 16% per annum between 1978 and 1993 (Lardy 1994, Table 2.1), and it imports and exports fossil fuels. In simulating the effects of carbon dioxide abatement it is therefore important to take into account general global trading activities and the fact that oil prices are determined at the world level. It is for this reason that, in modelling the costs of carbon abatement for China, we use a global model: OECD's **G**ene**R**al **E**quilibrium **EN**vironmental model (GREEN).

4.1 The model

GREEN contains twelve country/regional sub-models, each with fifteen productive sectors. Twelve cover fossil and backstop fuels, electricity, gas and water distribution while the remainder cover agriculture, energy intensive industry, and other industry and services. The representative consumer allocates income between four consumer goods which include fuel and power, and transport and communications. Governments collect carbon and other taxes and all revenue from any carbon tax is recycled to ensure revenue neutrality. GREEN derives its dynamics from fossil fuel depletion, productive capital accumulation and the putty/semi-putty specification of technology.

As GREEN was developed to measure the costs of abating carbon dioxide, the energy sectors are modelled in considerable detail. Coal reserves are assumed to be infinite within the model's time horizon of 2050. Crude oil and natural gas supplies are derived from resource depletion sub-models. The real world price of crude oil is endogenous but coal and gas prices are determined by the supply elasticities of their respective resource bases. Imports from different regions are treated as imperfect substitutes - the Armington assumption. This implies that each region faces downward sloping demand curves for its exports. The only exception is crude oil which is assumed to be homogeneous across regions indicating a unique world price. Natural gas and coal, on the other hand, are assumed heterogeneous across regions because their transportation costs are greater than that for oil. (Further details of GREEN can be found in van der Mensbrugghe 1994).

4.2 The scenarios

Industrialised countries must take most responsibility for the rising levels of carbon dioxide emissions. China, while recognising the dangers of global warming, is naturally reluctant to curb its rapid economic growth as, in common with other developing countries, a majority of its population - especially those living in remote rural areas - experience low living standards. Despite its heavy reliance on coal, China's current per capita emissions are only 2.27 metric tons compared with a world average of 3.9 in 1995 (World Resources Institute 1998, Table 16.1). With rising living standards these emissions will increase sharply. We explore two abatement scenarios: in the first, all regions of the world stabilise carbon dioxide emissions at 1990 levels. In the second scenario, we recognise the need for developing countries to have some leeway in curbing emissions and they are allowed to increase CO_2 emissions by 50% above 1990 levels. In order to achieve the same overall emission abatement target as the first scenario, industrialised countries have to cut emissions to 20% below 1990 emission levels. In all scenarios we have assumed that all Chinese fuel subsidies have been removed from 1995 and a tax (other than a carbon tax) introduced from 2005.

Unlike the situation with sulphur dioxide and nitrogen oxide, there are currently no satisfactory solutions to the removal and storage of carbon. Abatement therefore has to be achieved by decreased energy consumption and by fuel switching from the dirtier fuel, coal, to relatively cleaner oil and natural gas. Introducing a carbon tax provides an incentive to make these changes and has the beneficial secondary effect of simultaneously reducing emissions of SO_2 and NO_X. In our simulations, these secondary benefits arise solely from the changes induced by a carbon tax and do not reflect directly increased use of technologies controlling SO_2 and NO_X emissions. However, allowance for some autonomous improvement has been incorporated into the model. While China has no declared intention to introduce a carbon tax, it is already experimenting with other environmental taxes: water pollution charges have been implemented for some time and more recently taxes on sulphur emissions have been introduced on an experimental basis in a few provinces.

The baseline scenario simulates "business-as-usual" emissions of CO_2, SO_X and NO_X. If no attempt is made to abate CO2 emissions, by 2050 China's emissions would be nearly six times those at 1990. This would make China the largest emitter at 17.6% of the global total, overtaking the USA which would emit 16.2%. China's CO_2 emission growth rate is not as fast as for India and the dynamic Asian economies but, as expected, greatly exceeds the rate for industrialised countries.

Scenario 1 reports the costs of abatement if all regions were to stabilise carbon dioxide emissions at 1990 levels. That this would place a heavy burden on developing countries can be seen in Table 19.2. China would have to cut carbon dioxide emissions, relative to the baseline scenario, by about 80%. This would result in a 1.4% loss of GDP relative to baseline levels but there are significant similar offsetting benefits from reductions in SO_X and NO_X emissions.

Table 19.2: Emission reductions and GDP loss in 2050 - percentage change relative to base

Region	CO2 targets relative to 1990 levels		Global pollutant CO_2		Local/regional pollutants SO_x		NO_x		GDP loss relative to base	
	Scenario 1	Scenario 2	Scenario 1	Scenario 2	Scenario 1	Scenario 2	Scenario 1	Scenario 2	Scenario 1	Scenario 2
China		**50%**	**79.7**	**69.6**	**77.2**	**69.6**	**78.2**	**70.5**	**1.44**	**1.05**
India		50%	84.7	77.1	75.4	69.2	81.4	74.5	0.66	0.53
Dynamic Asian Economies		50%	83.4	75.1	59.8	54.4	68.5	61.7	1.14	0.94
Brazil		50%	80.4	70.6	61.0	56.1	60.3	54.6	1.55	1.58
Energy Exporting DCs	Stabilis-	50%	75.0	62.5	48.0	42.4	56.7	49.0	5.23	5.11
Rest of world	ation at	50%	77.2	65.8	67.6	61.9	70.7	64.2	1.55	1.25
USA	1990	-20%	50.5	60.4	41.1	47.0	59.9	67.6	0.38	0.55
European Union	levels	-20%	46.2	56.9	33.4	40.4	47.2	56.3	0.29	0.65
Other OECD		-20%	53.6	62.9	53.2	59.0	62.1	68.9	0.41	0.67
Former USSR		-20%	50.9	60.7	44.7	52.6	49.2	57.5	2.60	3.52
Central & E. Europe		-20%	46.5	57.2	45.3	55.5	46.6	57.3	1.01	1.31
World			64.9	64.9	54.6	55.4	62.6	63.3	1.39	1.44

Scenario 2 achieves the same 65% global reduction in carbon dioxide emissions but shares the costs more evenly. Developing countries are permitted to increase carbon dioxide emissions up to 50% above 1990 levels while industrialised countries have to constrain their emissions to 20% below 1990 levels. The cost to China in lost GDP relative to base is now 1%, just under three-quarters of the cost for scenario 1 - see Table 19.2. Other developing countries also benefit but, amongst industrialised countries, the European Union experiences a considerably greater loss of GDP relative to base at 0.65% compared with 0.29% in scenario 1. While the reductions in SO_X and NO_X are smaller than in scenario 1, they are still considerable.

4.3 *Discussion*

Our estimates of the GDP loss for China are smaller than those reported elsewhere - see, for example, OECD 1992 and Manne and Richels 1992. As model construction, underlying assumptions and scenarios differ, comparison is not easy (Clarke, Boero and Winters 1996). However, the OECD (1992) compares various models including GREEN and their fourth scenario stabilises emissions at 1990 levels (our scenario 1). The study reports China's GDP loss relative to baseline at 5.5% (GREEN) and 4% (Manne & Richels model) by 2050, higher than our results. As we adopt most of the same parameter values in our study, including those for AEEI (autonomous energy efficiency improvement) and inter-fuel substitution, the explanation of the differences must be found elsewhere. The data and some parameters in the version of GREEN which we are using have been revised in 1992-94 and, from checks, changes appear to have been most noticeable for China. In particular, labour productivity growth trends have been recalculated by OECD to allow for changes in real GDP trends.

The major difference between our study and that of OECD (1992) is our assumption that consumer fossil fuel subsidies terminate in 1995 and a tax (other than a carbon tax) is imposed on fossil fuels from 2005. The OECD studies assumes that Chinese fuel subsidies continue over the whole simulation period. This change would raise the price of fossil fuels, especially that of coal which has been the most subsidised fuel, and reduce consumption. However, this change cannot explain all the difference in our results and it seems likely that the main explanation must lie in data revision.

The business-as-usual scenario (BaU) suggests that without any attempt to abate, the proportion of energy from coal would rise from 77 to 84 per cent - see Table 19.3. Under the two abatement scenarios, oil is substituted for coal though, compared with most other countries, China would still be a relatively heavy user of coal. This move into oil, a much cleaner fuel, can only be accomplished if the oil market is fully liberalised in China and price and availability of oil will clearly be important factors in determining abatement costs. If China continues to use regulation of the oil market as a means of controlling inflation[3], abatement costs would be higher but, with a global model such as GREEN, it is impossible to incorporate such "on-off" regulatory measures into the scenarios.

As Horsnell (1997, p.33) observes, past projections of China's crude oil production have been overly optimistic and hopes for oil from the Tarim basin have yet to be fulfilled. Our scenarios take a similarly cautious position and assume a slight decrease in China's crude production over time on the basis that currently producing fields have passed their peak. This means that two-thirds of oil is imported by 2050 under the baseline scenario, rising to approximately 80% in the two abatement scenarios. The simulations show world oil prices rising sharply after 2030, with prices 126% higher than in 1990 by 2050 in scenario 2. Clean renewable energy does not make much of an impact until 2030 and even then, in China, provides only a very small proportion of energy consumed in China.

[3] For a more extensive discussion of China's oil market, see Chapter 3 in Horsnell 1997.

Table 19.3: Fossil fuel consumption (percentages) - China and USA

Scenarios	Year	Coal	Oil	Gas
China				
Base	1990	77.3	20.6	2.2
BaU	2050	84.3	11.4	4.4
Scenario 1	2050	29.1	59.6	11.3
Scenario 2	2050	39.7	49.9	10.4
USA				
Base	1990	28.7	44.0	27.4
BaU	2050	53.0	23.9	23.1
Scenario 1	2050	13.5	56.3	30.2
Scenario 2	2050	9.0	55.3	35.7

5 Conclusions

Our study suggests that the cost to China of abating carbon dioxide, measured in terms of GDP, will not be heavy, ranging from 1% to 1.4% by 2050 depending on abatement targets. Removing fuel subsidies has rectified many of the underlying distortions and the introduction of a carbon tax provides a further incentive to conserve energy use and switch to cleaner fuels. If no abatement measures are undertaken, by 2050 China will be the world's largest emitter of carbon dioxide and will rely even more heavily on coal as its major source of energy, imposing heavy environmental costs in terms of health and through acid rain. Our abatement scenarios show a switch from coal to oil, bringing China's fossil fuel proportions more in line with other countries. However, the scenarios assume that the government ceases to use the oil market to regulate the economy. If the market is not liberalised, then abatement costs to China would be heavier. Although the main objective of a carbon tax is to cut CO_2 emissions, it has the additional benefit of reducing other local pollutants and, as our study shows, these reductions are significant.

References

China's Agenda 21 (1994), China Environmental Science Press, Beijing.
China Environment Yearbook (1996), China Environment Yearbook Inc., Beijing.

Clarke Rosemary, Boero Gianna and L. Alan Winters (1996), 'Controlling Greenhouse Gases: A Survey of Global Macroeconomic Studies', *Bulletin of Economic Research*, 48 (4): 269-308.
Clarke Rosemary and T. Huw Edwards (1997), *The Environmental Effects of Deregulating the Japanese Oil Product Markets*, mimeo, University of Birmingham.
Clarke Rosemary and L. Alan Winters (1995), 'Energy pricing for sustainable development in China', in Ian Goldin and L. Alan Winters (eds.), *The Economics of Sustainable Development*, Cambridge University Press, Cambridge.
Ekins Paul (1996), 'How large a carbon tax is justified by the secondary benefits of CO_2 abatement?', *Resource and Energy Economics*, 18: 161-187.
Fesharaki F., Tang Chuanlong and Li Binsheng (1994), *China's Energy Pricing: Current Situation and Near Term Perspective*, China Energy: Short Memos No. II, Program on Resources, East-West Center, Honolulu, Hawaii.
Han Zhenjun (1997), 'China's energy industry blended into the great tide of market economy' Xinhua News Agency, *SWB** FEW/0524 WG/7 [30] 11 February 1998.
Horsnell Paul (1997), *Oil in Asia: Markets, Trading, Refining and Deregulation*, Oxford University Press for the Oxford Institute for Energy Studies, Oxford.
IEA (1989), *World Energy Statistics and Balances 1971-1987*, OECD, Paris.
IEA (1997), *Energy Statistics and Balances of Non-OECD Countries 1994-65*, OECD, Paris.
Johnson Todd M. (1995), 'Development of China's energy sector: reform, efficiency, and environmental impacts', *Oxford Review of Economic Policy*, 11 (4): 118-132.
Lardy Nicholas R. (1994), *China in the World Economy*, Institute for International Economics, Washington DC.
Lin Xiannuan (1992), 'Declining energy intensity in China's industrial sector', *Journal of Energy and Development*, 16 (2): 195-216.
Manne Alan S. And Richard G. Richels (1992), *Buying Greenhouse Insurance: The Economic Costs of CO2 Emission Limits*, MIT Press, Cambridge, Massachusetts and London.
Mumford J.L.K. *et al.* (1987), 'Lung cancer and indoor pollution in Xuan Wei, China', *Science*, 235: 217-220.
OECD (1992), 'Costs of reducing CO_2 emissions: evidence from six global models', *OECD Economic Studies*, 19: 15-47.
Sinton J.E. and M.D. Levine (1994), 'Changing energy intensity in Chinese industry', *Energy Policy*, 22 (3): 239-255.
Smith K. (1988), 'Air pollution: assessing total exposure in developing countries', *Environment*, 30 (10): 16-35.

SWB* (9 July 1997) FEW/0494 WG/4 [20], source Xinhua News Agency.
SWB* (25 March 1998) FEW/0530 WG/11 [43], source Xinhua News Agency.
SWB* (17 June 1998) FEW/0542 WG/15 [51], source Xinhua News Agency.
SWB* (22 July 1998) FEW/0547 WG/12 [45], source Xinhua News Agency.
Thomson Elspeth (1996), 'Reforming China's coal industry', *The China Quarterly*, 147: 726-750.
Todd Daniel and Jin Fengjun (1997), 'Interregional coal flows in China and the problem of transport bottlenecks', *Applied Geography*, 17 (3): 215-230.
Van der Mensbrugghe D. (1994), *GREEN: the reference manual*, Economics Department Working Paper 143, OECD, Paris.
Wang Haijiang (1995), 'The prospects of oil supply in China', *Energy exploration and exploitation*, 13 (6): 631-648.
Wells Gary J., Xu Xiping and Todd M. Johnson (1994), *China: Issues and Options in Greenhouse Gas Control: Valuing the health effects of air pollution*, World Bank Sub-report No. 8, World Bank, Washington DC.
World Bank (1996), *From Plan to Market: World Development Report 1996*, Oxford University Press, New York.
World Bank (1997), *Clear Water, Blue Skies*, The World Bank, Washington D.C.
World Resources Institute (1998), *World Resources 1998-99*, Oxford University Press, New York and Oxford.
Xie Zhijun and Michael Kuby (1997), 'Supply-side-demand-side optimisation and cost-environment trade offs for China's coal and electricity system', *Energy Policy*, 25 (3): 313-326.
Yang Ming and Yu Xin (1996), 'China's power management', *Energy Policy*, 24 (8): 735-757.

* SWB = Summary of World Broadcasts, British Broadcasting Corporation, Weekly Economic Report.

CHAPTER 20

INVESTMENT APPRAISAL IN THE TRANSPORT SECTOR IN THE UK - GETTING THE SIGNALS WRONG ON ENERGY AND THE ENVIRONMENT?

A.L. BRISTOW

Institute of Transport, University of Leeds, Leeds LS2 9JT, UK
Email: abristow@its.leeds.ac.uk

Keywords: environmental impact; externalities; measurement; model split; transport; UK.

1 Introduction

In the UK the transport sector is responsible for approximately 25% of CO_2 emissions, congestion is rising, the local environmental impacts of traffic are causing greater public concern and traffic levels are still rising. In such circumstances there is an increasing recognition that unrestricted traffic growth is unsustainable. In recent years considerable attention has been given to identifying ways in which traffic growth might be restrained and CO_2 targets met. The Government is committed to a 12.5% reduction in greenhouse gas emissions from 1990 levels by 2008-2112. The revised National Road Traffic Forecasts (DETR, 1997a) show that traffic is expected to rise by 28% between 1996 and 2011 with a continuation of existing polices, including the fuel price escalator. Local air pollution and noise are generating considerable concern, particularly in relation to the health effects of small particulates.

There has been an emphasis in research and policy development on the issue of pricing in transport and on taxation in particular to correct for the under pricing of road travel with respect to externalities, e.g., accidents, environmental impacts and congestion. There has been detailed consideration of fuel taxation at the national policy making level and of road pricing as a local tool. A range of other pricing measures have been discussed, e.g., differential Vehicle Excise Duty, company car taxation, parking charges, etc. The potential effectiveness of such policies has also been considered. However, so far, actual intervention has largely been limited to increases in the rate of fuel duty taxation on petrol and diesel, decreases in taxation of Compressed Natural Gas (CNG) and Liquid Petroleum Gas (LPG), and Vehicle Excise Duty (VED) concessions for "clean" Heavy Goods Vehicles (HGV) and buses.

An area which has received less attention concerns the appraisal methods used to aid decision making on investment and subsidy in the transport sector. It is clear that where, as in the UK, forms of appraisal vary considerably between modes, the outcomes may well be inconsistent with strategies to reduce traffic growth. Another important related issue is the weight given to environmental and resource impacts in appraisal, and their appropriateness. It is the intention of this chapter to review the various methods currently in use in order to examine these two critical issues. However, this is an area where change is occurring. The Department of Environment, Transport and the Regions (DETR) is currently working on new appraisal methods and has recently (1998a) issued new guidance for trunk road appraisal. These developments are considered.

2 The planning framework

Before moving on to look at appraisal methods in detail it is necessary to define, briefly, the planning context in which the methods are used and the sources of funding available.

Trunk roads are the responsibility of central government, administered by the Highways Agency. Local roads are the responsibility of Local Authorities and funded through grants from central government, the general Rate Support Grant (RSG) or the Transport Supplementary Grant (TSG) and credit approvals. Local public transport infrastructure is also the responsibility of local government. TSG can only be used for road and traffic schemes, so for specific funding, application has to be made for Section 56 Grant, which again can be supplemented by credit approvals. Revenue support for subsidised bus services comes through the RSG. Rail is directly supported through subsidy to Railtrack and through the Office of Passenger Rail Franchising. The Passenger Transport Executives are also funded to continue support for rail services in their areas. Other possible sources of funding include the private sector and EU grants.

Local Authorities submit their bids for capital funding for transport expenditure each year in a Transport Policies and Programmes (TPP) document. A recent development has been the adoption of the "package approach" to such bids, whereby a package of measures, across modes, is developed to meet the objectives of the area as a whole.

The bulk of public investment expenditure goes into trunk and local roads, and rail (£4,121 million and £2,000 million in 1995/6 respectively). Investment in local public transport schemes was £128 million in 1996/7. Funds allocated to "packages" were £78.7 million in 1996/7. It can be seen that rail investment is around half that on roads, while investment in other forms of public transport is a tiny fraction of total transport investment.

3 Appraisal methods for transport

This chapter concentrates on land based transport, as this is the sector responsible for the vast majority of transport related environmental pollution. The initial consideration is whether current appraisal methods succeed in producing a level playing field in the treatment of investments for private or freight road transport, public transport and non-motorised modes. In brief, large road schemes are subject to a Social Cost Benefit Analysis (SCBA), rail is subject to commercial criteria as is most bus industry investment, while local urban transport infrastructure, such as tramways and light rapid transit, is subject to a form of Restricted Cost Benefit Analysis (RCBA) and since 1993 a full CBA. The slower modes of cycle and walking are in the main neglected. The methods currently in use to assess land based modes are as follows.

3.1 Highways

All major trunk road schemes are subject to a formal appraisal using a form of Cost Benefit Analysis (CBA) known as COBA (Highways Agency et al, 1996). In urban areas the modeling and assessment is carried out using URECA, which has a similar format. For a typical road scheme the cost benefit analysis would include the impacts shown in Table 20.1.

Table 20.1: Impacts included in trunk road scheme appraisal CBA

Costs	Benefits
Capital and Maintenance	Time savings
Operating costs	Accident cost savings
Disruption to traffic during construction	Vehicle operating cost savings
	Maintenance and operating cost savings on relieved roads

Scheme costs are estimated from past experience and projected future costs. In most trunk road appraisals the main element of benefit is time savings. In COBA, savings during work time are estimated at the average wage rate by vehicle type. This then assumes that the wage rate reflects the marginal product of labour and that any time saved will be used productively. Non-work time savings are valued using a single "equity" value. This value was derived from willingness to pay studies (MVA/ITS/TSU). The decision to use an equity weight treats all time savings as having the same value, despite evidence that values vary according to, for example, income and journey purpose (Wardman, 1997).

Accident cost savings have two main elements, the direct measurable costs and those incurred by the injured, the dead and their families. The direct costs - damage to vehicles and property, emergency services, medical treatment, etc. - are valued at market prices (net of tax). The cost to the individual has been assessed through the use of willingness to pay studies looking at the risk of various outcomes (for injuries) and risk reduction (for fatalities) (Jones-Lee, 1987, Hopkin & Simpson, 1995).

Vehicle operating costs include all costs of operation paid by the owner/driver (except any drivers' wages which are included as time savings). Savings in operating costs may arise from shorter routes, better alignment and/or reduced congestion.

The environmental and social costs of road schemes are not monetised but are included in a table or framework. Where possible a quantified measure, eg noise measured as the change in decibels, is used. Elsewhere a descriptive or subjective measure is used, e.g., extent of visual intrusion. The weight to be given to such impacts is left to the decision maker. It should be noted that any mitigation measures, e.g., to comply with noise regulations, are included in the capital costs of the scheme. The consumption of resources is not considered, except through the mechanism of market prices. An exception is the treatment of land, where the environmental impact tables include one showing the temporary and permanent land take, by class of land.

New guidance on trunk road appraisal (DETR, 1998a) places the COBA and environmental assessment within a framework with five main objectives:

- environment;
- safety;
- economy;
- accessibility;
- integration.

The COBA results enter under safety and economy. The indicators used in the summary assessment table are expressed in money terms or on scales of impact. They are not additive. This development clarifies objectives, but does not indicate the weight to be placed on the different criteria.

3.2 Major public transport infrastructure

In this section the treatment of large scale passenger transport investment is discussed, excluding investment in heavy rail schemes, which is covered in the following section. The main source of funding is Section 56 Grant (deriving from Section 56 of the 1968 Transport Act which provided for specific grant funding). Section 56 Grant is administered under principles outlined in a 1989 circular (Department of Transport, 1989). Scheme costs are to be set against the following benefits:

- fare revenue;
- developer contributions;
- savings in support to tendered passenger transport services;
- non-user benefits, e.g., decongestion, environmental and/or development.

In common with roads appraisal a form of CBA is used in appraisal, in this case a restricted form, RCBA, outlined in the 1989 circular. Since 1993 a full CBA has been required alongside a financial appraisal and the RCBA. These additional requirements retain the principle that public transport schemes should pay their way and that any subsidy be justified by benefits to non-users.

The key difference between the RCBA and a trunk road CBA is the treatment of user benefits, which form the main element of road scheme benefits. For public transport schemes, there is an obligation on scheme promoters to maximise farebox revenue. User benefits are included in the appraisal in terms of fares paid, which cannot fully capture consumer surplus. Passenger transport investments, therefore, need to establish benefits to non-users, for example, through de-congestion benefits, in order to secure public funding.

Another difference lies in the treatment of accident cost savings. In a full SCBA all accident cost savings are included and valued. In the RCBA only accident cost savings to those not using the scheme are included, on the grounds that all benefits to users should be reflected in the fare paid.

Development benefits are not included in roads appraisal. There is uncertainty as to their significance in a dense highway network, and their inclusion runs the risk of double counting if it is merely the capitalisation of direct user benefits. In the case of Section 56 grants, the inclusion of development or regeneration benefits is justified by the role of large public transport schemes in regenerating inner city areas. However, any such benefits are not included in the CBA and there is no standardised methodology for including such effects, though there is guidance (HM Treasury, 1995).

The exclusion of user benefits, save those expressed in terms of fares is the most important issue. A number of arguments may be used to justify the payment of subsidy to passenger transport. The main economic arguments are:

1. The often low marginal costs of operation, whereby marginal cost pricing would require subsidy.
2. The benefits to existing users of frequency increases which are not captured in fares.
3. Diversion of car users, where car use is under priced.

The first two are based on benefits to users which are not reflected in fares. Further arguments in favour of subsidy may be based on objectives relating to accessibility and equity. Arguments against the payment of subsidy tend to be based on the risk of inefficiency in operation.

3.3 Bus operations

Approximately 80% of all bus services are operated commercially by private companies and are not covered by public appraisal. However, it should be remembered that these operators are in receipt of public support where concessionary fares are supported by the Local Authority. The combined total of concessionary fare reimbursement and fuel duty rebate, available to commercial and supported services, was £668 million in 1995/6, far in excess of the direct support to socially necessary services of £268 million. Yet Local Authorities have very little power of regulation over the commercial sector. The development of "quality partnerships" in some areas is a positive step which sees public bodies and private operators working together voluntarily seeking to improve the provision of local passenger transport services. These partnerships are likely to be strengthened as a result of the recent transport white paper (DETR, 1998b) and there may be moves towards exclusive franchises for certain routes through "quality contracts".

The remaining, unprofitable but socially necessary services are supported by local authorities. A variety of techniques is used to allocate resources (Bristow et al 1992). These include demand related financial criteria, standards and targets, priority scores and ranking. The use of CBA is extremely limited, partly because of the scale of costs involved in carrying out a CBA relative to the often small amounts of money used to support any particular service, and the perceived complexity of the technique. The issue of value for money in supported services is currently under review by the Audit Commission.

3.4 Walking and cycling

The non-motorised modes are largely excluded from formal investment appraisal. However, it is possible for local authorities to include cycle- or pedestrian developments within a TPP or specific package bid.

3.5 Railways

The rail industry is now almost completely privatised and has been transformed from a single organisation, British Rail, to a multitude of private companies including: Railtrack (responsible for infrastructure and signaling); the TOCs (25 Train Operating Companies); the ROSCOs (3 passenger rolling stock companies; and a number of companies covering track maintenance and renewal, rolling stock maintenance, parcels and freight, research and development, etc.

These companies operate on commercial criteria. Investment decisions are, therefore, made on commercial grounds. However, the rail industry is in receipt of approximately £2 billion in subsidy each year. There is then a need for an appraisal method to assess proposals involving subsidy. The body responsible for allocating support to rail operations is the Office of Passenger Rail Franchising (OPRAF). New objectives and guidance (Secretary of State for the Environment Transport and the Region 1997) issued to the Franchising Director in November 1997, state that the principal objectives are:

- to increase the number of rail passengers;
- to promote the interests of rail passengers; and
- to secure improvements in the quality of rail services and stations.

OPRAF have recently (OPRAF, 1997) issued guidance on the appraisal of support for rail passenger services. This document outlines a form of SCBA, which it is acknowledged may be revised in the light of the Labour Government's review on developing an integrated transport strategy (DETR, 1997b). This SCBA has similarities to that used in road scheme appraisal outlined earlier, but again with some critical differences.

1. The treatment of user benefits:
   "The Franchising Director expects fares alone to remain the most commonly used indicator of user benefits but he is willing to consider fare options which trade off other benefits against returns subject to affordability" (OPRAF 1997). This is an advance from the Section 56 guidelines which exclude user benefits. There is no commitment to consider such benefits, but the door is open for a case to be made.
2. Consideration of option values attached to rail services. These are considered solely in regard of existing services, whereas they might in some degree also apply to new ones.
3. The main decision criterion is NPV/k where k is net cost to OPRAF and NPV is Net Present Value. In the consultation document on the guidelines OPRAF proposed requiring new services to meet a higher threshold than existing services. This asymmetry in the decision criterion was intended to meet the original objective that change in service levels should be gradual and to recognise the transaction costs imposed on users as a result of service withdrawals. In response to consultation this has been amended to a disaggregation in the appraisal to show user gains and user losses separately. A ratio of gross user gains to gross user losses would then be used in decision making.
4. Taxes, which are normally netted out of an SCBA, are left in. This would be a reasonable approach where taxes might equally affect the costs and benefits to rail and competing modes. However, as public passenger transport is exempt from Value Added Tax, while taxes on road transport are high, this is not the case. In switching travellers from road to rail, there will be a tax loss to the exchequer and this could distort an appraisal which included road vehicle operating costs.

A similarity with the approach to highways appraisal is in the treatment of environmental impacts, where a framework style presentation is adopted.

3.6 Comparing the methods

A discussion of the various methods of appraisal has also contained a comparison against the roads appraisal, which is the most comprehensive of the public methodologies. There are a number of areas where methods should be improved in order to provide a consistent and comprehensive assessment. These include:

- the inclusion of user benefits in all appraisals;
- consistent treatment of safety;
- developing appropriate modelling and appraisal techniques to handle non-motorised modes;
- improved treatment of the environment and resources.

In the remainder of this chapter the focus will be on improving the treatment of the environment and resources. The methods that include environmental costs do so in a descriptive way and do not attempt to incorporate them into a CBA. In this case the weight to be placed on these effects relative to the NPV of the scheme is solely the responsibility of the decision maker. None of the methods consider energy use or that of other resources, save land, except through market pricing.

4 A way forward?

In this section the main options by which environmental and resource issues could be more effectively incorporated into appraisal are discussed. These include valuation of impacts, the setting of limits and the use of Multi-Criteria Assessment techniques that can handle conflicting objectives.

4.1 Valuation of environmental impacts

There has been a considerable amount of progress on the valuation of environmental impacts in recent years, such that it may be possible to bring at least the main environmental impacts into a CBA. The total environmental costs of the transport sector are acknowledged to be large, though there is disagreement on precisely how large (see Table 20.2).

Table 20.2: External costs of road transport in the UK per year (£ bn.)

	Pearce et al	**Mauch & Rothengatter**	**RCEP**	**Maddison et al**
Year	1991	1991	1994/5	1993
Impacts				
Accidents	4.7 to 7.5	13.3	5.4	2.9 to 9.4
Noise	0.6	3.4	}	2.6 to 3.1
Air pollution	2.4	6.2	} 4.6 to 12.9	19.7
Climate change	0.4	4.1	}	0.1
Totals	8.1 to 10.9	27.0	10.0 to 18.3	25.3 to 39.3

Sources: Mauch & Rothengatter, 1995, Maddison et al, 1996 and Royal Commission on Environmental Pollution (RCEP), 1994.

In part, the differences may be explained by differences in detailed methods and approach. Pearce et al and Maddison et al estimated the marginal external costs of climate change over the next two to three hundred years, producing low estimates. Mauch and Rothengatter set targets based on the precautionary principle and estimated the costs of meeting these targets in a much shorter time scale. The Royal Commission reported a range of available estimates. Other differences may be explained by changes in knowledge. The high estimate for air pollution impacts of Maddison et al incorporates new evidence on the health effects of small particulate matter that was not available to earlier researchers.

A wide range of studies has been carried out to estimate the value of environmental impacts based on behaviour or surveys of individuals. These too have produced a range of estimates (see Tinch 1995 and Perkins, 1997 for an overview). However, there have not been many studies in the UK. A recent study in Edinburgh, using Stated Preference techniques, investigated the values households place on transport related noise and air pollution. The results suggest that a 10% improvement in noise levels would be worth around 32 pence per week, while a similar change in air pollution would be worth around 81 pence (Wardman et al, 1997). A number of interesting issues have been identified including an asymmetry in values, with a deterioration from the current situation being valued more highly than a similar improvement.

A further complication in attempting to identify values for use in appraisal is the way impacts of additional pollutants vary according to location (Eyre et al, 1997, Watkiss & Collings, 1997). This has not prevented a number of European countries from adopting environmental values for use in appraisal, most commonly for local air pollution, noise and climate change (Nellthorp et al, 1997).

Although the impacts of CO_2 and other gases contributing to climate change is uncertain, in Europe there is a commitment to a reduction in CO_2 emissions. It therefore seems reasonable to base values on the costs of meeting those targets. Stabilisation of emissions in the EU implies a shadow price of 50 ECU per ton of CO_2 and this would need to double to achieve a 15% reduction by 2015 (Perkins, 1997). Taking the value of 50 ECU per ton, Perkins gives a figure of 5.6 ECU per 1000 car kilometres and 3.4 ECU per 1000 ton kilometre for trucks (the respective figures for rail are 3.0 and 1.1 ECUs). In this case, where there is uncertainty about long run impacts, adopting a precautionary approach and incurring costs to meet targets is an acceptable approach. It is then possible to use such figures in appraisal.

4.2 Resource consumption

Where it is difficult to value environmental impacts it may still be possible to develop more appropriate ways of treating them in appraisal. Land use is an important example, as it includes issues of landscape and nature conservation, although certain types of landscape and nature reserves are given protection from development in theory, e.g., National Parks, Areas of Outstanding Natural Beauty, Sites of Special Scientific Interest, etc. The practice can be rather different. Land in rural areas that is protected from commercial development has a very low market value, which could be seen as an attraction for infrastructure projects where the market price of land applies. Given the density of population, an already dense transport network, and the difficulty of escaping from the influence of motorised traffic, it may be that the point has been reached at which beautiful landscapes and SSSIs should be given complete protection from development. This can become part of a sustainability objective:
"to pass on to future generations a portfolio of landscape qualities at least as good as current generations enjoy" (Bowers et al 1991).

One way in which this becomes a strong, yet not absolute measure, would be to argue that if any part of a top grade landscape is to be damaged then either it must be fully restored after the construction process, e.g., by constructing a tunnel, or a degraded landscape elsewhere must be upgraded as a replacement. This would enforce replacement or preservation costs as the appropriate cost for valued landscape and SSSIs.

4.3 Multi-criteria approaches

Another way of addressing the problem is to conclude that the objectives that a transport system is required to fulfill are varied and sometimes conflicting. The new guidance on trunk road appraisal recognises this, identifying five explicit objectives. It may also be difficult to assess some of the direct benefits using conventional measures of journey cost and time saving, where a sustainable system will involve greater use of the slower non-motorised modes, cycling and walking. It is however debatable whether cycle is slower than car in congested urban conditions. In the light of such issues, use of Multi-Criteria Analysis techniques (MCA) may be appropriate. The derivation of weights for use in appraisal can be used as an exercise in public participation, to identify objectives and strategies. A risk with MCA is that it becomes a substitute for work to identify the costs and benefits clearly, and falls back on the use of subjective measures. However, an MCA which is clearly rooted in an assessment of costs and benefits and which includes a CBA, may be more acceptable. The new guidance on trunk road appraisal steps away from reliance on CBA and towards an MCA approach (with CBA results central to it) but no explicit weights.

5 Conclusions

The question remains as to whether it is possible to develop a truly multi-modal assessment method that is capable of including and assigning the correct weight to all the possible impacts of transport schemes. There are still a number of critical issues in transport appraisal, some more tractable than others.

1. The treatment of user benefits should be consistent across all modes. This requires an estimation of changes in consumer surplus for all travellers.
2. The treatment of accident costs should be consistent across all modes. Where a value exists for injury and death, it should be applied to all such occurrences.
3. The treatment of taxation should be the same across modes.
4. Further consideration needs to be given to the treatment of environmental impacts. For some impacts there is a growing body of evidence on values derived from individual preferences (noise), impacts on health (local air pollutants) and consensus on reduction (CO_2). These values may be brought into appraisal, in an experimental way at first to assess their significance and to investigate the sensitivity of results. However, there are a range of environmental impacts that cannot be valued at present.
5. Impacts on pedestrians and cyclists tend to be neglected in appraisal, except perhaps when commenting on the degree of severity. There is a need to model behaviour and to value changes in journey time or quality.
6. Resource consumption needs to be considered in appraisal. In regard to energy consumption, if air pollutants are given the correct values, then the signals on energy will be improved. The only case then for a further value would be one based on reducing the rate of consumption and hence exhaustion of a non-renewable resource. There is a case for using replacement or preservation cost when dealing with quality landscapes or nature reserves.

The White Paper (DETR, 1998b) suggests that a new approach to appraisal will be introduced for all modes based on five criteria, integration, safety, economy, environment and accessibility. Guidance to this effect has already appeared for trunk roads. Such an approach applied to all transport projects would be a major step towards consistency in the treatment of different modes in appraisal, and will answer some of the points above. However, research is still required to establish appropriate models and methods to assess benefits and costs across modes, including pedestrians and cyclists, and which reflect environmental impacts and resource consumption.

References

Bowers J.K., Bristow A.L. and P.G. Hopkinson (1991), *The treatment of landscape in public investment appraisal*, School of Business and Economic Studies, University of Leeds. Report to the Countryside Commission.

Bristow A.L., Mackie P.J. and C.A. Nash (1992), 'Evaluation Criteria in the Allocation of Subsidies to Bus Operations', in *Transport Investment Appraisal and Evaluation Criteria, Proceedings of Seminar C*, 20th PTRC Summer Annual Meeting, held at UMIST, September 1992. PTRC Education and Research Services Ltd, London.

Department of the Environment, Transport and the Regions (1997a), *National Road Traffic Forecasts* (Great Britain) London.

Department of the Environment, Transport and the Regions (1997b), *Developing an Integrated Transport Policy: An invitation to contribute*, London.

Department of The Environment, Transport and the Regions (1998a), *Guidance on the New Approach to Appraisal*, London.

Department of the Environment, Transport and the Regions (1998b), *A New Deal for Transport: Better for Everyone*, Cm 3950,HMSO, London.

Department of Transport (1989), *Section 56 Grant for Public Transport. Circular 3/89*, Department of Transport, London.

Eyre N.J., Ozdemiroglu E., Pearce D.W. and P. Steele (1997), 'Fuel and location effects on the damage costs of transport emissions', *Journal of Transport Economics and Policy*, Vol 31, pp 5-24.

Highways Agency, Scottish Office, Welsh Office, Department of the Environment for Northern Ireland (1996), *Economic Assessment of Road Schemes: COBA Manual, Volume 13, Design Manual for Roads and Bridges*, HMSO, London.

Highways Agency, Scottish Office, Welsh Office, Department of the Environment for Northern Ireland (1993), *Environmental Assessment, Volume 11, Design Manual for Roads and Bridges*, HMSO, London.

HM Treasury (1995), *A Framework for the Evaluation of Regeneration Projects and Programmes*, London.

Hopkin J.M. and H.F. Simpson (1995), *Valuation of Road Accidents*, Transport Research Laboratory, Report 163, Crowthorne.

Jones-Lee M (1987), *The value of transport safety*, Policy Journals, Newbury, UK.
Maddison D., Pearce D., Johansson O., Calthrop E., Litman T. and E. Verhoef (1996), *The True Costs of Road Transport, Blueprint 5*, Earthscan, London.
Mauch S.P. and W. Rothengatter (1995), *External Effects of Transport*, International Union of Railways (UIC), Paris.
Nellthorp J., Bristow A.L. and P.J. Mackie (1997), 'Valuing the Costs and Benefits in European Transport Appraisal', *Proceeding of Seminar E, Transportation Planning Methods, Vol 1*, pp. 187-200, PTRC European Transport Forum, Annual Meeting, London.
Office of Passenger Rail Franchising (OPRAF) (1997), *Appraisal of Support for Rail Services: Planning Criteria: An Interim Guide*, London.
Perkins S. (1997), 'Bringing Monetisation into Environmental Assessment: The OECD Approach', paper to the conference, *Determining Monetary Values of Environmental Impacts*, October, London.
Royal Commission on Environmental Pollution (RCEP) (1994), *Transport and the Environment*, 18th Report, Cm2674. HMSO, London.
Secretary of State for the Environment, Transport and the Regions (1997), *Objectives, Instructions and Guidance for the Franchising Director*, London.
Tinch R. (1995), *The valuation of environmental externalities: full report*, report to the Department of Transport, UK.
The MVA Consultancy, Institute for Transport Studies, University of Leeds & Transport Studies Unit, University of Oxford (1987), *The Value of Travel Time Savings*, Policy Journals, Newbury, UK.
Wardman M (1997), *A review of evidence on the value of travel time in Great Britain*, Institute for Transport Studies, University of Leeds, Working Paper 495.
Wardman M., Bristow A.L. and F.C. Hodgson (1997), 'Valuations of noise, air quality and accessibility: evidence for households and businesses', paper to PTRC European Transport Forum, Annual Meeting, London.
Watkiss P. and S. Collings (1997), 'The ExternE Transport Project. Methodologies and their Application in Appraising Public Transport Systems', paper to the conference, *Determining Monetary Values of Environmental Impacts*, October, London.

CHAPTER 21

ELECTRICITY LIBERALISATION, AIR POLLUTION AND ENVIRONMENTAL POLICY IN THE UK

PETER J G PEARSON

T H Huxley School of Environment, Earth Sciences & Engineering
Imperial College of Science, Technology & Medicine,
48 Prince's Gardens, London SW7 2PE, UK
Email: p.j.pearson@ic.ac.uk

Keywords: carbon dioxide; electricity; liberalisation; pollution; sulphur dioxide; UK.

1 Introduction

This chapter examines the relationships between air pollution, environmental policy and the restructuring and privatisation of the public electricity supply industry (ESI) in Great Britain in 1990. The paper begins by examining briefly the energy and environmental expectations that might be associated with a restructured and privatised ESI. It then considers what happened to plant capacity and fuel use after privatisation, and explores the consequences for carbon and sulphur emissions associated with the 'dash for gas' and the increased contribution of nuclear electricity. The various revisions in official energy/emissions scenarios associated with these changing emissions profiles are reviewed, and their implications for environmental policy and its presentation are explored. The chapter concludes by reflecting on the intended and unintended effects of restructuring on present and future environmental policies and outcomes.

2 Expectations from a liberalised ESI

This section outlines some of the energy and environmental expectations that might be expected to be associated with a liberalised ESI. The first is the adoption of commercial/private rather than governmental/social time-horizons, discount rates and attitudes to risk. Electricity generation and pollution abatement technologies would, therefore, tend to be relatively more attractive to market-oriented, profit-seeking enterprises if they exhibited: (a) relatively low front-end capital costs; (b) smaller efficient scales; (c) shorter construction times; and (d) less commercially risky environmental profiles. A key feature of these attributes is that they offer enhanced flexibility in decision-making, at a point in time and over time. Technologies with most or all of these attributes would be expected to figure prominently in new investment plans, especially if the new industry structure offered commercial or regulatory incentives to invest in new capacity. On these criteria, CCGT plants look attractive, relative to other larger scale, less flexible technologies.

With freedom of entry into a liberalised ESI, the arrival of new entrants, unencumbered by the baggage of past choices, could stimulate the diffusion of a range of more efficient new technologies and fuels, some of which might have enhanced environmental performance. In terms of emissions of carbon dioxide (CO_2), sulphur dioxide (SO_2) and nitrogen oxides (NO_x), for example, combined cycle gas turbine (CCGT) plant performs well against the other fossil-fuelled capacity (mostly coal and some fuel oil) that it has displaced. Nevertheless, it does not necessarily follow that those technologies and fuels that possess commercially attractive characteristics will always be less environmentally damaging than the alternatives.[1] The fact that a technology has a less privately risky environmental profile does not necessarily ensure less social risk. Much will depend on the expected regulatory situation and on the methods and extent of internalisation of environmental externalities.

A second feature of successful liberalisation is that ESI efficiency may rise. Greater technical efficiency, through more efficient technologies and improved operating efficiency, provides the opportunity for reduced fuel use - and hence pollutant emissions - per kWh generated and supplied. Enhanced efficiency should also imply lower electricity costs and, to the extent that they are passed through to consumers, falling electricity prices. In this case, on the one hand, declining prices stimulate electricity consumption and hence raise emissions. On the other hand, they might also encourage substitution towards electricity and away from less efficient and probably more polluting uses of other fuels, especially for industrial and commercial end-uses. The net effect on emissions would then depend on the interplay of these influences.

A third feature of liberalisation is that the freer choice of technologies, fuels and pollution abatement technologies open to the liberalised industry could, in some circumstances, lead to lower pollution abatement costs per unit (e.g., by switching to gas instead of installing flue gas desulphurisation (FGD) equipment). It would not necessarily lead to higher abatement *levels*, however. The outcomes would also depend on the regulatory authorities' choice of environmental policy instruments and on the stimuli they offered to innovation in pollution control technologies and practices.

A fourth feature of liberalisation is a new relationship between environmental regulator and regulated ESI, with the regulator dealing with several private companies rather than one state-owned company. Also, in the case of the UK the

---

[1] It has been suggested, for example, that in some circumstances orimulsion could have been the fuel of choice.

industry is now regulated by two regulators: the environment regulator, the Environment Agency (EA), and the industry regulator, OFFER. There are possibilities both for co-operation and for conflicts of interest and jurisdiction between the agencies. Furthermore, as the behaviour of the gas industry regulator, OFGAS, has shown in the past, there may be a need for more explicitly formulated environmental policy and a clearer delineation of who is to bear the costs of environment-related policies (e.g., gas/electricity shareholders, employees and consumers, or the general public purse). Environmental policy considerations do not appear to have formed any significant part of the objectives that underlay privatisation – and the 1989 Electricity Act does not refer directly to environmental issues. OFFER's duties require it to take the environmental effects of generation, transmission and supply 'into account', and it is supposed to promote efficiency and economy in the transmission and use of electricity. Apart from this, it cannot be said to have been assigned an active role in environmental protection.

A fifth feature of liberalisation concerns whether private companies will engage in different levels and patterns of investment in R&D from those under state-ownership (especially in fledgling technologies which are some distance from commercialisation). If there were significant differences between the private and social returns to R&D expenditure (e.g., falling costs from 'learning by doing' with new technologies), then privatisation could result in lower expenditures than before. In particular, of course, the market offers little incentive to introduce commercially unprofitable but environmentally benign technologies. The UK privatisation led in practice to considerable reductions in electricity-related R&D financed by the UK ESI, including the closing of most of the industry's large and well-regarded laboratories (Chesshire, 1996, 36). The R&D programme had grown to cover generation, utilisation, fundamental science (including combustion and emissions abatement technologies) and environment. Chesshire notes, however, that, 'Latterly a hesitant CEGB was forced to expand R&D on the environment, particularly to address issues such as acid rain' (*ibid.* 36). He also observes that the CEGB had often been criticised for failing to commit enough resources into renewable technologies.

In fact, the UK liberalisation did lead to explicit support for renewable fuels and technologies. However, this appears to have happened fortuitously, largely as a side-effect of support for nuclear-generated electricity, e.g., through the Fossil Fuel Levy in England and Wales (originally set at 10% of consumers' bills, reduced to 3.7% in November 1996 and to 2.2% in April 1997). The Renewables Orders - the Non Fossil Fuel Obligation (NFFO) in England and Wales, the NI-NFFO in Northern Ireland, and the Scottish Renewables Obligation - have offered a small but significant stimulus to the development of some near-competitive renewable technologies (Mitchell, 1996, 1999, DUKES, 1998).

A related issue is the opening up of the entire supply market to competition, in stages through 1998 and beyond. In a liberalised market, companies are free to offer differentiated products, including energy (efficiency) services and various forms of 'green electricity'. If substantial numbers of consumers turn out to be willing to pay a premium for the environmental characteristics of the electricity they use, liberalisation would stimulate the production and use of more environmentally benign technologies (Fouquet, 1998, 1999).

3 Plant capacity and fuel use after liberalisation

This section examines briefly what happened to plant capacity and fuel use, in order to provide a basis for understanding the evolution of emissions after ESI restructuring in 1990. By the end of 1997, 33 'major power producers' were involved in generation in the UK, operating just over 68 MW of the total capacity of 72.5 MW (*DUKES* 1998, 153). Table 21.1 shows percentage shares in UK electricity plant capacity for 1989-1997.

Table 21.1: Percentage shares in UK electricity plant capacity, 1989-1997

	1989	*1990*	*1991*	*1992*	*1993*	*1994*	*1995*	*1996*	*1997*
Conventional steam	77	75	74	73	71	65	61	57	56
CCGT	0	0	0	0	2	8	12	17	17
Nuclear	12	15	15	16	17	17	18	18	18
GT & oil engines	5	5	4	4	4	3	3	2	2
Hydro	6	6	6	6	6	6	6	6	6
Other renewables	0	0	0	0	0	0	1	1	1
Total	100	100	100	100	100	100	100	100	100
Total capacity (MW)	70348	74557	73545	70535	67499	68523	68741	73248	72498

Source: *DUKES 1993, Table 51, 97 (1989 data), DUKES 1994, Table 51, 109 (1990), DUKES 1995, Table 50, 105 (1991-92), DUKES 1997, Table 62, 147 (1993), DUKES 1998, Table 6.5, 163 (1994-96).*

Broadly, between 1990 and 1997 a number of old thermal coal and oil plants were decommissioned, so that conventional steam capacity fell by about 15,000 MW. Also some existing plants were upgraded and more combined heat and power (CHP) was introduced, with concomitant effects on the efficiency of fuel use. Nuclear capacity rose slightly, as Sizewell B was completed and plants achieved intended ratings. CCGT capacity surged in the 'dash for gas' (Surrey, 1996), from less than 100 MW in 1990 to a little over 13,000 MW by 1997. Renewables other than large hydro and pumped storage achieved penetration for the first time, stimulated by four NOFFO Renewables Orders for England and Wales, as well as by two Renewables Orders in Northern Ireland and in Scotland.

Table 21.2 shows shares in electricity generated for 1989-1997. In particular, between 1990 and 1997 it shows the major fall in the share of the conventional steam stations, from 77% to 43%, the rise of CCGT from zero to over one-quarter, the rise in the nuclear share from 21% to 28%,[2] and the small but increasing penetration of CHP.

Table 21.2: Percentage shares in total electricity generated, UK, 1989-97

	1989	*1990*	*1991*	*1992*	*1993*	*1994*	*1995*	*1996*	*1997*
Conventional steam	75	77	76	72	63	58	56	51	43
CCGT	0	0	0	1	7	12	15	19	26
Nuclear	23	21	22	24	28	27	27	27	28
GT & oil engines	0	0	0	0	0	0	0	0	0
Hydro	2	2	2	2	2	2	2	1	2
Other renewables	0	0	0	0	0	1	1	1	1
Total	100	100	100	100	100	100	100	100	100
of which from CHP	-	3	3	4	4	4	5	5	6
Total electricity generated (GWh)	314585	319739	322863	321043	323102	324978	334047	347386	345342

***Source: calculated from** DUKES 1993, Table 48, 93 (for 1988-89), DUKES 1995, Table 47, 100 (1990-91), DUKES 1997, Table 59, 142 (1992-95), DUKES 1998, Table 6.2 (1996-97).*

4 Air pollution emissions after liberalisation

This section considers what happened after liberalisation to the emissions of three key air pollutants from power stations: carbon dioxide, sulphur dioxide and nitrogen oxides. First, however, to view them in perspective, in 1996 UK power station emissions of CO_2 were about 27% of total UK emissions from all sources, while SO_2 was about 65% and NO_x about 22% (*DUKES* 1998, Tables A.2, A.6 & A.9).

For power station atmospheric emissions, the evolution of levels and shares in fossil and non-fossil fuel inputs is central. Table 21.3 shows what happened to these shares. Between 1990 and 1997, although the total input of all fuels hardly changed, coal's share fell from a dominant 65% to 38%, while oil fell from 11% to 2%, less than one fifth of its 1990 share. The beneficiaries were natural gas, with a striking increase from 1% to 28%, and nuclear, up from 21% to 30%.

[2] There were striking improvements in the average performance of the AGRs in England and Scotland in the early years of liberalisation, with significant increases in annual load factors. These improvements, the increases in capacity and the guaranteed market for nuclear electricity through the Fossil Fuel Levy, explain the increased share in total generation. There is some debate about how much of the improvement in productivity would have happened in the 1990s anyway, even in the absence of the stimulus from liberalisation (MacKerron, 1966).

Table 21.3: Percentage shares in fuel input for electricity generation, UK, 1989 to 1997

	1989	*1990*	*1991*	*1992*	*1993*	*1994*	*1995*	*1996*	*1997*
Coal	65	65	65	61	53	50	48	43	38
Oil	9	11	10	11	8	5	5	5	2
Natural gas	1	1	1	2	9	13	17	21	28
Nuclear	24	21	23	24	29	29	28	29	30
Other	2	2	2	2	2	2	2	2	2
Total	100	100	100	100	100	100	100	100	100
Total fuel input (mtoe)	75.27	76.34	76.87	76.57	75.40	73.71	75.12	76.6	76.08

Source: *calculated from DUKES 1998, Table 6.8, 165.*

Given that the increased nuclear and gas fuelled generation was displacing mostly coal fired generation (as well as some oil use), we would expect, inter alia, carbon dioxide, sulphur dioxide and nitrogen oxide emissions to fall significantly. Nuclear generation uses zero-carbon, non-fossil fuel, while gas produces just over half of coal's CO_2 emissions per gigajoule, less than half of coal's NO_x emissions and virtually no SO_2, as Table 21.4 indicates.

Table 21.4: Fossil fuel emission factors and ratios, UK, 1996

	Emission factors (grams/GJ)			*Ratios*		
	Natural gas	*Oil*	*Coal*	Gas/Oil	Gas/Coal	Oil/coal
CO_2 (weight of C)	14000	19000	25000	0.74	0.56	0.76
SO_2	0	520	850	0	0	0.61
$NO_{\diamond}$	51	120	270	0.43	0.19	0.44

Source – emission factors: *DUKES 1998, Table A.1, 237. Calculated from the estimated UK total emissions from stationary combustion processes and total energy consumed in these processes.*

Table 21.5 shows what happened to emissions, electricity generated and fuel used between 1990 and 1996. It indicates the significant declines in each of the selected pollutants; for example, between 1990 and 1997, while electricity generated rose by 9%, CO_2 fell by 20%, SO_2 fell by 51% and NO_x by 43%.

These declines were associated particularly, of course, with the changing nature of plant capacity and fuel inputs, indicated in Tables 21.1 and 21.3. They were also influenced by general improvements in the efficiency of power stations, and especially the greater fuel efficiency of CCGT generation compared with displaced conventional steam generation.[3] Overall, fuel input per unit of electricity generated fell by nearly 8% between 1990 and 1997. Enhanced control of nitrogen and sulphur emissions (much of which was planned before the restructuring of the ESI) also played a part. This included the installation of low NO_x burners at coal-fired

[3] Over the period 1993-97, the thermal efficiency of CCGTs was about 25% greater than that of conventional steam stations (DUKES, Table 6.7, 164).

stations and the progressive introduction from 1993 of flue-gas desulphurisation on the 6GWe of coal-fired generating capacity at the Drax and Ratcliffe-on-Soar stations.

Table 21.5: Power station emissions (million tonnes), electricity generated (GWh) and Fossil Fuel Used (mtoe), UK, 1989-96

	1989	1990	1991	1992	1993	1994	1995	1996
CO_2 (weight of carbon)	52	54	53	50	45	44	44	43
SO_2	2.641	2.723	2.535	2.434	2.089	1.764	1.589	1.318
$NO_{\diamond}$	0.773	0.781	0.683	0.672	0.577	0.524	0.494	0.449
Electricity generated (GWh)	314585	319739	322863	321043	323102	324978	334047	347386
Total fossil fuel used (mtoe)	56.2	58.8	58.1	56.6	52.4	51.0	52.3	52.9

Source - emissions: *Department of the Environment (1998), Digest of Environmental Statistics No.20, London. TSO (CO2 expressed in terms of weight of carbon emitted; to convert to CO2, multiply by 44/12).*

Source – electricity, fossil fuel: *DUKES 1993, Table 48, 93 (1988-89), DUKES 1995, Table 47, 100 (1990-91), DUKES 1997, Table 59, 142 (1992), DUKES 1998 (1993-96), DUKES, 1998, Table 6.8, 165.*

The reductions can also be seen in terms of indices of emissions per unit of electricity generated. Table 21.6 shows the greatest decline to have taken place in the SO_2 intensity of electricity (after an initial more rapid fall in NO_x until 1993), followed by the nitrogen intensity and then the carbon intensity

Table 21.6: Index of emissions per unit of electricity generated, UK, 1988-96 (1990 = 100)

	1989	*1990*	*1991*	*1992*	*1993*	*1994*	*1995*	*1996*
CO_2.	98	100	97	92	82	80	78	76
SO_2	99	100	92	89	76	64	56	46
NOx	101	100	87	86	73	66	61	55
Electricity generated	98	100	101	100	101	102	104	109

Source: *calculated from Table 21.5*

Table 21.7 shows indices of emissions per unit of fossil fuel used. The declines in these intensities reflect the switch to gas and the rising use of emissions control equipment, noted earlier.

Table 21.7: Index of emissions per unit of fossil fuel used, UK, 1988-96 (1990 = 100)

	1989	1990	1991	1992	1993	1994	1995	1996
CO_2	101	100	99	96	93	94	92	89
SO_2	101	100	94	93	86	75	66	54
NOx	103	100	88	89	83	77	71	64
Total fossil fuel used	96	100	99	96	89	87	89	90

Source: *calculated from Table 21.5*

5 Revisions in official energy/emissions scenarios

5.1 Fuel shares and CO_2 emissions scenarios

The changes in ESI plant capacity after 1990 prompted significant changes in official forecasts for fuel shares in the ESI. These changes had commensurate implications for emissions and for those charged with formulating and implementing UK and ESI environmental policy. As an illustration, Table 21.8 shows two sets of ESI fuel use share forecasts for the year 2005. The first forecast was published in 1990 in the Department of Trade and Industry's (DTI) *Energy Paper 58* (EP58), while the second appeared in 1995 in *Energy Paper 65* (EP65). The EP65 forecast shows coal use falling by more than 50%, against increases of more than 50% in renewables, natural gas and imports (from EdF), and smaller but significant increases in petroleum and nuclear. These changes led to striking revisions in the DTI's emissions forecasts.

Table 21.8: Forecasts of ESI fuel use percentage shares in 2005 (EP58 and EP65)

	Renewables	*Coal*	*Petroleum*	*Nat. Gas*	*Nuclear*	*Imports*	*Total*
EP58: 2005	1	52	7	20	19	1	100
EP65: 2005	2	25	10	36	25	2	100

***Sources:** EP65 (1995), Table C3, CL scenario, p. 132; EP58 (1990), Table 4.11, CL scenario, p. 25.*

Figures 21.1 and 21.2 below show three sets of CO_2 emissions scenarios published by the DTI. The first forecast appeared in EP58 in 1990, the year of ESI privatisation, the second was in EP59 in 1992, while the third appeared in EP65 in 1995. Figure 21.1 is for total UK emissions from all sources, while Figure 21.2 shows ESI emissions (the EP65 graph shows the CL Scenario - 'central growth, low oil prices').

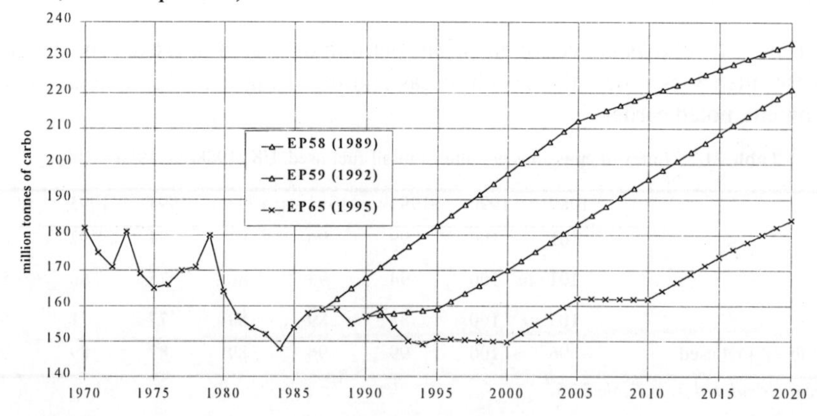

Figure 21.1: UK total carbon emissions scenarios to 2020 (MtC)

Given the then UK Government's UNFCCC commitment to reach 1990 CO_2 emission levels by the year 2000, the changing forecasts show how the UK's anticipated ability to achieve the target improved strikingly over the five years from privatisation. By 1995 EP65 was forecasting year 2000 emissions well below the 1990 level. As Figure 21.2 indicates, moreover, most of the differences were expected to come from the ESI.

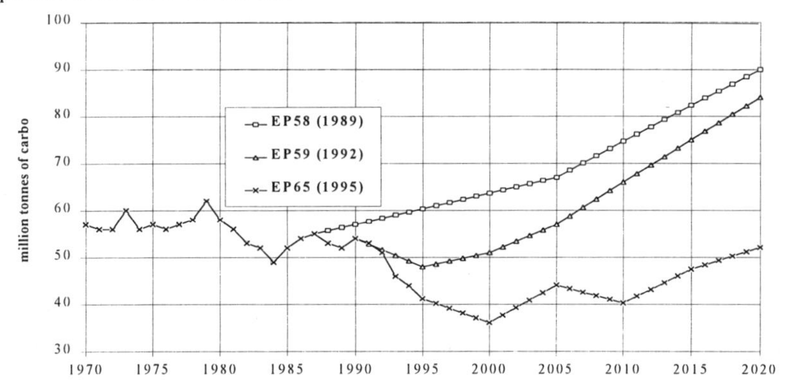

Figure 21.2: UK ESI carbon emissions scenarios to 2020 (MtC)

In EP59 in 1992 the UK had forecast a need to cut emissions by about 10 MtC per year against projected emissions for 2000, in order to meet its Rio commitments. Table 21.9 shows from where the overall savings were to come. Onc of the key measures was the Energy Savings Trust (EST), to be financed by the government and from 1994 by levies of £1 per customer in the electricity and gas markets (domestic consumers and small businesses). The EST was originally intended to achieve savings of 2.5 MtC by 2000.

Table 21.9: UK climate change programme 1994

Sector	*Reduction in Emissions by 2000 (MtC)*
Energy consumption in the home	4
Energy consumption by business	2.5
Energy consumption in the public sector	1
Transport	2.5
Total	10

Note: *EST was assumed to account for 2.5 MtC; VAT at 17.5% on domestic fuel and power was assumed to account for 1.5MtC.* ***Source:*** *EP 65 (1995), Table 3.18, 48.*

In the event, as Figures 21.1 and 21.2 showed, and Table 21.10 illustrates, by 1995 even the 'central growth, low oil prices' CL scenarios were suggesting that UK emissions would be below the 1990 level, *without* the first Climate Change Programme (CCP) and its planned 2.5 MtC EST savings.

Table 21.10: UK CO_2 emissions in 2000, from EP65 (million tonnes of carbon)

	EP59 Projections[1]	*EP65 Projections*			
	CL Scenario[2]	*CL Scenario*		*CH Scenario*	
		Without CCP[3]	*With CCP*	*Without CCP*	*With CCP*
1990 (base year)	158.3	158.3	158.3	158.3	158.3
2000	169.4	157.8	149.7	157.2	148.4

***Notes:** (1) Adjusted to same basis as EP65; (2) CL – central growth, low oil price assumptions; CCP – Climate Change Programme.* ***Source:** EP65 (1995) Table 7.1.*

By the time that the UK government published its second report under the UNFCCC (Climate Change, 1997), several months before Kyoto, it was comfortably projecting that total UK CO_2 emissions to lie from 4-8% below 1990 levels by 2000. In identifying 'market liberalisation' as an instrument of policy, the report credited the switch from oil and coal to gas and the use of more efficient CCGTs (17 MtC), and improvements in the productivity of the nuclear sector (2.9 MtC), with saving about 20 MtC a year by 2000. This amounted to more than half of the Climate Change Programme's projected total savings of 35.2 MtC per year, against projections of 193 MtC per year in 2000 in the absence of the Programme's 'principal carbon-savings measures'.

In addition, the 'market stimulation' of the Renewables Orders was 'working towards' 1,500 MW of renewable capacity (2 MtC) and the government set a target of 5,000 MW of CHP capacity by 2000 (3.5 MtC). The Energy Efficiency Best Practice Programme was to achieve savings of about 4.5 MtC in industry (including 1.5 MtC towards the CHP target). The Energy Savings Trust's programme for domestic and small business users was scaled down by four-fifths, to 0.5 MtC, after resistance to the levies from the electricity and gas regulators, particularly gas. The second tranche of value-added tax on domestic electricity and gas supplies was also abandoned, after political opposition (the first tranche,at 8%, had been introduced in April 1994).[4]

5.2 *SO_2 emissions scenarios*

The DTI's SO_2 emissions scenarios from EP65 showed ESI and total UK levels falling sharply after 1990, as Figure 21.3 indicates. They suggested that there would be no serious difficulty in meeting the United Nations Economic Commission for Europe's (UNECE) Second Sulphur Protocol requirements (to meet sulphur emission reduction targets of 50% by the year 2000, 70% by 2005 and 80% by 2010 from a 1980 baseline) with existing flue gas desulphurisation capacity. In fact, by the end of 1996 the UK had attained a 59% reduction in total SO_2 emissions from

[4] VAT on domestic fuels was reduced to 5% in September 1997.

1980 baseline levels, 9% ahead of the UNECE target level for the year 2000 (DETR, 1998, 23).

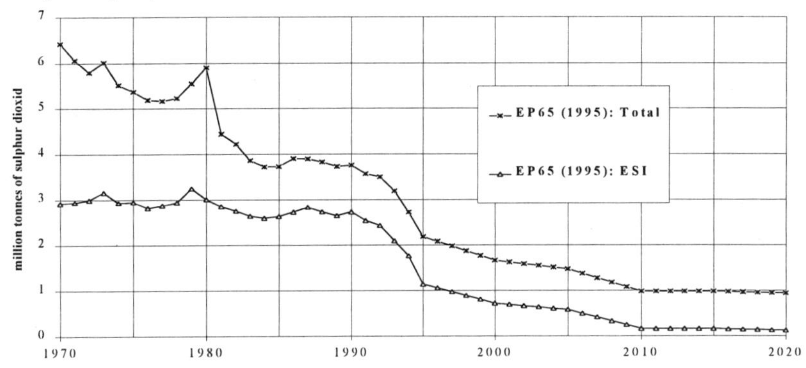

Figure 21.3: UK sulphur dioxide scenarios to 2020 (million tonnes of sulphur dioxide)

The EP65 scenarios also suggested that there should be no problems in meeting the requirements of the EC's 1988 Large Combustion Plant Directive (88/609/EEC). The LCPD set national targets (which vary by sector) for the reduction of sulphur and nitrogen emissions to the year 2003.[5] In fact, by 1996 SO_2 emissions from large combustion plants in the UK were 57% below the 1980 baseline level, 17% ahead of the 1998 EC target level (DETR, 1998, 24).[6] Table 21.11 shows how by 1995 the ESI was expected to be well inside the limits. Indeed, the DTI noted in this table that, although the columns did not indicate figures for 1998 and 2003, there was little point in troubling about whether or not the UK would meet its LCPD limits in those years.

Table 21.11: EC large combustion plant directive: EP65 CL scenario for SO_2 (million tonnes)

	1990	*1995*	*2000*	*2005*
ESI emissions	2.722	1.140	0.720	0.580
ESI limit	2.906	2.339	1.572	1.208
% reduction in expected ESI emissions on 1980	-9%	-62%	-76%	-81%
Total LCPD plant emissions	2.952	1.350	0.912	0.772
Total LCPD plant limit	3.338	2.695	1.858	1.452
% reduction in expected LCPD plant emissions on 1980	-24%	-65%	-77%	-80%

***Note:** target dates for LCPD are 1998 & 2003 - but emissions are below limits.* ***Source:** EP65 (1995), Table E4, 141.*

[5] These targets require reductions in total SO_2 emissions from existing combustion installations with an annual capacity of greater than 50 MW thermal, of 20% by the end of 1993, 40% by 1998 and 60% by 2003, taking 1980 emissions as the baseline.

[6] By 1996, NO_X emissions from large combustion plants in the UK were 44% below those estimated for 1980, 29% below the 1993 EC target and 14% below the 1998 EC target (DETR, 1998, 26).

Although the CL scenario forecasts for 1995 turned out to be somewhat optimistic, the realised figures were still well within the limits (see Table 21.5). As in the case of CO_2 emissions, successive governments that had in the late 1980s and early 1990s been concerned about meeting targets for sulphur and nitrogen emissions, now found their burden lightened. The December 1996 UK sulphur dioxide strategy document (Department of the Environment, 1996a) said that: 'New emissions projections which take account of the review of the electricity supply industry in England and Wales, and of assumptions about progressive upgrading of other industrial combustion plant in the UK, indicate that the UNECE targets for 2000 and 2005 will be met without the need for additional measures. […] Forward projections also indicate that the targets in the National Plan, which gives effect to the European Community Directive on large combustion plant, will be met by a significant margin.'

6 Conclusion

The UK ESI, whether privatised or not, would always have needed to pay attention to externally agreed limits on carbon, sulphur and nitrogen. The dash for gas, the improvement in nuclear performance and, to a lesser extent, other improvements in the efficiency of plant, pollution control and management made it significantly easier to meet those targets[7]. This was not, however, entirely costless: the dash for gas implied real resource costs, both in terms of the early bringing on of new capacity and in the social costs imposed by the abrupt decline in the UK coal industry's market, while the Fossil Fuel levy imposed significant nuclear-related costs on electricity consumers rather than the Treasury and taxpayers (MacKerron, 1996).

In the case of the CCGTs, it was to some extent fortuitous that the commercial considerations of a liberalised industry focused at that time on a technology that happened also to yield lower carbon, sulphur and nitrogen emissions. There seems little reason to suggest that commercial considerations can be relied upon automatically to select not only commercially viable but also environmentally benign technologies and fuels (if that were so, moreover, there might be little need for environmental policy).

This does not mean, however, that the circumstances of a liberalised industry will necessarily make it more difficult to implement environmental policy than would those of a state-owned industry. In some ways, it could be easier. As has been suggested, for example, the opportunities for new market entrants to

[7] Of course, some of these outcomes would have happened even in the absence of privatisation and liberalisation; ideally more should be done to compare the actual outcomes with a set of plausible counterfactual scenarios.

experiment with new products (such as energy services or 'green electricity') and systems, such as decentralised systems, drawing on a range of new technologies, seem promising in a world where governments lack the reputation for picking technological or environmental 'winners'. Nevertheless, in the presence of environmental market failure, there is still a role for governments to set environmental targets and to select the instruments with which to stimulate their achievement. Ironically, the UK, an innovator in applying market principles to energy, has been much slower to adopt market-based environmental policy instruments, which have not as yet sat comfortably with the UK's traditional modes of implementing environmental policy. Recently, however, there have been signs of change (Marshall, 1998).

The early years of the UK ESI restructuring and liberalisation were associated with a rapid improvement in emissions of several key pollutants. It can be argued that this encouraged the Environment Secretary of the day, John Gummer, to take a more proactive environmental stance than he and the government would otherwise have done, both nationally and internationally.[8] In this sense, liberalisation may – unintentionally - have changed the path and targets of future UK environmental policy. It can also be argued, however, that the authorities, freed from the immediate constraints of meeting international obligations, yielded to the temptation to relax the pursuit of proactive environmentally benign policies (e.g., energy-saving policies and R&D investment, including investment conducive to the control or sequestration of CO_2 emissions). The outcome could then be that that when international environmental constraints, including those arising from the Kyoto Protocol,[9] threaten once more to bite in the first two decades of the next century,[10] the catch-up could become more difficult and costly than it might otherwise have been.

References

Chesshire J (1996), 'UK Electricity Supply under Public Ownership', Ch. 2 in Surrey, J (ed.) (1996), *The British Electricity Experiment*, Earthscan, London.

---

[8] See, e.g. the foreword to Climate Change (1997) and the statement in Department of the Environment (1996b): 'We have buried the old myth that Britain is the Dirty Man of Europe'.

[9] The December 1997 Kyoto Protocol specified legally binding commitments by most industrialised countries to reduce greenhouse gas emissions to a fraction of their 1990 levels by 2008-2012. The UK's commitment, as part of the European Union's reductions, implies a 12.5% reduction on 1990 carbon dioxide emissions, while the UK's own expressed target is a more demanding 20% cut by 2010.

[10] See, for example, Grimston (1998).

Climate Change (1997), *Climate Change: the UK Programme*, Cm 3558, The Stationery Office, London.

Department of the Environment (1996a), *Reducing National Emissions of Sulphur Dioxide: A Strategy for the United Kingdom*, Department of the Environment, London.

Department of the Environment (1996b), 'Cleaning up acid rain – United Kindom well ahead of the field', News Release 571, London, 17 December.

DETR (Department of the Environment, Transport and the Regions) (1998*), Digest of Environmental Statistics No. 20*, TSO, London.

DTI (Department of Trade and Industry) (1988), *Energy Paper 58. An Evaluation of Energy Related Greenhouse Gas Emissions and Measures to Ameliorate them*, HMSO, London.

DTI (Department of Trade and Industry) (1992), *Energy Paper 59. Energy related carbon emissions in possible future scenarios for the United Kingdom*, HMSO, London.

DTI (Department of Trade and Industry) (1995), *Energy Paper 65. Energy Projections for the UK*, HMSO, London.

DTI (Department of Trade & Industry)(1997), *The Energy Report*, Vol. 1, HMSO, London.

DUKES (1998), *Digest of United Kingdom Energy Statistics 1998*, DTI (Department of Trade & Industry), The Stationery Office, London. (Also DUKES 1993-1997, annual publication, published by HMSO until 1997.)

Fouquet R. (1998), 'The United Kingdom demand for renewable electricity in a liberalised market', *Energy Policy*, 26(4), 281-93.

Fouquet R (1999), 'Lessons for the United Kingdom from previous experiences in the demand for renewable electricity,' *Ch. 17 in this volume.*

Grimston M. (1998), 'The greenhouse challenge – just how difficult is it going to be?' *Nuclear Energy*, 37(3), 163-169.

MacKerron G. (1996), 'Nuclear Power under Review', Ch. 7 in Surrey, J (ed.) (1996), *The British Electricity Experiment*, Earthscan, London.

Marshall Lord (1998), *Economic Instruments and the Business Use of Energy*, A Report by Lord Marshall, HM Treasury, London.

Mitchell C. (1996), 'Renewable Generation – Success Story?' Ch. 8 in Surrey, J (ed.) (1996), *The British Electricity Experiment*, Earthscan, London.

Mitchell C. (1999), Renewables in the UK –how are we doing?', *Ch. 15 in this volume.*

Surrey J. (ed.) (1996), *The British Electricity Experiment*, Earthscan, London.

CHAPTER 22

RISK ASSESSMENT AND EXTERNAL COST VALUATION: HOW USEFUL IS THE 'ANALYTICAL FIX' IN THE ENVIRONMENTAL APPRAISAL OF ENERGY OPTIONS?

ANDREW STIRLING

Science Policy Research Unit, University of Sussex, Falmer, East Sussex BN1 9RF, UK
Email: a.c.stirling@sussex.ac.uk

Keywords: electricity; environmental appraisal; environmental valuation; external cost; risk assessment; uncertainty.

The pressures to demonstrate a robust and rigorous basis for the environmental appraisal of energy options are now stronger than ever. The stakes are as high where investment decisions are dominated by Government policy or regulation as where they are influenced more through the structures of commercial markets and consumer demand. However, the available analytical tools for environmental appraisal are suffering something of a 'crisis of confidence'. After a period of ascendancy in the 1980's, comparative risk assessment has been cumulatively undermined by criticism of its neglect of the scope and complexity of environmental performance. There is currently an energetic bid to establish environmental cost-benefit analysis as a replacement approach to analysis for regulation and public policy intervention. However, though the issues may be concealed by a new vocabulary, essentially the same questions may be raised.

Based on insights gleaned in the field of risk analysis, this paper examines some of the difficulties encountered in attempts to characterise the broad environmental effects of energy options as monetary 'externalities'. A series of important general factors in environmental appraisal are identified: such as the divergent forms and distribution patterns of the effects of different options; the varying degrees of autonomy on the part of those affected by different types of impact; the treatment of uncertainty and variability and the framing and presentation of analysis. These factors are then discussed in relation to a number of highly influential recent environmental valuation studies. It is shown that, if anything, neoclassical environmental economics in practice actually adds to the deficiencies displayed by comparative risk analysis.

The results obtained by more than thirty major environmental valuation studies in the electricity sector are then reviewed. Attention is focused on the variability in the literature and on the resulting ambiguity in the possible rank-orderings of different options. The results obtained by different studies are found to vary by factors ranging up to several orders of magnitude. The literature as a whole might be taken to accommodate several radically different environmental rank orderings of generating technologies. In addition, it is argued that the values derived for energy externalities may reflect the constraints imposed by existing market prices as much as they do the relative magnitudes of different environmental effects.

The paper concludes by pointing to alternative practical approaches to the social appraisal of environmental impacts. Looking at recent experience in the USA, Germany and Switzerland, it is argued that other techniques are available which offer greater transparency and rigour and which, through structured public participation, avoid futile and politically polarising attempts to impose an 'analytical fix' on the essentially political business of comparative environmental appraisal of energy options.

1 Introduction

The pressures to demonstrate a robust and rigorous basis for the environmental appraisal of energy options are now stronger than ever. The stakes are as high where investment decisions are dominated by Government policy or regulation as where they are influenced more through the structures of commercial markets and consumer demand. However, the available analytical tools for environmental appraisal are suffering something of a 'crisis of confidence'. After a period of ascendancy in the 1980's, comparative risk assessment has been cumulatively undermined by criticism of its neglect of the scope and complexity of environmental performance. There is currently an energetic bid to establish environmental cost-benefit analysis (environmental valuation) as a replacement approach to analysis for regulation and public policy intervention. The present paper seeks to establish the degree to which environmental valuation suffers from the same problems as its predecessor and to explore the implications both for energy and environment policy and for the future conduct of environmental appraisal of energy options.

The chapter is divided into three parts. The first section summarises the results of a detailed review of the theoretical and methodological issues raised in any assessment of the practical utility of environmental valuation studies. A series of important general factors in environmental appraisal are identified, namely: the divergent forms and distribution patterns of the effects of different options; the varying degrees of autonomy on the part of those affected by different types of impact; the treatment of uncertainty and variability and the framing and presentation of analysis. These factors are illustrated by drawing examples relating to a number of highly influential recent environmental valuation studies.

The second section reviews the results obtained by more than thirty studies of the external environmental costs associated with the principal electricity supply options in industrialised countries. Attention is focused on the degree of variability in the results obtained throughout the literature as a whole, on any resulting ambiguity in the possible rank-orderings obtained for different options and on the possible influence of circumstantial factors such as the prevailing values of market prices for electricity.

Based on a survey of recent experience and emerging practice in North America and northern Europe, the final section of this paper examines the potential for alternative approaches to the social appraisal of the environmental impacts of energy options. In particular, an illustration is provided of the kind of result that might be obtained by the pursuit of a multi-criteria approach to appraisal, conducted within a more pluralistic participatory framework.

2 Some key theoretical and methodological issues

Throughout the 1970's and early 1980's, the dominant analytical approach to the environmental appraisal of energy options was provided by comparative risk assessment. Over this period, there accumulated an increasingly substantive body of criticism[1]. Serious questions were raised concerning the degree to which the aggregated numerical values derived in risk assessment satisfactorily reflect the scope, complexity, disparity and subjectivity of environmental performance.

The environmental effects of different energy options may differ radically in the ***forms*** which they take. For instance, some may be more manifest as risks of death, others as injury or disease. They may differ in the immediacy or latency of their impacts. The effects of some options may be concentrated in a few large events, whereas others may spread across a larger number of smaller incidents. Effects of different options may vary in the degree to which they are reversible. The determination of the relative importance of these different dimensions must inevitably be a highly circumstantial and subjective matter. The aggregated numerical values obtained in risk assessment and environmental valuation compress these different dimensions onto a single yardstick, adopting implicit assumptions about their relative importance. Some of the specific practical questions which are raised in considering the potentially divergent forms of environmental effects of different electricity supply options are summarised in Box 1.

Box 1: Questions relating to the form of environmental effects

Severity: Do the options differ in the ratios of risks of death to risks of injury or disease which they pose? How much illness or how many serious injuries equate in severity with one death? (E.g., offshore wind and wave vs biomass).

Immediacy: Are the effects associated with different options equally immediate in their manifestation or do they differ in the degree of latency between the initial commitment of a burden and the eventual realisation of an effect? For instance, are some risks manifest as injuries and others as disease? (E.g., rooftop solar arrays vs nuclear power).

Gravity: Are the risks associated with some options dominated by low probabilities of large impacts, while those of other options are characterised predominantly as high probabilities of relatively low impacts? To what extent are impacts the result of single or repeated events? (E.g., nuclear vs coal).

Reversibility: Are the effects associated with different options all equally reversible after they have been committed? (E.g., nuclear and fossil fuels vs wind).

---

[1] For instance, specifically with regard to energy technologies in the decade to 1985: Budnitz and Holdren, 1976; Holdren et al, 1979; Cohen and Pritchard, 1980; Watson, 1981; Holdren, 1982; Rowe and Oterson, 1983; Kayes and Taylor, 1984.

The environmental effects of different energy options also differ in terms of their ***distribution*** across space, through society and beyond. This raises issues concerning the 'fairness' of the distribution of impacts across different groups and the way this correlates (or not) with the distribution of the benefits arising from the operation of the investments concerned. Particularly intractable difficulties emerge in contemplating the distribution of environmental effects through time, and the balance between burdens which fall on human and non-human life. Where the patterns in the distribution of the environmental effects of different energy options vary along these dimensions, further serious questions must be raised about the value of discrete numerical results, such as those delivered by risk assessment and environmental valuation. Some of the detailed implications are summarised in Box 2.

Box 2: Questions relating to the distribution of environmental effects

Spatial Distribution: Are the effects associated with different options identical in their spatial extents? Is it better that impacts of a given magnitude be geographically concentrated or dispersed? (E.g., wind vs fossil fuels).

Balance of Benefits and Burdens: To what extent is the social distribution of the environmental burdens caused by each option balanced by the distribution of associated benefits? (E.g., distributed vs centralised).

Fairness: To what extent do the distributions of burdens imposed by the different options act to alleviate or compound pre-existing patterns of privilege or social disadvantage? To what extent should exposure to other (unrelated) risk-inducing agents be taken into account in the assessment of the acceptability of incremental burdens? (E.g., urban waste-to-energy vs domestic PV).

Public or Worker Exposure: To what extent do different options impose different distributions of risks across workers and the general public? (E.g., offshore wind vs oil).

Intergenerational Equity: Do the effects associated with certain options present risks to future generations to a degree not associated with others? What is the appropriate discount rate, if any? (E.g., nuclear and fossil vs renewables).

Human or Non-human: Do the options differ in the degree to which their impacts affect the well-being of humans and non-human organisms? (E.g., biomass vs gas).

As illustrated in Box 3, the environmental effects of different energy options also impact differently on the ***autonomy*** of those affected. Exposure to the effects of some technologies is more voluntary than is the case for others. Likewise, different effects vary in their familiarity and the degree to which they are controllable. Finally, serious, complex and pervasive issues are raised in considering the trust that should be placed in the communities and institutions associated with the operation of the different options, and the appraisal results which they obtain.

Box 3: Questions relating to the autonomy of those affected

Voluntariness: Do the environmental effects of different options vary in the degree to which exposure may be considered to be 'voluntary' prior to the commitment of an impact? (E.g., do-it-yourself home insulation vs centralised coal).

Controllability: Once committed, are the impacts associated with different options all equally controllable from the point of view of the individuals or communities who stand to be affected? Do certain effects require efforts at control which are perceived to pose a threat to democratic institutions or processes? (E.g., nuclear vs wind).

Familiarity: Do the effects associated with different options differ in terms of the degree to they are familiar to individuals, communities and established social institutions? Do responses to the different effects involve equally disruptive changes to normal routines and attitudes? (E.g., nuclear vs biomass).

Trust: Do options differ in terms of the degree of trust enjoyed in the wider society by the institutions and communities charged with evaluating and managing their associated risks? Does the appraisal of certain options tend to be more a specialised undertaking than that of others? (E.g., nuclear vs biomass).

Turning to the characteristics of analysis, rather than the environmental effects themselves, the results obtained in environmental appraisal are, obviously, highly sensitive to the selection of primary quantitative ***indicators***. Although final results are (in environmental valuation) expressed as monetary values or (in risk assessment) mortality or morbidity probabilities, these represent conversions and aggregations over a wide variety of basic indices. The different primary metrics employed with each individual environmental effect may vary radically in the degree to which they capture the full character of that individual effect and the fidelity with which they track its dynamics. Some effects are intrinsically much more readily quantifiable than others, compounding the potential for incoherence between the approaches adopted to different effects both within individual studies, and between different studies.

Some of the specific issues raised in contemplating the selection of indicators are summarised in Box 4. The general problems of fidelity, resolution and coherence raised there are also important issues on a more restricted canvas, in relation to the choice of particular valuation methodologies. Figure 22.1 provides a schematic summary of the way in which different studies tend to emphasise different methodologies.

Box 4: Questions relating to the choice of indicators
Quantifiability: Are the effects associated with different options all equally quantifiable? How has appraisal avoided a disproportionate emphasis on the more quantifiable aspects - and thus an overemphasis of the impacts of the associated options? (E.g., nuclear waste vs aesthetic landscape impacts).
Fidelity: What is the fidelity of the models employed to track the relationship between the magnitudes of particular burdens and the scale of their associated effects? Are there any non-linearities, or even 'non-monotonicities' in the dose-response function? (E.g., acid rain, ozone and low level ionising radiation).
Resolution: How well does each disaggregated performance indicator resolve the full character and scope of the individual effect which it is intended to represent? (E.g., radioactivity vs 'relative biological effectiveness')
Coherence: How coherent is the classificatory scheme adopted in any particular study with respect to the full range of environmental effects? Are there gaps or overlaps between the different classes of effect which are recognised for the purposes of analysis? (E.g., emissions, burdens, or effects).

Some valuation studies obtain their results largely through the pursuit of a 'mitigation cost' approach, based on an assessment of the costs incurred in alleviating environmental damage once committed[2]. Other studies mix results obtained through application of mitigation cost techniques (e.g., to certain atmospheric effects) with values obtained by the use of 'hedonic market' and 'contingent valuation' methods (e.g., to certain water effects) which assess values, respectively, by examining prevailing property or wage markets or responses to questionnaires[3]. Still other analysts favour the use of 'abatement cost' techniques, which take the costs of controlling pollution at source as a proxy indicator for the social costs of the environmental impacts thereby avoided[4]. A final group of studies, is based mainly on a fourth methodology: the 'damage function' technique[5]. This involves the 'bottom-up' assessment of the costs associated with each physical dose-response relationship. There is a considerable literature concerning the relative merits and deficiencies of these different techniques[6]. The differing characters of these approaches and the fact that the results obtained for specific effects are found

---

[2] E.g., Hohmeyer, 1988.

[3] E.g., Ottinger et al, 1990.

[4] E.g., Tellus, 1991.

[5] E.g., Externe, 1995. Although Externe does also use contingent valuation for appraising some effects, such as the ecosystem impacts of hydroelectricity (Externe, 1995: volume 6)..

[6] E.g., Peterson et al, 1988; OECD, 1989; Hohmeyer and Ottinger, 1990.

often to vary between methods[7], are suggestive of particularly significant difficulties of resolution, fidelity and coherence in environmental valuation.

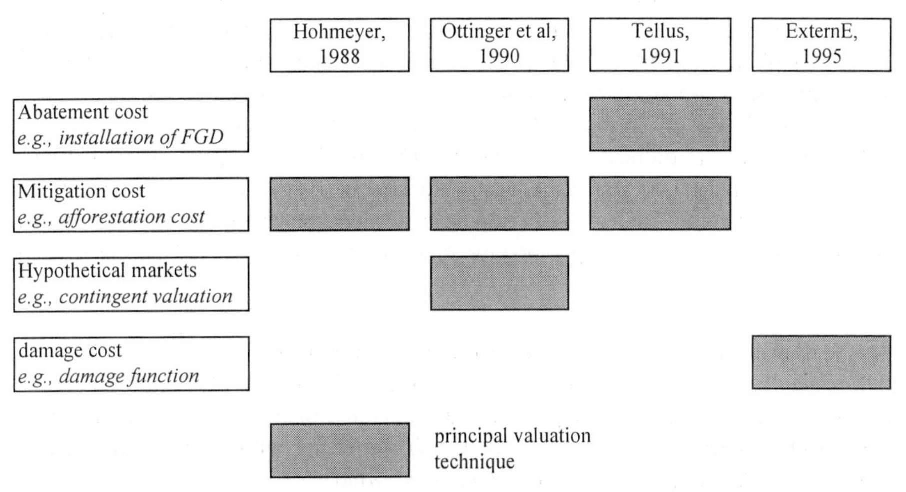

Figure 22.1: Schematic illustration of the use of different valuation methodologies

One of the most important issues in the general field of environmental appraisal concerns the analysis of ***uncertainty***[8]. In studies of individual classes of environmental effect, it is typical that upper and lower bounds to the ranges of results expressed in individual studies may differ by several orders of magnitude[9]. Against this background, it might be thought reasonable that many risk assessment studies express their numerical results only to one[10] or two[11] significant figures. It is curious, then, that - over the years - a significant number of risk studies evidently feel such confidence in their results that they employ as many as three significant figures[12]. Yet the treatment of uncertainty in the environmental valuation literature is even more optimistic. Despite the inherent intractability of their task, authors of energy externality studies often seem to feel they can justify levels of precision which are at least as great, and sometimes greater, than any professed elsewhere in the environmental appraisal literature. For instance, the 1990 Pace University study

---

[7] As exemplified by the divergent results obtained by variants of the contingent valuation method (Peterson et al, 1988) and (in the specific context of energy risks) by the discussion of the value of statistical life in DTI, 1992.

[8] De Jongh, 1988; Perrings, 1991; Talcott, 1992.

[9] E.g., Comar and Sagan, 1976 ; Holdren et al, 1980; Ferguson, 1981; Fritzsche, 1989.

[10] E.g., Ferguson, 1981; Ball et al, 1994.

[11] E.g., Fritzsche, 1984; Inhaber, 1978.

[12] E.g., Comar and Sagan, 1976; Cohen and Pritchard, 1980; Holdren et al, 1980; UNEP, 1985; IAEA et al, 1991.

for the US Department of Energy (and other agencies) (Ottinger et al, 1990) and a report by Voss et al for the German electricity industry in 1989 (Voss et al, 1989) give some results to three significant figures. The pioneering study by Hohmeyer for the European Commission in 1988 presents results to a daunting four significant figures (Hohmeyer, 1988; 1990). Likewise, the stated ranges of variation of valuation studies are similarly indicative of higher confidence than that enjoyed by other approaches to environmental assessment. Hohmeyer's ranges are as narrow as factor ten at most (Hohmeyer, 1988; 1990; 1992), whilst Pace (Ottinger et al, 1990) and a study for the UK Department and Industry in 1992 (DTI, 1992) present no ranges at all in some final results[13]. Some of the key aspects of uncertainty which are concealed in this ostensibly precise mode of presentation are summarised in Box 5.

Box 5: Questions relating to the treatment of uncertainty
Ignorance: How important is the element of surprise? Do some effects involve complex, novel or highly contingent mechanisms more than others? Are there are large discrepancies in the degree of established experience with particular options or effects? (E.g., genetically engineered biomass vs wind).
Data Quality: Are the performance data for the different options all of comparable quality and pertinence? Do some derive from ex post and other from ex ante studies? (E.g., energy forestry vs offshore wave).
Aetiology: Are the effects of all options equally "direct" in their manifestation, or do some involve complex, contingent or synergistic interactions with other agents or activities to a greater extent than others? What possible interactions might exist between the effects associated with different options in an overall portfolio? (E.g., 'business as usual' vs 'all-renewable' future).

The sixth and final set of 'dimensions of variability' in the environmental appraisal of energy options concerns the fundamental underlying assumptions adopted in the ***framing and presentation*** of the analysis. Some of the key specific questions raised here are set out in Box 6. In short, assumptions concerning the specific operational circumstances of the different options, their developmental trajectories and the 'system boundaries' set for the purpose of analysis may all have a determining influence on the nature of the results obtained. Further issues may be

---

[13] The 1995 Externe study for the European Commission adopts a more sophisticated approach to the treatment of data quality, specifying the degree of confidence associated with the values obtained for different disaggregated effects and itself avoiding summing over categories (Externe, 1995). However, even this study nevertheless presents its results as discrete values rather than as ranges or sensitivities. Where qualifications are buried in the more theoretical passages of such studies, they are all too easily lost in derivative work which treats the results obtained as if they *were* meaningfully additive (e.g., ESD et al, 1995).

raised in considering the degree to which any individual set of results constitute a 'complete' account of the issues pertaining to any individual decision, and the way in which those factors which are included in analysis are articulated with those which are not in subsequent interpretation of results.

Box 6: Questions Relating to the Framing and Presentation of Appraisal

Specificity: How site-specific is the performance data for the different options? How sensitive are results to assumptions about the operational characteristics of the individual options and of the system in which they are embedded? (E.g., combined heat and power, dual electric/irrigation hydro, integration of intermittents).

Trajectories: How long a historic data series is appropriate as a basis for the appraisal of current options? How robust are assumptions concerning the likely future behaviour of those at risk? Are different options on different 'learning curves' in terms of the potential for future improvements in performance? (E.g., radioactive waste, photovoltaics).

System Boundaries: How systematically does analysis address the resource chains and facility life cycles associated with the different options? How far back into the wider economy should analysis regress in assessing energy and material inputs? (E.g., material inputs to renewables, uranium mining for nuclear).

Completeness: How complete is the scope of appraisal with respect to the full range of relevant environmental and health effects and all pertinent cross-cutting dimensions of these effects? How might changes in the scope of appraisal alter the apparent rankings of the different options? (E.g., global warming, nuclear proliferation).

Articulation: How are the results of analysis to be articulated with wider considerations and the subsequent decision making process. At what point does the domain of analysis end and that of politics begin? (E.g., are results to be regarded as 'real', 'true' or 'full'?).

To illustrate the potential importance of these 'dimensions of variability' in determining results, examples may be given concerning just two such factors in recent environmental valuation studies of energy options: the treatment of system boundaries and the completeness in the scope of analysis. Focusing on three major Government-sponsored analyses of energy externalities, Figure 22.2 displays schematically the extent to which the apparently neat numerical values derived in environmental valuation may conceal the crucial fact that different studies address different stages in the 'fuel cycles' associated with individual options and in the 'life cycles' of associated[14] Hohmeyer's 1988 study (Hohmeyer, 1988) and its subsequent updates (Hohmeyer, 1990, 1992) are essentially restricted to the electricity generation stage (omitting mining or drilling, fuel processing, storage and transport and waste management) and to the operational phase (omitting inputs of energy and materials, and the impacts of the construction and decommissioning processes). The 1990 Pace study is almost as restricted in scope, addressing waste management burdens and decommissioning (for some options but not others).

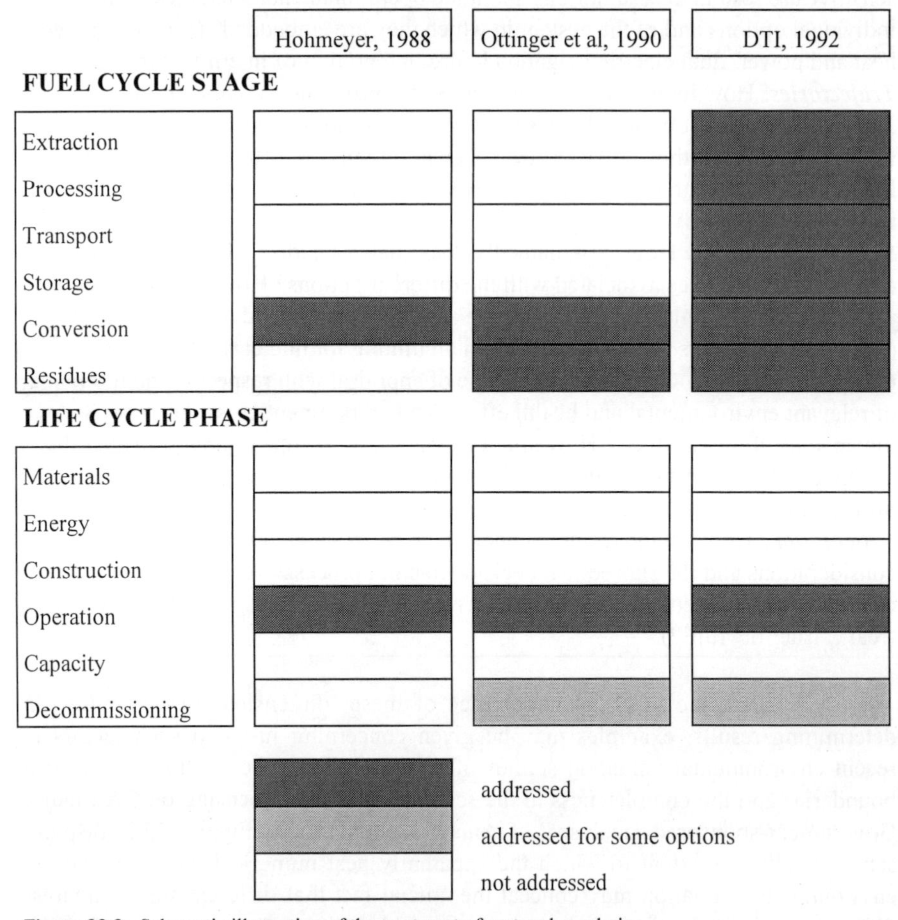

Figure 22.2: Schematic illustration of the treatment of system boundaries

[14] The fact that neither the chains of transactions and transformations involved in fuel use nor those of facilities themselves are actually 'cycles' does not seem to have inhibited the use of these misleadingly cosy terms.

The 1992 study for the UK DTI (DTI, 1992) adopts wider system boundaries, including some reference to fuel extraction, processing, transport and storage and waste management, but also omits material and energy inputs and construction and decommissioning impacts. With the exception of the 1995 Externe report, it is notable that the system boundaries set in valuation studies tend to be much narrower than those which have for some time been conventional in the comparative risk assessment of energy options [15]. Even the Externe report, however, omits energy and material inputs to construction of fossil and nuclear facilities, while including these for some other options [16]. Crucial underlying assumptions on system boundaries are not conveyed in aggregated numerical results. As a result, the environmental valuation literature as whole is vulnerable to serious difficulties of interpretation.

A similar picture emerges with respect to the completeness of environmental valuation studies. With reference to just four studies, Figure 22.3 illustrates schematically the way in which different analyses include and exclude different categories of effect. Hohmeyer's 1988 study for the European Commission (Hohmeyer, 1988) (and its subsequent revisions (Hohmeyer, 1990, 1992)) exclude aesthetic effects, thereby omitting a factor widely regarded as the most serious single environmental impact of wind power. The 1990 Pace University study (Ottinger et al, 1990) does address aesthetic impacts, but omits to account for occupational safety risks, another potentially important effect in assessing wind power. Although relatively comprehensive in scope, the major 1995 Externe study (Externe, 1995), also conducted for the European Commission, omits to include global warming, despite the fact that this is addressed in both the other earlier studies mentioned. The Externe study also fails to consider the possibility of environmental damage due to terrorist attacks or sabotage at nuclear power stations, factors which are elsewhere often viewed as significant[17]. All three studies exclude any attention to the environmental implications of nuclear proliferation, although efforts in this regard were made in an early US valuation study in 1982 (Shuman and Cavanagh, 1982) and are also commented on in a 1994 survey by the US Office of Technology Assessment (OTA, 1994). Despite including some of the most thorough and systematic studies in the field, each of these reports, in different ways, may therefore be judged to be seriously incomplete. If environmental valuation results are taken at face value, then this important factor is entirely missed.

When taken together, it is an uncomfortable but undeniable fact that the adoption of different but equally reasonable assumptions or conventions on

[15] E.g., Inhaber, 1978; Ferguson, 1981; IAEA et al, 1991; Ball et al, 1994

[16] Externe, 1995: volume 2: 429. An interesting discussion of energy inputs to different generating options with possibly differing implications may be found in Mortimer, 1991.

[17] E.g., Hirsch et al, 1986.

potentially any one of the different dimensions of environmental performance discussed in this section might radically alter the apparent environmental merit order of different electricity generating options. Environmental valuation studies seem no more to avoid this problem than did comparative risk assessment before them. Indeed, based on the examples provided here, it may be difficult to escape the conclusion that, by combining additional methodological complexity with apparent presentational simplicity, environmental valuation can make the problem worse rather than better.

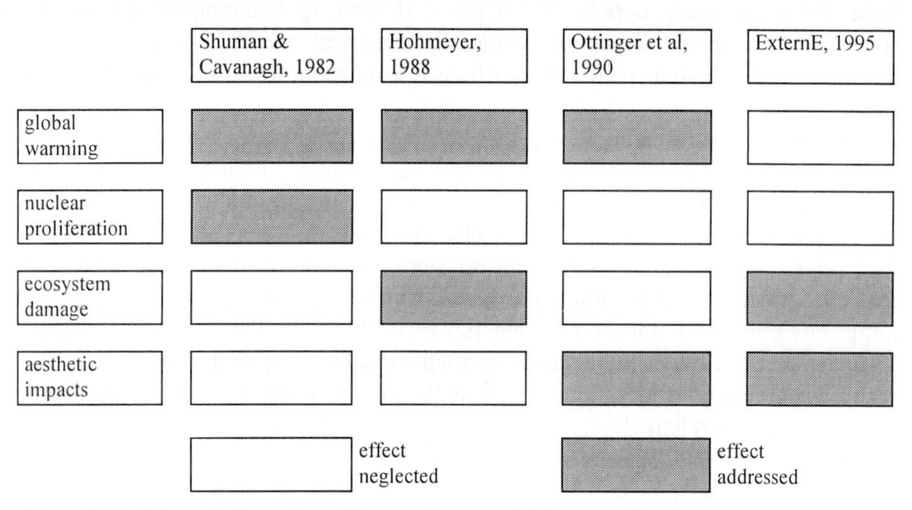

Figure 22.3: Schematic illustration of the completeness of different studies

3 A survey of environmental valuation results

The theoretical and methodological issues discussed in the last section are borne out by examination of the practical results obtained in the environmental valuation literature over recent years. Based on a recent survey by the author of studies conducted in industrialised countries (Stirling, 1997b), Figure 22.4 displays the degree of variability evident in the literature as a whole in relation to just one option - new coal power[18]. The highest and lowest externality values for new coal vary by

[18] The results are expressed in US dollars at 1995 prices. They are, in the order displayed in Figure 4, those of: Ramsay, 1979; Shuman and Cavanagh, 1982; ECO Northwest, 1987; EPRI, 1987; Hohmeyer, 1988, Chernick and Caverhill, 1989; Shilberg, 1989; CEC, 1989; Friedrich et al, 1990; Koomey, 1990; Hohmeyer, 1990; Bernow and Marron, 1990; Ottinger et al, 1990; Bernow et al, 1990; Hagen et al, 1991; Koomey, 1991; Stocker et al, 1991; DTI, 1992; Hohmeyer, 1992; Cline, 1992; Ferguson, 1992; Hohmeyer et al, 1992; Externe, 1993; Friedrich et al, 1993; Eyre and Jones, 1993; Fankhauser, 1993; Pearce, 1993; Lazarus et al, 1993; Meyer et al, 1994; Eyre, 1995; Externe, 1995 and Tol, 1995.

more than four orders of magnitude - far exceeding the range expressed in any individual study[19]. There is no categorical trend evident over time, nor even a consistent relationship between the results of those studies which include and exclude global warming[20].

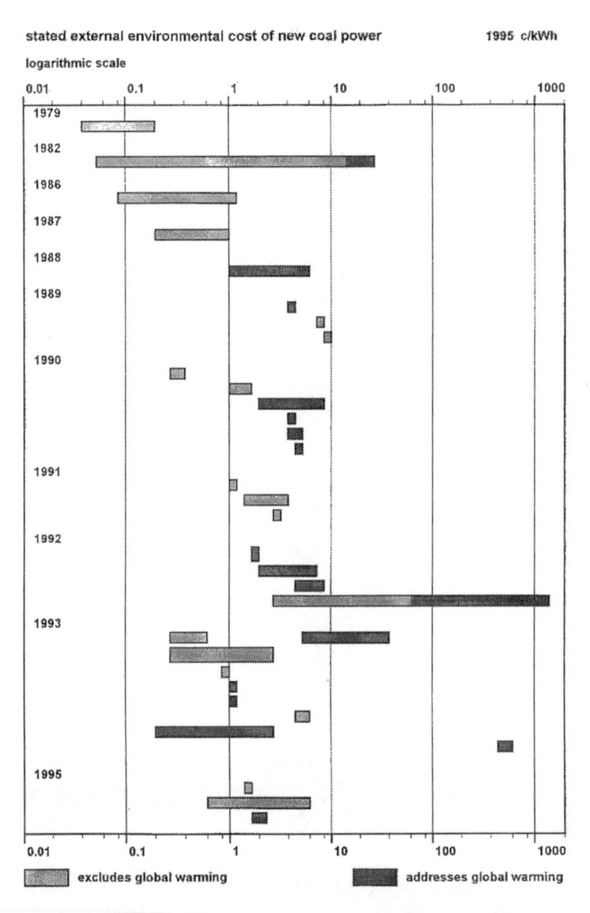

Figure 22.4: Variability in the monetary valuation results obtained in the literature for new coal power

[19] Where an individual study acknowledges variability or uncertainty by stating a range of values, this is represented in Figure 4 by a horizontal bar. One of the single most important dimensions of variability is addressed by showing the inclusion or exclusion of consideration of global warming effects in the shading of these bars.

[20] Values including and excluding attention to global warming overlap across an interval which is some two and a half orders of magnitude wide. Some of the lowest values obtained in the literature as a whole involve some consideration of global warming, while some of the highest overall values actually exclude this effect.

When attention turns to a comparison of the externality results obtained for a range of different electricity supply options, the ambiguity of the overall picture is further compounded. Based on the same survey, Figure 22.5 displays the externality values derived in the literature as a whole for eight key generating options. Again, the picture is dominated by enormous variability. Indeed, since the lowest values obtained for the worst ranking option (coal) are lower than the highest values obtained for the apparently best ranking options (wind), the overall picture would accommodate any conceivable ranking order for these eight options[21].

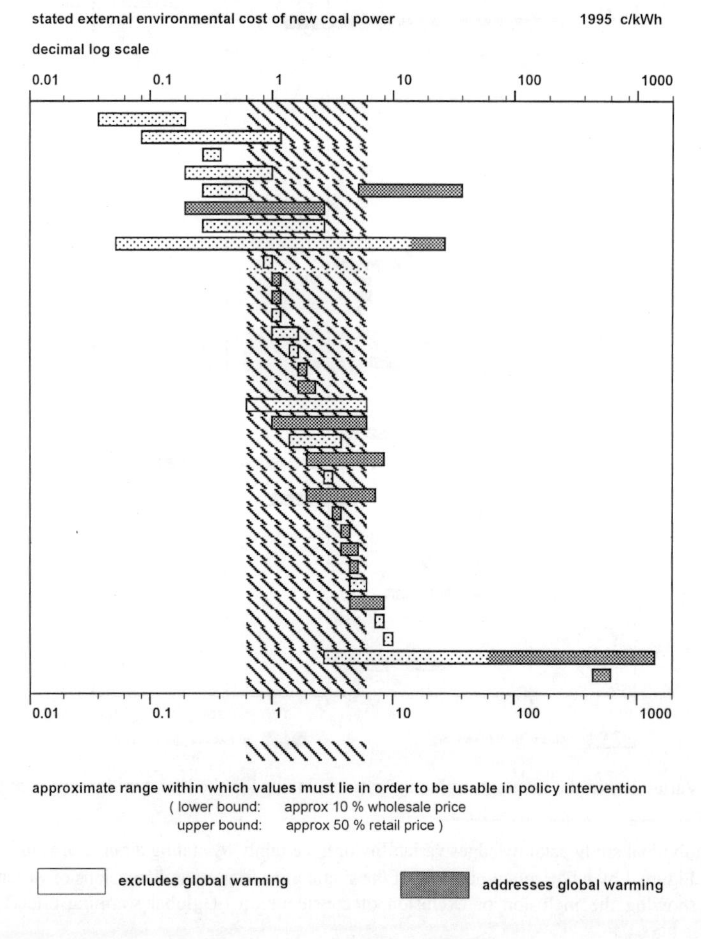

Figure 22.5: The 'Price Imperative' in the environmental valuation of generating options

[21] Individual studies show results at the high end of the overall range for some options but lower in the distributions for others

A final observation that is prompted by this survey of results is that, despite the enormous variability, there is a marked tendency for the results obtained for the generating option which dominates current electricity supply systems (coal) to congregate in the region of existing market prices. Based on the data displayed chronologically in Figure 22.4, Figure 22.6 displays this effect[22], which might be termed the 'price imperative'[23]. This phenomenon may be related to the intended use of environmental valuation results as values for 'externality adders' in regulation or as 'Pigovian' environmental taxes. If the externalities derived for the dominant option were found to lie well below existing market prices, then their inclusion in retail prices would have virtually no impact on consumer behaviour. On the other hand, if externalities were found to lie too high above market prices, then they might be feared to be 'unrealistic' in relation to any politically credible environmental tax. In other words, environmental externality results display the property (unique in environmental appraisal) that if they are to be usable in practical policy instruments, the results obtained for the dominant option would *have* to fall somewhere near the bounded region in Figure 22.6.

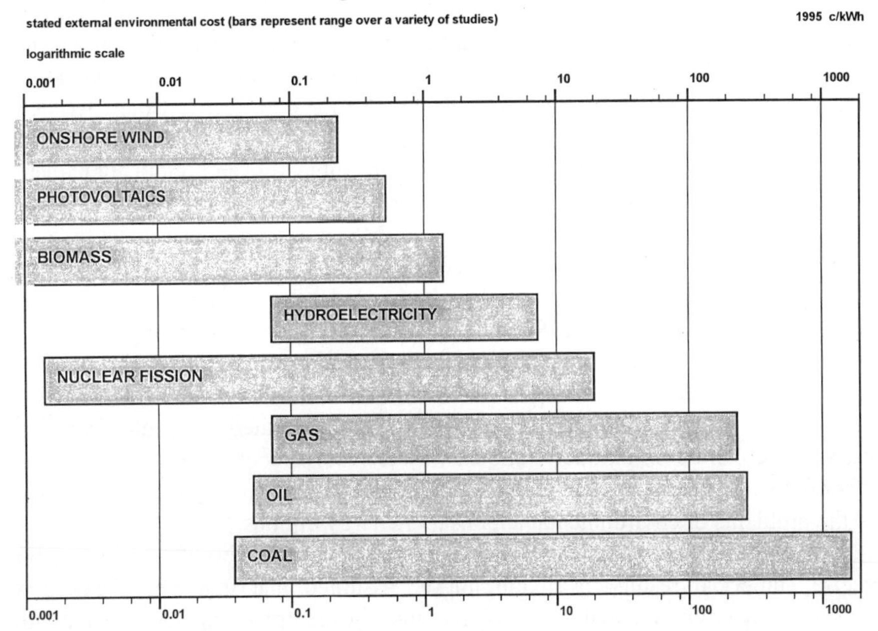

Figure 22.6: Ambiguity in the ranking of electricity supply options in the monetary valuation literature

---

[22] The shaded region of the graph is bounded at the lower end by a value which is approximately 10 per cent of the typical wholesale price of electricity in OECD countries and at the upper end by a value which is about half the typical retail price.

[23] This is discussed in more detail in Stirling, 1997b.

It is evident from the variability of results displayed in Figures 22.4 and 22.5 that the 'dimensions of variability' discussed in the last section can have an enormous impact on the practical results. Since a variety of positions on these different dimensions may all be equally 'rational', there is little prior reason to regard as 'definitive' the results obtained under any individual set of assumptions. Consequently, it is apparent from the ambiguity of the picture in Figure 22.5 that the environmental valuation literature taken as a whole fails to provide an unequivocal basis even for the *ranking* of electricity supply options. One of the most basic tasks in environmental appraisal is to achieve some notion of the overall ranking of different options under different assumptions. Since many of the dimensions of variability discussed in this section remain implicit in much environmental valuation, serious questions over whether the associated methodologies are of any practical policy use at all.

4 Some practical implications for environmental appraisal

Over recent years, the notion that different forms of environmental and health effect may be fruitfully compared in objectively determinate quantitative terms has fallen under serious doubt. Bodies such as the US National Research Council (NRC, 1996) and the British Royal Society (Royal Society, 1992:9) and Treasury (HMT, 1996) have come to acknowledge the intrinsically subjective and political character of this business. Whilst specialists may often reasonably claim greater authority with respect to the assessment of the likely probabilities or physical magnitudes of *individual* effects, it is increasingly recognised that expert judgements are as essentially subjective as any other when it comes to the relative prioritisation of *different* effects (Bradbury, 1989; Stirling, 1995). Indeed, in many ways, the predicament experienced by environmental cost-benefit analysis and risk assessment are simply a manifestation of a general insight which has been well established in theoretical terms in welfare economics and social choice theory for many years [24]. Just as there can be no uniquely rational way to resolve contradictory perspectives or conflicts of interest in a plural society, so there can be no single 'analytical fix' for the problems of environmental appraisal of energy options.

Although undermining the 'objective' status and credibility of ostensibly precise numerical results, it should not be assumed that this newly emerging consensus on environmental appraisal requires the complete abandonment of the discipline and clarity of quantitative techniques. The implication is simply that they be treated as 'tools' rather than as 'fixes'. Once we are prepared to relinquish the aspiration to single definitive 'results', the key features of a more realistic approach

---

[24] As expressed in the famous 'Arrow Impossibility' (Arrow, 1963). See discussion in Stirling, 1997b and 1998.

to the environmental appraisal of energy options seem quite readily identifiable. For instance, by treating environmental performance as a vector (rather than a scalar) quantity, straightforward ***multi-criteria*** techniques permit a more systematic approach to the multi-dimensional character of environmental effects [25]. Likewise, numerous tools exist for the substitution of single values expressed to several significant figures with systematic ***sensitivity analysis***. Finally, it may be that the problem of divergent assumptions, values and uncertainty might also be addressed by the adoption of a rigorous approach to ***diversification*** - focusing on portfolios as a whole, rather than on the 'first-past-the-post' identification of the 'best' individual options[26].

Either way, it is clear that an essential but hitherto neglected input to environmental appraisal is the transparent inclusion of divergent public perspectives and value judgements. In this light, the need for active public participation in the analysis underlying environmental regulation is not simply a question of democratic accountability and political legitimacy. It is a fundamental matter of analytical rigour[27]. In response to this emerging new climate in environmental appraisal, a large array of new techniques and procedures are under development in North America and northern European countries for enabling the efficient inclusion of divergent public values at the outset in environmental appraisal[28]. Although valuable experiments have been conducted in many areas[29], such techniques have for the most part yet to be seriously pursued on a large scale as a means to inform real policy decisions concerning the environmental regulation of energy options.

So what would the environmental appraisal of energy options actually look like, were it to be based on comprehensive and systematic sensitivity analysis under a multi-criteria framework addressing portfolios rather than individual options? Despite the present dearth of major empirical exercises of this sort, a schematic exercise undertaken by the author and reported in more detail elsewhere (Stirling, 1997a) may have some illustrative value. Accordingly, Figure 22.7 displays as a set of pie charts the implications for the UK generating mix of adopting a range of perspectives concerning the relative importance of different appraisal criteria. This stylised and purely illustrative exercise models the appraisal of three groups of UK

---

[25] Vatn and Bromley (1994) express concerns in this regard, but in fact these relate more to the reification of a single set of weightings. The trick is to treat the weighting of attributes as a heuristic for the systematic exploration of trade-offs, with results expressed as sensitivity analysis rather than as discrete values.

[26] As proposed, for instance, by Stirling (1994a; 1994b; 1996a; forthcoming).

[27] Cf: Stirling, 1998.

[28] Webler et al, 1991, 1995; Renn et al, 1993. See Renn et al (1995) for a comprehensive survey of techniques including citizen's advisory panels, planning cells, citizen's juries, 'study groups', mediation and regulatory negotiation.

[29] Interesting examples in this regard include Renn et al, 1986; Hope et al at, 1988.

generating options (nuclear, fossil fuels and renewables) under three major classes of environmental criteria (land use, air pollution and 'nuclear issues'). In addition, account is taken of the economic performance of the different options under prevailing market conditions, and of the possibility of deliberately retaining some diversity in the generating mix as a whole. Based on a systematic set of permutations, the eighty one different pie charts each represent an electricity supply mix which would be 'optimal' for the UK under a particular set of weightings on the various appraisal criteria. To the extent that it employs real technical performance data under each criterion and to the extent that the overall range of weighting schemes might be held to accommodate a large portion of the present energy debate, this hypothetical exercise might be viewed as a very rough first order approximation of what the results of a real empirically-based appraisal might look like, were it to be undertaken as part of a Royal Commission, public inquiry or Green Paper process, or were it to be substituted for one of the many major officially-sponsored valuation studies reviewed in this paper.

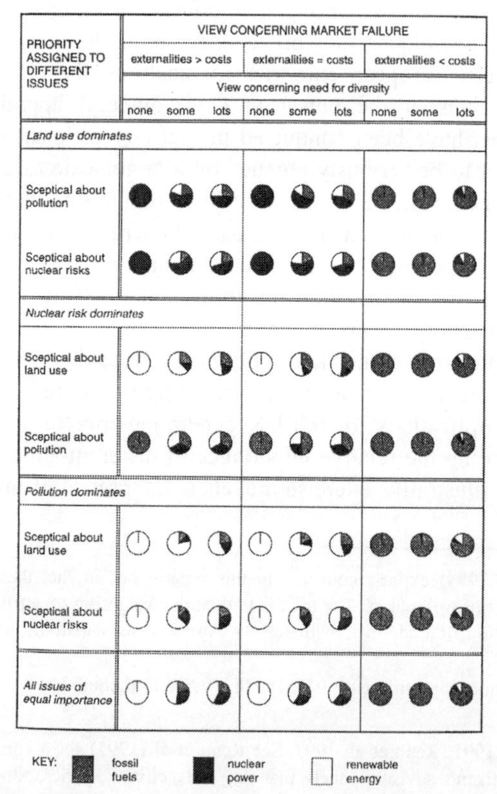

Figure 22.7: An illustrative multii-criteria 'sensitivity map', based on a hypothetical exercise

Although just a schematic reflection of a hypothetical exercise, Figure 22.7 nevertheless serves to illustrate a number of key differences between this sort of approach, and one based on orthodox risk assessment or environmental valuation.

First, this type of exercise is predicated on an inclusive ***participatory*** appraisal process rather than on technical analysis conducted exclusively by specialists. Instead of being based on a single position concerning the many dimensions of variability discussed in this paper, this type of appraisal accommodates in parallel a potentially unlimited range of disparate positions.

Second, a multi-criteria framework offers a far more ***transparent*** way of dealing with the dimensions of variability in appraisal. Where results are presented as ostensibly precise discrete numerical values, aggregated under familiar metrics such as monetary value or mortality, attention is drawn away from the fundamental determining importance of the issues discussed in this paper. Under a multi-criteria approach, these factors are all more readily highlighted as the key determining factors in analysis.

Third, and perhaps most importantly, the results are presented as a systematic 'map' of ***sensitivities***, rather than as a single prescriptive set of values. Essentially subjective value judgements concerning the relative merits of the disparate forms and distributions of the various effects, variations in the autonomy of those affected, divergent choices of indicators, differences in the treatment of uncertainty and inconsistencies in the framing of analysis are all represented as different 'regions' on this map. In this way, the central matters of subjective value judgement are effectively removed from analysis and placed firmly in the domain of politically accountable decision making, where they belong.

References

Arrow K.J. (1963), '*Social Choice and Individual Values*', 2nd edition, Wiley, New York.

Ball D.J., Roberts L.E.J. and A.C.D. Simpson (1994), '*An Analysis of Electricity Generation Health Risks - a United Kingdom Perspective*', Centre for Risk Assessment, University of East Anglia, 1994.

Bernow S. and D. Marron (1990), '*Valuation of Environmental Externalities for Energy Planning and Operations, May 1990 Update*', Tellus Institute, Boston, May.

Bernow S., Biewald B. and D. Marron (1990), 'Environmental Externalities Measurement: Quantification, Valuation and Monetization', in Hohmeyer & Ottinger.

Bradbury J. (1989), 'The Policy Implications of Differing Conceptions of Risk', *Science, Technology and Human Values*, Vol.14, No.4, Autumn.

Budnitz R.J. and J.P. Holdren (1976), 'Social and Environmental Costs of Energy Systems', *Annual Review of Energy*.

California Energy Commission (1989), '*Energy Technology Status Report*', June.
Chernick P. and E. Caverhill (1989), '*The Valuation of Externalities from Energy Production, Delivery and Use, Fall 1989 Update*', Chernick and Caverhill Inc., Boston.
Cline W. (1992), '*The Economics of Global Warming*', Institute for International Economics, Washington DC.
Cohen A.V. and D.K. Pritchard (1980), '*Comparative Risks of Electricity Production: a critical survey of the literature'*, UK Health and Safety Executive Research Paper 11, HMSO.
Comar C.L. and L.A. Sagan (1976), 'Health Effects of Energy Production and Conversion', *Annual Review of Energy.*
De Jongh P. (1988), 'Uncertainty in EIA', in P Wathern ed, *Environmental Impact Assessment: Theory and Practice*, Unwin Hyman.
Pearce D., Bann C. and S. Georgiou (1992), '*The Social Cost of the Fuel Cycles*', report to the UK Department of Trade and Industry by the Centre for Social and Economic Research on the Global Environment, HMSO, September.
ECO Northwest et al (1987), '*Generic Coal Study: Quantification and Valuation of Environmental Impacts*', report commissioned by Bonneville Power Administration, January.
EPRI (1987), Results cited but not referenced in Hohmeyer, 1990.
ESD et al (1995), Energy for Sustainable Development, ZEW, IARE, IER, Coherence, FhG, '*SAFIRE Final Report'*, final report for the European Commission DG XII, Corsham, November.
Externe (1993), UK Energy Technology Support Unit, CEC, DG XII, et al, '*Assessment of the External Costs of the Coal Fuel Cycle*', draft position paper prepared for the CEC/US Joint Study on Fuel Cycle Costs, Harwell, February.
Externe (1995), European Commission '*Externe: externalities of energy*' Volumes 1 - 6, EUR 16520 - 16525 EN, Brussels.
Eyre N. (1995), 'The External Costs of Wind Energy and What They Mean for Energy Policy', paper to the *Third International Conference on External Costs*, Ladenburg, May.
Fankhauser S. (1993), '*Global Warming Damage Costs: some monetary estimates*', CSERGE GEC Working Paper 92-29, University of East Anglia, Norwich.
Ferguson R.A.D. (1981), *Comparative Risks of Electricity Generating Fuel Systems in the UK*, UKAEA, Peter Peregrinus.
Ferguson R.A.D. (1991), results cited in Hohmeyer, 1990 and HoC, 1992 and confirmed by personal communication, November.
Friedrich R. and U. Kallenbach (1990), 'External Costs of Electricity Generation' in Hohmeyer & Ottinger.
Friedrich R. and A. Voss (1993) 'External Costs of Electricity Generation', *Energy Policy*, February.

Fritszche A.F. (1989), 'The Health Risks of Energy Production', *Risk Analysis*, Vol.9, No.4.
Hagen D. and S. Kaneff (1991), '*Application of Solar Thermal Technology in Reducing Greenhouse Gas Emissions - Opportunities and Benefits for Australian Industry*', ANUTECH Pty.
Hirsch H. (ed.) (1986) '*International Nuclear Reactor Safety Study: design and operational features and hazards of commercial nuclear power reactors in the world*', report prepared by Gruppe Ökologie for Greenpeace International, September.
HMT (1996) Inter-Departmental Liaison Group on Risk Assessment, '*The Setting of Safety Standards: a report by an interdepartmental group and external advisers*', HM Treasury, London, June.
Hohmeyer O. (1988), '*Social Costs of Energy Consumption: external effects of electricity generation in the Federal Republic of Germany*', prepared for DG XII of the European Commission by the Fraunhofer Institut fuer Systemtechnik und Innovationforschung, Springer Verlag, Berlin.
Hohmeyer O. (1990), 'Latest Results of the International Discussion on the Social Costs of Energy - How Does Wind Compare Today?', presented at the *1990 European Wind Energy Conference*, Madrid, October.
Hohmeyer O. (1992), 'Renewables and the Full Costs of Energy', *Energy Policy*, April.
Hohmeyer O. and R. Ottinger (eds.) (1990), '*External Environmental Costs of Electric Power Production and Utility Acquisition Analysis and Internalization: Proceedings of a German-American Workshop*', Fraunhofer ISI.
Hohmeyer O. (1992), 'Renewables and the Full Costs of Energy', *Energy Policy*, April.
Holdren J. (1979), 'Energy: calculating the risks', *Science*, Vol.204.
Holdren J., Morris G. and I. Mintzer (1980), 'Environmental Aspects of Renewable Energy Sources', *Annual Review of Energy*, Vol5.
Holdren J. (1982), 'Energy Hazards: What to Measure, What to Compare', *Technology Review*, February.
IAEA et al (1991), International Atomic Energy Agency et al, '*Senior Expert Symposium on Electricity and the Environment: key issues papers*', IAEA, Vienna.
IAEA (1992), International Atomic Energy Agency, '*Methods for Comparative Risk Assessment of Different Energy Sources*', IAEA-TECDOC-671, Vienna.
Inhaber H. (1978), '*Risk of Energy Production*', AECB-1119/REV-1, (Canadian) Atomic Energy Control Board.
Jones M., Hope C. and R. Hughes (1988), '*Examining Energy Options Using a Simple Computer Model*', Cambridge University Department of Management Studies, Cambridge.

Kayes R.J. and P.J. Taylor (1984), '*Health Risks of Nuclear and Coal Fuel Cycles in Electricity Generation: a critique*', Political Ecology Research Group, PERG RR-13, Oxford.
Koomey J. (1990), '*Comparative Analysis of Monetary Estimates Associated with the Combustion of Fossil Fuels*', paper presented to conference of New England Public Utility Commissioners Environmental Externalities Workshop.
Koomey J. (1991), '*Comparative Analysis of Monetary Estimates Associated with the Combustion of Fossil Fuels*', paper presented to conference of New England Public Utility Commissioners Environmental Externalities Workshop.
Lazarus M. et al (1993), '*Towards Global Energy Security: the next energy transition - an energy scenario for a fossil free energy future*' Stockholm Environment Institute, Greenpeace International, Amsterdam, February.
Meyer H.J., Morthorst P.E. and L. Schleisner (1994), '*Assessment of Environmental Costs: External Effects of Energy Production*, submitted to 18th IAEE International Conference, revised version, July 1995, Riso.
Mortimer N.D. (1991), 'Energy Analysis of Renewable Energy Sources', *Energy Policy*, Vol.17, No.4, May.
NRC (1996), Stern P. and H. Fineberg, '*Understanding Risk: informing decisions in a democratic society*' US National Research Council Committee on Risk Characterisation, National Academy Press, Washington DC.
OECD (1989), Pearce D. and A. Markandya, '*Environmental Policy Benefits: Monetary Evaluation*', OECD, Paris.
OTA (1994), US Congress Office of Technology Assessment, '*Studies of the Environmental Costs of Electricity*', OTA, Washington DC, September.
Ottinger R.L., Wooley D.R., Robinson N.A., Hodas D.R. and S.E. Babb (1990), '*Environmental Costs of Electricity*', prepared for NYSERDA and US DOE by Pace University Center for Environmental Legal Studies, Oceana Publications.
Ottinger R. (1990), 'Getting at the True Cost of Electric Power', *The Electricity Journal*, July.
Pearce D. (1993), '*The Economic Value of Externalities from Electricity Sources*', Paper presented at Green College Seminar , University of Oxford, April.
Perrings C. (1991), 'Reserved Rationality and the Precautionary Principle' in Costanza (ed).
Peterson G.L., Driver B.L. and R. Gregory (eds.) (1988), '*Amenity Resource Valuation: integrating economics with other disciplines*', Venture, Philadelphia.
Ramsay W. (1979), '*Unpaid Costs of Electrical Energy: health and environmental impacts from coal*', Resources for the Future, John Hopkins University Press.
Renn O., Webler T., Rakel H., Dienel P. and B. Johnson (1993), 'Public Participation in Decision Making: a three step procedure', *Policy Sciences*, Vol.26, p.189.
Renn O. (1995), '*Fairness and Competence in Citizen Participation*', Kluwer.

Rowe W.D. and P. Oterson (1983), '*Assessment of Comparative and Non-comparative Factors in Alternate Energy Systems*', Commission of the European Communities.
Royal Society (1992), '*Risk: analysis, perception, management*', Royal Society, London.
Shilberg G.M., Nahigian J.A. and W.B. Marcus (1989), '*Valuing Reductions in Air Emissions and Incorporation into Electric Resource Planning: Theoretical and Quantitative Aspects*', JBS Energy Inc.
Shuman M. and R. Cavanagh (1982), '*A Model Conservation and Electric Power Plan for the Pacific Northwest, Appendix 2: Environmental Costs*', Northwest Conservation Act Coalition, Seattle, November.
Stirling A. (1994a), 'Diversity and Ignorance in Electricity Supply Investment: addressing the solution rather than the problem', *Energy Policy*, Vol.22, No.3, March.
Stirling A. (1994b), 'Technology Choice for Electricity Supply: putting the money where the mouth is?', D.Phil thesis, *Science Policy Research Unit*, University of Sussex, September.
Stirling A. (1995), 'The Nirex Multi-Attribute Decision Analysis as a Justification for the Siting of the Rock Characterisation Facility at Sellafield', Proof of Evidence to *Public Local Planning Inquiry on the Application by NIREX to Site a Rock Characterisation Facility at Sellafield*, Cumbria, September.
Stirling A. (1996a), 'Optimising UK Electricity Portfolio Diversity', chapter in G. MacKerron and P. Pearson, (eds), '*The UK Energy Experience: a model or a warning?*', Imperial College Press, March.
Stirling A. (1997a), 'Multicriteria Mapping: mitigating the problems of environmental valuation?', in J. Foster, 'Valuing Nature: economics, ethics and environment', Routledge.
Stirling A. (1997b), 'Linits to the Value of External Costs', *Energy Policy*, Vol.25, No.5.
Stirling A. (1998), 'Risk at a Turning Point', *Journal of Risk Research*, Vol.1, No.2.
Stocker L., Harman F. and F. Topham (1991), '*Comprehensive Costs of Electricity in Western Australia*', Canberra.
Talcott F. (1992), 'How Certain is that Environmental Risk Estimate?', *Resources*, No 107, Spring.
Tellus Institute (1991), '*Valuation of Environmental Externalities: sulfur dioxide and greenhouse gases*', Report to the Massachusetts Division of Energy Resources, December.
Tol R. (1995), results cited in Externe.
UNEP (1985), United Nations Environment Programme, '*Comparative Data on the Emissions, Residuals and Health Hazards of Energy Sources*', Environmental Impacts of the Production and Use of Energy, Part IV, Phase I, UNEP.

Vatn A. and D.W. Bromley (1994), 'Choices without Prices without Apologies', *Journal of Environment Economics and Management*, Vol.26.

Voss A., Friedrich R., Kallenbach E., Thoene A., Rogner H.-H. and H.-D. Karl (1989), *Externe Kosten der Stromerzeugnung Studie im Auftrag der VDEW*, Frankfurt.

Watson S. (1981), 'On Risks and Acceptability', *Journal of the Society for Radiological Protection*, Vol.1, No.4, 1981 p21.

Webler T., Levine D., Rakel H. and O. Renn (1991), 'A Novel Approach to Reducing Uncertainty: the Group Delphi', Technological Forecasting and Social Change, V39, pp235-51.

Webler T., Kastenholz H. and O. Renn (1995), 'Public Participation in Impact Assessment: a social learning perspective', Environmental Impact Assessment Review, V15 p443-463.

327-43

CHAPTER 23

SUSTAINABILITY AND NUCLEAR LIABILITIES

(UK)

GORDON MACKERRON[1]
Science Policy Research Unit, University of Sussex,
Mantell Building, Brighton BN1 9RF, UK
Email: gmackerron@mistral.co.uk

Q42
L94

MIKE SADNICKI[1]
50 Paines Lane, Pinner, Middlesex HA5 3DA, UK
Email: sadnicki@aol.com

Keywords: British Energy; electricity; Magnox Electric; nuclear liabilities; privatisation; sustainability.

Nuclear liabilities are the costs of dealing all the unwanted products of the nuclear age - radioactive wastes, spent reactor fuel and redundant nuclear structures. These have been accumulating in the UK over the past 50 years and now amount, on official estimates, to some £41.8 billion (undiscounted). There is a conventional wisdom that nuclear generation, alone among the energy sources, accounts for and makes financial arrangements to meet all its costs, including liabilities.

Our recent paper[2] demonstrates that this 'wisdom' is false. Past arrangements for the payment of future nuclear liabilities have failed. This is primarily because nuclear generation was and is uneconomic, so that provisions invested internally in new nuclear reactors did not generate the required rate of return. A large funding gap has developed. Slight improvements in the funding arrangements were instituted by the 1996 privatisation in the form of the British Energy Segregated Fund - but the liabilities *covered* by this Fund are only a small proportion of the total ostensibly in the private sector. In addition, most of the liabilities left in the public sector are unfunded.

The consequence is to raise serious questions about the compatibility of nuclear liability arrangements with the commitment of successive UK Governments to the principle of sustainability in environmental policies. Sustainability - "development which does not compromise the ability of future generations to meet their own needs"[3] - proves easy to define but difficult to implement.

1 Introduction

This chapter summarises arguments about sustainability and nuclear liabilities in the UK in six steps:

---

[1] **Gordon MacKerron** is Head of the Energy Programme.
Mike Sadnicki is an Independent Operational Research Consultant.

[2] M.J. Sadnicki and G. MacKerron (1997). The present chapter draws heavily on this report. We also draw from a later paper by M.J.Sadnicki (1997).

[3] This is the so-called 'Brundtland' definition from World Commission on Environment and Development (1987, page 43).

1. Are current official estimates of nuclear liabilities complete and adequate?
2. Have nuclear liability funding arrangements been adequate in the past?
3. Was the 1996 Privatisation allocation of nuclear liabilities equitable between the public and private sectors?
4. What proportion of nuclear liabilities is securely funded?
5. How far are current arrangements compatible with Government policy on sustainability?
6. How far are present liability management policies consistent with cost-effectiveness?

2 Are official estimates of nuclear liabilities complete and adequate?

After the failure to privatise any of the civilian nuclear sector as part of the general UK electricity privatisation in 1989/90, a second attempt proved more successful in 1996. The bulk of the nuclear generating capacity, AGRs plus the PWR at Sizewell B, were sold into private ownership under the umbrella of a new company, British Energy. The older and more expensive Magnox stations remained in the private sector as Magnox Electric.

Sadnicki and MacKerron (1997) estimate total nuclear liabilities as £41.8 billion, based on the official estimates given in the most recently published Reports and Accounts. Figure 23.1 shows how this total is made up.

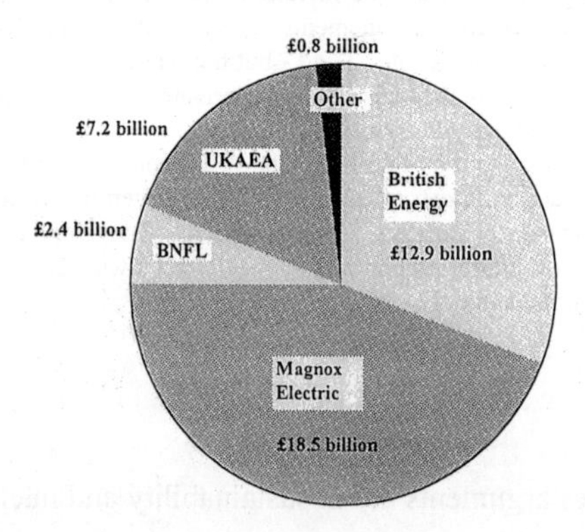

Figure 23.1: UK undiscounted nuclear liabilities 1997. Official total: £41.8 billion.

The present chapter confines itself to the liabilities attached to the two main generators, British Energy and Magnox Electric[4]. Of the total of £41.8 billion, the

[4] Magnox Electric has now been absorbed into BNFL. However, this has had no discernible impact on liabilities for Magnox reactors.

two generators are responsible for over 75% of the total: BE has £12.9 billion, and Magnox Electric £18.5 billion.[5]

2.1 *Omissions from official liability estimates*

The £31.4 billion of future expenditure anticipated by the two nuclear generators are very large sums indeed. However this does not guarantee that the sums are large enough. First, there appears to be one category of liability - for dealing with plutonium - that seems to be almost ignored in official liability estimates. By 2020, the UK will on present plans have accumulated over 100 tonnes of separated plutonium, and there is no plausible way in which this material will ever be used directly. This in turn requires the development of a management strategy for plutonium which recognises this material as a waste rather than a potentially useful material. Sadnicki and MacKerron (1997) estimate such a programme at some £2.3 billion undiscounted, while present policy appears to allow only some £200 million for dealing with plutonium.

2.2 *Reasons why nuclear liabilities might escalate*

Present liability management strategies emphasise complex and capital-intensive technological options, such as reprocessing spent fuel, and early deep disposal of wastes. In other words, much of the cost within the £41.8 billion consists of the kinds of projects which have historically been subject to much appraisal optimism and severe cost escalations. Sadnicki and MacKerron (1997) provide an analysis of past nuclear projects within the UK, and show that the *minimum* level of escalation has been around 30%, with many projects in the 100% to 200% range. Nuclear liability projects are commonly large, unique, complex and sometimes of a hole-in-the-ground type - in other words, the kinds of project which historically are subject to the largest degree of optimism/escalation. Sadnicki and MacKerron (1997) show how, if earlier experience in UK nuclear projects were to be repeated over the range of expected future liability projects, the total of £41.8 billion could easily rise to around £70 billion (Figure 23.2).

---

[5] The UK convention is to include within the liability total not only all those costs which are unavoidable because the physical basis for the liability already exists (e.g. spent fuel from past nuclear operation) but also avoidable costs of dealing with nuclear products that are not yet in existence (e.g. spent fuel from all expected *future* operations of reactors). This means that the total of £41.8 billion is an exaggeration of the genuinely unavoidable elements of nuclear liability costs. However, for other reasons, which can be briefly termed 'appraisal optimism' the £41.8 billion is probably a large underestimate of the cost of dealing with all liabilities, at least as long as current policies remain in force.

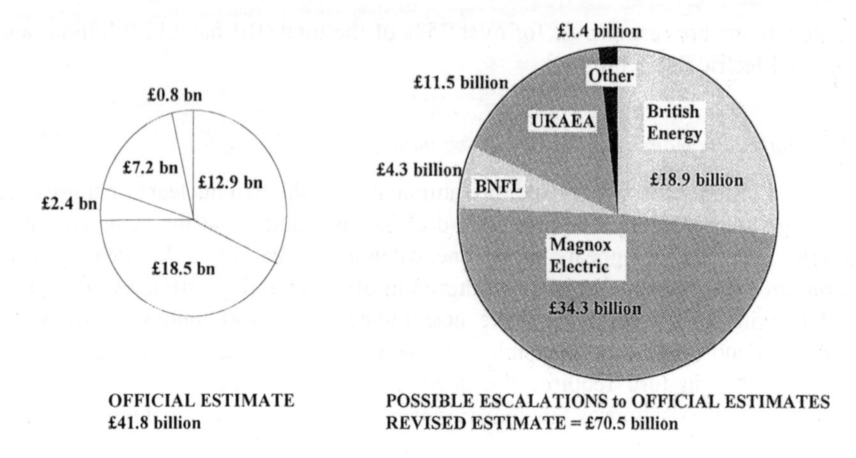

Figure 23.2: UK undiscounted civil nuclear liabilities (authors' estimates of possible escalations)

3 Are current funding arrangements adequate?

For the purposes of this discussion, it will be assumed that £41.8 billion does represent an adequately large estimate of liability costs. Do current funding arrangements give assurance that the £31.4 billion of generator liabilities will be met?

There are two major ways in which attempts can be made to fund liabilities:

- the *internal unsegregated route* (sometimes misleadingly referred to as a 'fund'), where provisions for liabilities are internally re-invested within the business;
- the *external segregated fund*, in which an annual contribution in cash is handed over to an external body (usually a trust) which invests in low-risk and relatively liquid assets.

The external segregated fund clearly offers much more assurance that provisions will provide adequate sums of cash to meet liability costs. This is especially true in the UK case, where the internal re-investment in the internal 'fund' was principally in further nuclear construction, which has historically earned very low and often negative returns. The internal segregated route was always used by nuclear generators in the UK until 1996. As a consequence, UK nuclear liabilities are seriously unfunded. The provisions shown in the balance sheet are not necessarily matched by cash, or by the likelihood that the assets described in the balance sheet will generate the required cash.

The 1996 re-structuring improved the liability management regime to a limited extent:

- British Energy set up an external segregated fund to deal more securely with some long-term liabilities;
- Magnox Electric was given a 'dowry' of up to £3 billion of unspent cash from the former Nuclear Electric, and this constitutes a quasi-segregated fund to help the company meet its very large long-term liabilities[6].

However these improvements were marginal in relation to the need. In order to establish the extent of unfunding under the post-privatisation arrangements, we look at the liabilities of the two main generators in turn. A number of critical issues are raised. The first is the equity of the allocation of liabilities between private and public sectors at the time of the 1996 privatisation.

4 Was the 1996 privatisation allocation of liabilities equitable - between public and private sectors?

In principle, there is no problem: AGR and PWR liabilities went with British Energy into the private sector, while Magnox liabilities stayed with Magnox Electric. However there are questions of timing (and of the re-categorisation of liabilities, ignored here) that cannot be fully explored given the limited state of public information, but which nevertheless deserve some discussion.

The timing question concerns large variations in the rates at which liabilities for Magnox and AGR reactors have been discharged in the last few years. These rates are shown in Figure 23.3.

In the two years immediately before the privatisation of the AGRs, the rates at which AGR liabilities were discharged accelerated from an average of £240 million in the three years up to 1993/4, to £801 million in 1994/5 and £642 million in 1995/96. Thus the public sector paid for the discharge of almost £1.5 billion of liabilities from reactors in the two years just before they entered the private sector. In 1996/97 however - the first year of private ownership - BE's liability discharge fell back to £346 million, and is expected by the company to stay at around £300 million annually for the next three years. There has clearly been some re-allocation of the expected timing of BE payments in the recent past. In early 1996, the discounted value of BE liabilities was £7.0 billion, but by mid-1996 this had fallen to £5.6 billion (using the same discount rate), while the undiscounted value fell by only £300 million.

---

[6] This 'dowry' has been transferred directly to BNFL together with the Magnox stations.

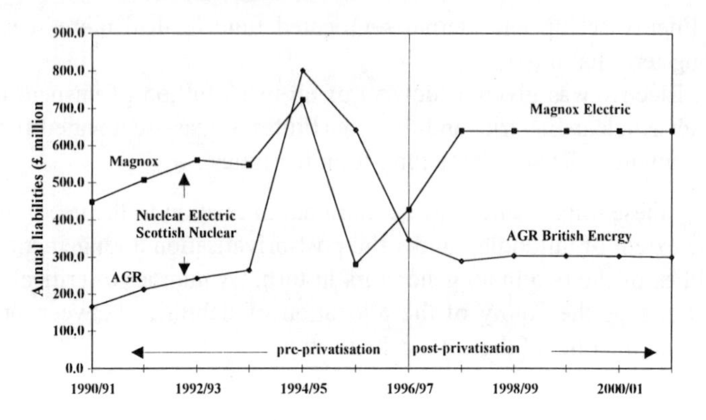

Figure 23.3: AGR and Magnox liability payments (actual and projected 1990-2002)

The Magnox pattern of liability payments has however taken a very different course. In 1995/96, Magnox liability payments were £280 million. Since then they have started to rise: to £427 million in 1996/97, and an expected average of over £600 million over the following five years.

What does this amount to? AGR liability payments rose very sharply in the two years before privatisation, and then fell equally sharply after the private sector took on ownership. In the Magnox case, liability payments were moderate before privatisation, and have now risen to much higher levels. These various trends may have good explanations, but such explanations have not yet been forthcoming. All that can be observed in the meantime is that the public sector has paid very heavily for liabilities - for AGRs just before privatisation and for Magnox liabilities since 1996.

5 What proportion of liabilities is currently securely funded?

5.1 British energy

Figure 23.4 shows the British Energy liabilities year by year, from 1997 to the final year of liabilities, beyond 2150. In the early years, the liabilities are at a level of up to £300 million a year, mainly reprocessing. Reprocessing finishes soon after 2020. After that the annual bill settles at around £50 million a year, mainly the storage, management and ultimate disposal of wastes. From time to time peaks associated with the various stages of decommissioning can be seen. For example, Stage III decommissioning is shown, mainly after the year 2140.[7] The exact total pattern depends on the closure pattern of the stations. The overall total is £12.9 billion.

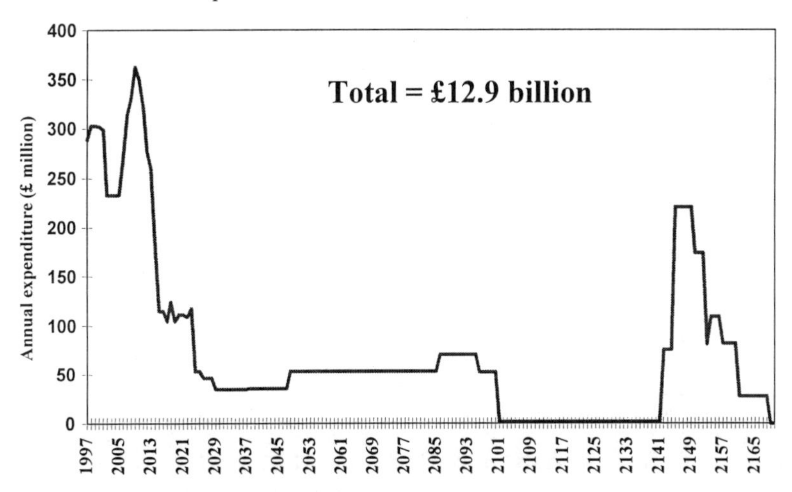

Figure 23.4: British Energy annual nuclear liabilities (1997-2168)

The curve of total British Energy liabilities can be divided into three components:

- liabilities probably covered by operational income;
- liabilities covered by the British Energy Segregated Fund;
- liabilities not securely covered by any means.

5.2 Liabilities probably covered by operational income

The first item is those liabilities which we can be reasonably certain that British Energy will meet from the Profit and Loss account; that is, the cash will come from day to day operating income. From information in British Energy Reports and Accounts, a detailed estimate can be made of such liabilities. The total liabilities in each year are plotted in Figure 23.5. The total in Figure 23.5 is about £4.0 billion, out of the overall total shown in Figure 23.4 of £12.9 billion.

---

[7] Stage 1 of decommissioning is the defuelling and immediate work necessary on closedown of a reactor. On present plans Stage 2 involves constructing a 'safestore' after about 30 years, Stage 3 (the most expensive part of decommissioning), involves dismantling the reactor and returning the site to alternative uses. This is expected to be delayed by 135 years beyond closedown, but in the Segregated Fund calculations it is assumed that the delay will be 70 years.

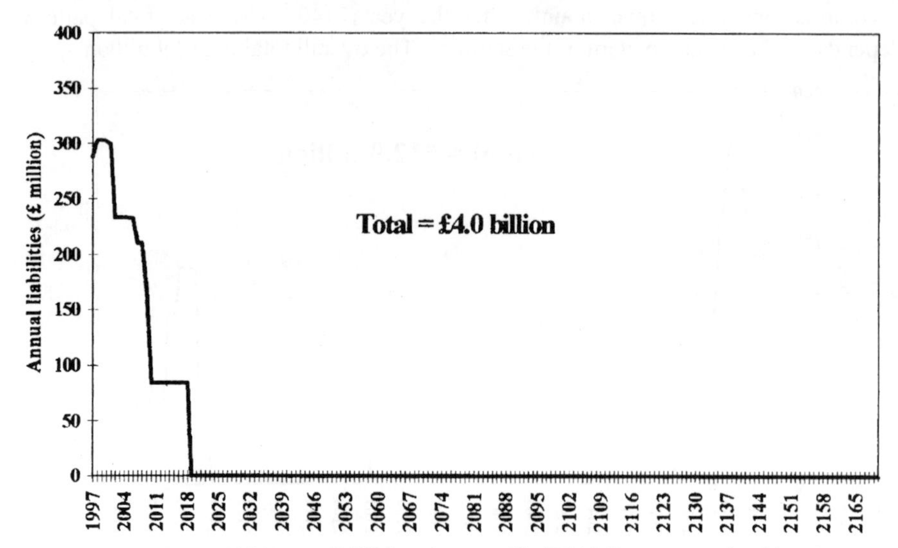

Figure 23.5: British Energy liabilities - 1: covered by British Energy operational income

5.3 Liabilities covered by the British energy segregated fund

The second source of coverage of British Energy liabilities is the Segregated Fund. At the time of privatisation in 1996, it was recognised that the internal unsegregated route was inadequate for liabilities in the private sector. There was much publicity concerning the setting up of a properly constituted external Segregated Fund.

The Segregated Fund has three sources of cash:

- The taxpayer contributed an initial endowment of £228 million in 1996.
- British Energy will make an annual contribution, £16 million to 2001, for each year in which all its stations continue to operate (on current plans up to about 2006). The annual contribution will then decline as power stations close. The cumulative total, by the time the last power station closes around 2035, is likely to be between £250 million and £300 million.
- The final and largest source of cash income is the interest assumed to accumulate from Fund surpluses.

However, when the full arrangements for the BE Segregated Fund were published in 1996, it became clear that the only liabilities brought into the coverage of the Segregated Fund are to be Stages II and III of decommissioning. The total liabilities thus covered by the Fund are plotted in Figure 23.6. The total of these liabilities in Figure 23.6 is £3.7 billion, out of the overall total shown in Figure 23.4 of £12.9 billion.

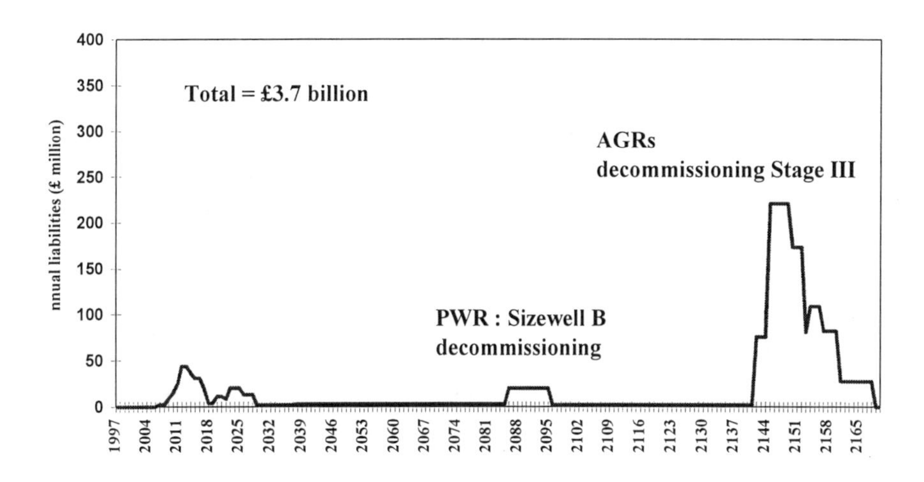

Figure 23.6: British Energy liabilities - 2: covered by British Energy's segregated fund

5.4 Liabilities not securely covered by any means

The Segregated Fund does not include the following post-operational liabilities:

- AGR post-closedown spent fuel management, especially of the final core (currently intended to be met predominantly by reprocessing);
- AGR Stage I decommissioning;
- decommissioning of AGR-related facilities at BNFL;
- disposal of AGR intermediate level waste (ILW), and storage and disposal of high level waste (HLW);
- post-closedown PWR liabilities which are not included in the Segregated Fund;
- plutonium disposition costs.

These liabilities total about £5.3 billion.[8] Figure 23.7 shows the phasing of the £5.3 billion. It demonstrates how the liabilities falling to the taxpayer occur throughout the next century, with particularly high levels in the first 25 years, when the AGR stations are closing.

[8] This is the total in the context of the 'official' estimates. As noted earlier in Section 1, these may prove to be significant underestimates. This is particularly true for plutonium disposition liabilities, which seem to be hardly considered at all in current official estimates.

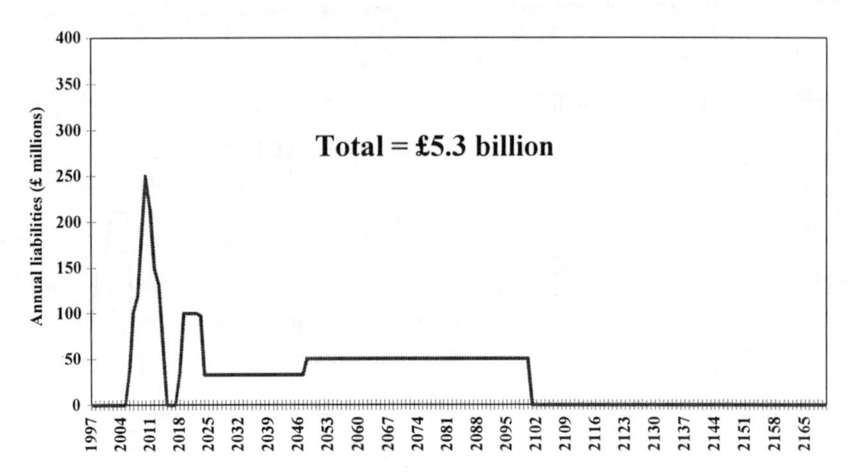

Figure 23.7: British Energy liabilities - 3: not covered by any secure arrangement

British Energy have said about these liabilities that they "intend to ensure that in due course there are sufficient investments in place to provide funding" (British Energy 1996, p. 73). This shows recognition of the problem but provides nothing to reassure taxpayers that the company will be able to generate over £5 billion of funds during the next two or three decades.

Generating an extra £5 billion above normal profits would be difficult for any company to achieve. However it may be particularly difficult for BE. This is first because it is committed to a level of annual dividend payout of around £100 million as a first call on profits, and second - more fundamentally - the profitability of the stock of reactors is unlikely in the longer term to be much better than marginal, especially now that most forecasts of future electricity prices are firmly downwards.

5.5 A possible explanation for the exclusions

An appropriate question is: why were the particular liabilities as shown in Figure 23.6 included in the coverage of the British Energy Segregated Fund, and the liabilities shown in Figure 23.7 excluded? The vital factor here is the way the compound interest works.

While the best solution for taxpayers would undoubtedly be to extend the coverage of the British Energy payments to the Segregated Fund to cover the omitted liabilities, the problem is that the contributions would need to be very large indeed. Because the omitted liabilities generally occur much sooner than the decommissioning liabilities included in the Fund, and because it is assumed that the Fund will generate interest at 3.5% (real) indefinitely, the discounted value of the omitted liabilities is proportionally much higher.

This is best demonstrated by an illustrative calculation as shown in Table 23.1.

Table 23.1: British Energy Segregated Fund - required annual contributions from 1997 onwards

	Liabilities covered (£ billion)	Annual contribution (£ million)
As constituted in 1996	3.7	16
Illustrative calculation	8.9	135

Source: first row: Privatisation Prospectus, 1996; second row: authors' calculations

The first row summarises the current Segregated Fund arrangements as constituted in 1996. The annual contribution is £16 million a year for the coverage of £3.7 billion. In the second row, the assumption is made that the excluded liabilities of £5.3 billion (as shown in Figure 23.7) are also included within the coverage of the Segregated Fund. They are added to the £3.7 billion, bringing the total coverage to £8.9 billion. The crucial fact is that the required annual contribution does not increase in a corresponding proportion. If it did, the revised required annual contribution would be a factor of about 2.4 higher, at almost £40 million a year. In fact calculation shows that the required annual contribution is over 8 times higher, at £135 million a year.

The reason for this is that the £5.3 billion liabilities which were originally excluded from the Fund generally arise far earlier in time than the £3.7 liabilities which were originally included. The compound interest does not have time to accumulate in the former case; consequently the required annual contributions while the power stations are still operating are proportionately much higher.

If the required annual payments to the Segregated Fund had not been at the relatively low level of £16 million a year, the AGR and PWR power stations would have been much less attractive to potential buyers in 1996. For example, if the required annual payments had been £135 million a year, this would have been greater than the dividends of £100 million a year which the company was predicted to pay, and indeed did pay out in its first year.

It clearly seems unrealistic to expect the company to start paying out such large sums as £135 million a year immediately. However the arithmetic is inexorable; and if limited further contributions are to be available to the Fund, then either new, non-nuclear investments will have to be very profitable indeed, or a very high long-term UK electricity price will be necessary for the surpluses to be made. None of these possibilities offers much reassurance for taxpayers that the private sector will definitely be able to pay for its liabilities in full.

5.6 Summary of British energy liabilities

Figure 23.8 shows the conclusions for British Energy liabilities. Of the £12.9 billion, some 31% - £4.0 billion - is probably covered by operational income. A further 28% - £3.7 billion - is covered by the Segregated Fund. A further 41% -

£5.3 billion - is not securely covered, and a significant proportion of this is likely to fall to the future taxpayer.

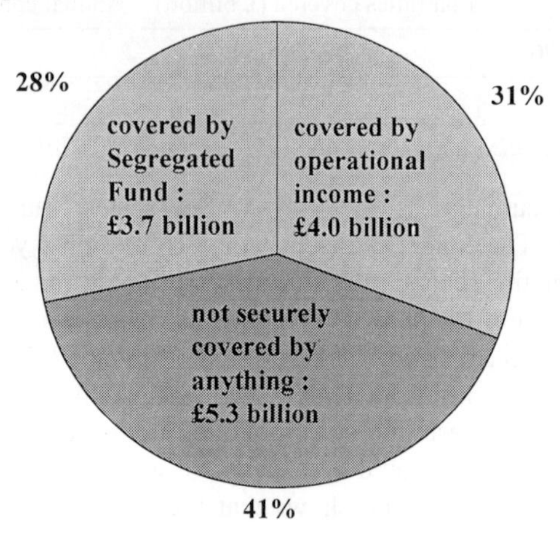

Figure 23.8: British Energy - Division of £12.9 billion liabilities

The previous Minister for Energy, Tim Eggar, said in a famous statement before the 1996 privatisation that the AGR and PWR liabilities would follow the assets[9]. The reality is that many of them will probably not follow the assets. Instead, the likelihood is that a significant proportion of them will fall to the future taxpayer. These considerations add further weight to the concerns expressed in Section 3 that the allocation of liabilities between private and public sectors as a result of the 1996 privatisation was not equitable.

5.7 Magnox Electric liabilities

The second main generator, Magnox Electric, remained in the public sector after privatisation. The situation with Magnox liabilities is substantially more difficult than that with BE. This is not surprising: Magnox stations could not be privatised in 1996 precisely because the burden of liabilities is so great in relation to prospective income. The value of Magnox liabilities (undiscounted) has been rising over the last two years, and has now reached £18.5 billion (against £17.0 billion in 1995).

Sources of cash to meet the £18.5 billion liabilities are as follows:

[9] "...the assets that go into the private sector will be followed by the attendant liabilities. It would be quite wrong for the liabilities to remain in the public sector while the assets went into the private sector. The Government's position on that is absolutely clear." (Hansard 1996, column 850).

- Magnox Electric received a cash endowment of more than £3 billion. This is mainly Fossil Fuel Levy, accumulated over the period 1990 to 1997 which was not spent on liabilities or on internal investment by Nuclear Electric.
- Under present plans all the Magnox stations will be closed by 2008. Their continued operation rests on the contention that future income exceeds future avoidable costs. It is likely that before the annual liabilities are counted - for example, before reprocessing operating costs are considered - the Magnox stations will be contributing a small positive cash surplus. It is unlikely that this cash surplus will amount to much more than £1 billion over the next decade.

There have been various attempts to prove that the Magnox cash resources are almost enough to meet the liabilities - in other words, to try to demonstrate that the funding gap is almost closed. However, such attempts always involve discounted liability figures, which can give misleading messages. Nuclear liabilities are always best analysed in terms of the undiscounted totals, if no real segregated fund exists.

Sadnicki and MacKerron (1997) carried out trial calculations to balance cash available against the likely profile of liability discharge. Figure 23.9 shows the true picture once the undiscounted Magnox liability figures are considered. The Magnox liabilities, mainly reprocessing charges, are initially very high, at over £600 million a year. The cash is likely to be exhausted by 2006, at which point some £13.0 billion of liabilities will remain to be paid for.

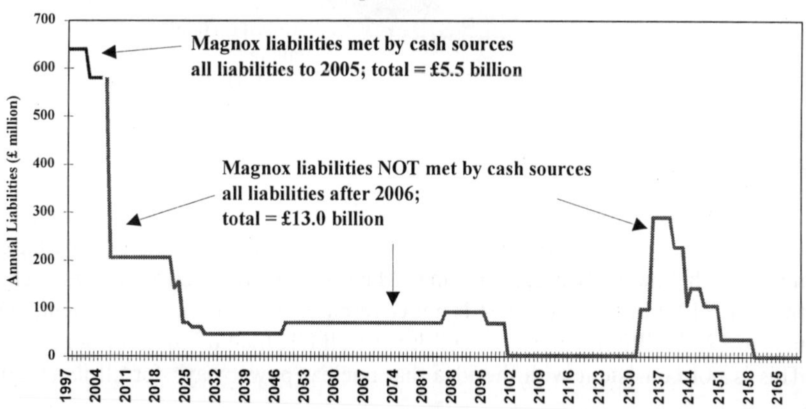

Figure 23.9: Magnox Electric - £18.5 billion liabilities in the public sector

The Government has undertaken to pay Magnox Electric up to £3.7 billion (escalated at 4.5% above the rate of inflation) when the company's own cash runs out. However this is simply to say that future taxpayers will need to fund Magnox liabilities. In any case it is clear that far more than £3.7 billion (escalated) will be needed.

5.8 Conclusions for UK civil liabilities: who will pay?

Overall then, there are some £13.0 billion of Magnox liabilities that will have no funding source other than taxpayers, starting from the middle of the next decade, plus £5.3 billion of BE liabilities for which no funding arrangement currently exists, and for which the ability of the private sector to fund is highly doubtful. A total of up to £18.3 billion of generators' liabilities could therefore fall for future taxpayers to pay, with a minimum of some £13.0 billion. In addition, liabilities of £9.6 billion belonging to UKAEA and BNFL will also remain in the public sector.

The overall conclusions for the total UK civil liabilities of £41.8 billion are therefore as shown in Figure 23.10.

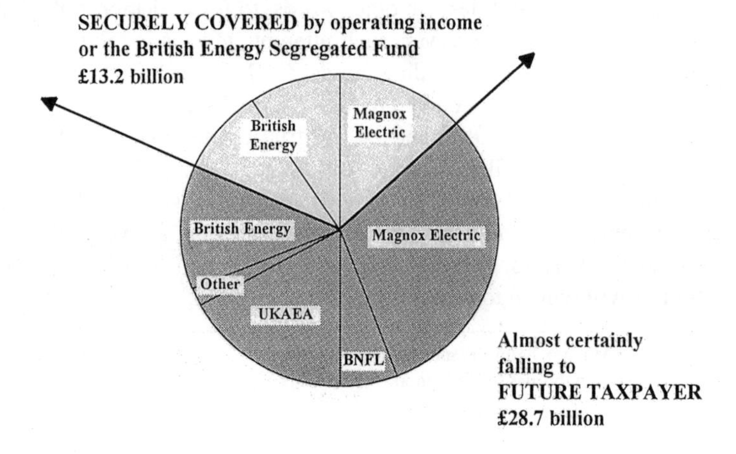

Figure 23.10: UK undiscounted liabilities: £41.8 billion

The lighter regions of Figure 23.10 are those which have been identified as securely covered. This coverage will be by one of: British Energy operational income; British Energy Segregated Fund; Magnox Electric cash surpluses. These regions total £13.2 billion. The rest is not covered by any secure arrangement. This amounts to £28.7 billion, nearly 70% of the overall UK total of £41.8 billion.

This is not consistent with the idea that nuclear power pays for all its external costs.

6 Sustainability

How far do liability management policies meet criteria of sustainability? Defining sustainability in an operational sense is difficult, but some components are clear. One important sub-set of the idea of sustainability is the *polluter pays principle.*

Evidently nuclear power is not paying for all its 'pollution' and to that extent current policy is not compatible with sustainability.

However there are other important inter-generational issues connected with sustainability. One implication of sustainability in the inter-generational sense is the simple idea that the generation responsible for creating an environmental problem should not leave it to future generations to deal with it. In the nuclear liability area this becomes complex. First, deliberate decisions have been made in the past to pass some problems directly to the future. Two clear examples were the 1982 decision to wait for 50 years before deciding on HLW disposal policy; and the ongoing policy to wait up to 135 years before dismantling radioactive cores. It is true that radioactivity becomes less harmful over time, and that this can be used as at least a partial defence of decommissioning delay (though not of waiting 50 years to decide on HLW). But even if we allow that future generations may have to deal with some radiation-related problems, we should not fail to endow them with sufficient funding to carry out tasks, like decommissioning, earlier than we now envisage.

Unfortunately current policy breaches such criteria. Decommissioning is the clearest case. There are two aspects to this:

- On current policy, funds will not have accumulated sufficiently to allow Stage 3 decommissioning until 70 years after reactor closedown. In other words we are assuming a positive fund accumulation rate into the 22nd century.
- In addition, early decommissioning is more expensive in undiscounted terms, because the higher radioactivity levels require use of more remote handling machinery and less human access.

On present policies therefore, the funds with which we endow the next generation for decommissioning will be entirely inadequate for any early move towards dismantling reactor cores. Even on this very weak sustainability criterion (in that the work of decommissioning will still be left to future generations) present policy falls short of adequate.

7 Technology choice

The picture painted so far seems unrelievedly gloomy: well over two-thirds of the official total of £41.8 billion of liabilities will fall to taxpayers; £41.8 billion is probably in itself a large under-estimate of the final cost; on sustainability grounds present policies fail on several counts, including inadequate funding by the present generation; and in the public/private split, the public sector appears to have come off badly. However, it is important that in the search for cost-effective solutions to liability problems, ways are found to contain these massive costs.

Fortunately, there are several promising potential ways forward. Traditional policies for liabilities, formulated in the days of 'cost-plus' when extra costs could automatically be passed on to consumers, have tended to favour highly capital-intensive solutions liable (as argued above) to large escalations. There are several alternative technical choices which promise large reductions in cost, at least over time.

These alternatives include:

- Progressive abandonment of fuel reprocessing and its substitution by policies of longer-term storage and direct disposal (Barker, Sadnicki and MacKerron, 1999). This has the large political benefit of avoiding the creation of more separated plutonium.
- A more measured approach to waste disposal. The Nirex proposal for ILW facilities at Sellafield involved too early a commitment to an underground research laboratory and threatened to lead to large cost escalation. The value of information effect (MacKerron and Sadnicki, 1997) meant that it would be cheaper (even ignoring discounting) to wait for the results of further research before committing to a deep repository.

8 Conclusions

This chapter has given only a very brief summary of a much larger research project. However it has demonstrated that the notion of nuclear power accounting realistically for all its costs, and then making secure funding arrangements to meet them, is entirely wrong, both in public and private sectors. An advance in recent years has been the setting up of a private sector segregated fund for at least a few long-term liabilities, and the earmarking of some £3 billion as a quasi-segregated fund for the public sector.

However such new beginnings need to be built on if the funding of nuclear liabilities is not to fall heavily on future generations. In particular, the case for properly segregated funds in the public as well as the private sector is a strong one (and an exceptional one, given the understandable aversion of Treasuries to hypothecation). But the issues are wider than this. The poor state of the funding situation, the lack of sustainability in present policies, and the need to review and amend technology choices for liability management all point to the need for a fundamental review of nuclear liability policy.

References

Barker F., M.J. Sadnicki and G. MacKerron (1999), *THORP: the case for contract renegotiation*, Friends of the Earth, London, June.

Sadnicki M.J. (1997), 'Nuclear liabilities: who is going to pay?' presented to the National Conference of the Nuclear Free Local Authorities, Kirkaldy, Fife, 23 October.

Sadnicki M.J. and G. MacKerron (1997), *Managing UK Nuclear Liabilities*, STEEP Special Report No. 7, SPRU, University of Sussex, October.

World Commission on Environment and Development (1987), *Our Common Future*, OUP.

CHAPTER 24

RETAIL MARKET LIBERALISATION AND ENERGY EFFICIENCY: A GOLDEN AGE OR A FALSE DAWN?

NICK EYRE

Eyre Energy Saving Trust, 21 Dartmouth Street, London SW1 9BP
Email: nicke@est.co.uk

Keywords: demand-side management (DSM); energy efficiency; liberalisation; natural monopoly; regulation; retail markets.

1 Introduction

It has long been known that the efficiency with which energy is used is not only far below the level which is technologically possible, but also lower than the level which would be economically optimum. Energy markets therefore exhibit economic inefficiency in providing energy services (warmth, illumination, etc.), because too little is invested in energy efficiency on the customer side of the meter and too much on new supply. Economic inefficiency implies that some of the conditions for a perfect market are not fulfilled in real energy markets. These deviations from ideal market conditions have been characterised as "barriers to energy efficiency". There have been various analyses of these barriers, but little attention to the underpinning characteristics of energy markets. Yet, in the context of the fundamental structural changes encompassed by market liberalisation, the size and/or nature of the barriers may change.

In recent years, concern has grown about the environmental effects of energy, and particularly anthropogenic climate change. Energy related carbon dioxide emissions are the dominant cause of climate change. International agreements already accept that these should be reduced. Critical international negotiations to define legally binding commitments took place in Kyoto in December 1997. In the absence of economically feasible "end of pipe" technologies to abate carbon dioxide emissions, energy efficiency has been accepted as a critical policy response capable of reducing emissions by up to 30% with no net economic costs (Watson et al, 1996). The effects of changes on market structure on energy efficiency are therefore of key importance to the world's biggest environmental debate.

There are also longer-standing social arguments for energy efficiency. In the UK, over 7 million households live in homes which require more than 10% of household income to provide reasonable comfort levels. Supply market liberalisation may exacerbate the problem, as poorer customers often are more expensive to supply (see e.g., Waddams, 1996). This "fuel poverty" can much better be addressed by capital expenditure on the energy efficiency of the housing stock than by income support (Boardman, 1991), and therefore energy efficiency is a key component of the social dimension to market liberalisation.

The objective of this chapter is to address the effects of retail market liberalisation on energy efficiency in the UK. Section 2 outlines the direct effects of liberalisation on energy efficiency. Section 3 looks in more detail at the barriers to energy efficiency and Section 4 at the extent to which they will be affected by liberalisation. Section 5 considers some implications for public policy - particularly economic regulation. Section 6 draws some conclusions, which, at this stage of the liberalisation process and debate, are necessarily preliminary.

2 Market liberalisation and energy efficiency

Four direct outcomes of market liberalisation are likely to affect energy efficiency:

- effects of price changes resulting from competition;
- re-regulation of residual natural monopolies;
- the loss of existing regulatory supported mechanisms for energy efficiency; and
- new opportunities for energy efficiency from liberalisation of supply.

The first two of these are considered only briefly. The emphasis is on the latter two, which are more affected by the longer term structural effects of liberalisation.

2.1 Price effects

Competition driven reductions in "supply costs" of gas and electricity can only be small as, for a typical residential customer, supply costs form only 7% of the final price for electricity and 13% for gas. Substantial future price reductions therefore rely on supply liberalisation placing further cost pressures on upstream activities - power generation and gas production (Burns and Carter, 1996).

Demand is rather price inelastic, especially in the short term. The UK Government's energy model estimates the long term household sector price elasticity is only -0.19 (DTI, 1997). Moreover, responses to price rises and reductions may not be symmetric as energy efficiency gains are largely technical rather than allocative, and therefore, once made, may not be lost as prices fall (Grubb, 1995). Thus, household energy market liberalisation may have only a small effect on demand even if price reductions are quite substantial. Also, increased demand is not synonymous with decreased energy efficiency. Much of any rise in demand will be in poorer households, using falling prices to raise comfort standards with little change in the efficiency with which energy is used. In short, price reductions may be the most obvious impacts of supply liberalisation on energy efficiency, but they are unlikely to be as large as often supposed. Other, structural effects in the energy industry are probably more important.

2.2 *Re-regulation of natural monopoly*

Liberalisation of supply will leave the natural monopoly networks as the focus of regulatory activity. But the economic interactions between monopoly and competitive activity make the regulation of distribution an important consideration. In the electricity sector, the incumbent suppliers (the PES) will retain distribution businesses. There are some disadvantages of this arrangement for energy efficiency. The current distribution price control relates 50% of allowable revenues to sales volume. As the distribution business is the main profit centre for the PES, total company profits would be reduced by the supply business undertaking (even very profitable) energy efficiency investment in its own area (Ilex, 1997). The co-ownership of supply and distribution, coupled to a volume related distribution price control, gives a strong disincentive to energy efficiency. Market liberalisation will not, in itself, remove this regulatory imperfection. Incumbent suppliers can be expected to maintain a strong position in former franchise markets at least for many years. Only separate ownership and/or no volume element in the price control will remove the disincentive.

Even without the complications of co-ownership of distribution and supply, there are important effects on energy efficiency of distribution monopoly regulation. Charges for distribution to domestic consumers are not fully cost reflective either in terms of geographical location or contribution to peak load. Energy efficiency programmes yield benefits to the distribution system which are not recognised in charges. Reductions in demand for space heating and lighting are correlated with peak demand and therefore improve distribution system efficiency (Ilex, 1997). In electricity, end use energy efficiency also reduces system losses with benefits to all consumers. Regulatory practice needs to be reviewed to ensure it reflects the benefits of more diverse and decentralised operations.

It can be concluded that the existing industry structures and regulatory practices provide disincentives to energy efficiency. Whilst supply market liberalisation provides a clear context in which to examine the issues, it is not clear how effectively they will be addressed.

2.3 *Regulatory driven energy efficiency programmes*

The most successful attempts by utilities to stimulate energy efficiency have been Demand Side Management (DSM) programmes in the USA under the system of Integrated Resource Planning (IRP). This is essentially a regulatory-driven process, requiring utilities to invest in energy efficiency instead of new supply where this is the most economic option for society. In its original form it was used in vertically integrated utilities, but it is applicable in any regulated supply monopoly. The use of IRP implies acknowledgement that regulators have a duty to intervene in the balance of supply and demand side investment, where this inefficient, so that the supply of end-use energy efficiency (commonly known as "Negawatts") is treated equitably. These demand side management programmes in the USA save electricity at an average cost of 3 c/kWh (Boyle, 1996).

Utility funded DSM in the UK was introduced into the regulated franchise markets largely in response to lobbying from environmental groups. In neither gas nor electricity was it fully-fledged IRP as the ideological commitment to deregulation took precedence. But both the E-factor in the British Gas supply price control and the Standards of Performance (SoP) schemes of the PES allowed for approved energy efficiency schemes to be funded by a small premium on prices. The E-factor schemes have been almost non-existent. The SoP programmes for domestic consumers have been more significant and save power at an average cost of 1.7 p/kWh (EST, 1997a) - less than a quarter of the cost of supply to the same consumers. But the scale of investment shifted from new supply to energy efficiency (about £50 million per year) is very small - only a few per cent of what is economically justified (EST, 1997b).

Despite the low volume of regulator-driven DSM in the UK, there are successes which could be built upon. The schemes which have been run have been economic. And there is an institutional framework: recognition of the issues in the regulators, a specialist agency (the Energy Saving Trust), supplier capability and support amongst NGOs and local authorities.

However, these models cannot be continued indefinitely into the liberalised market, without modification, as supply price controls from which they are funded will be phased out. An obligation on suppliers to operate energy efficiency programmes for their customers would need significant regulatory oversight and probably constitute a barrier to market entry. An alternative would be to fund energy efficiency schemes, either out of the distribution price control or a levy on all suppliers. The latter approach is already used to support renewable energy, and the former has been proposed as a transparent and non-discriminatory approach (Corry et al, 1996). An independent agency to operate the fund established would be the best option for implementation. It could create a competitive market in energy efficiency goods and services (Eyre, 1996). This would add to competition in the sector as the end-use efficiency market would almost certainly have more participants than "pure supply". Energy Service Companies (ESCOs) operating in both markets to provide efficient energy supplies would have the benefit of economies of scope.

Supply liberalisation therefore changes the nature of any regulatory-driven approach to energy efficiency investment, but it certainly does not prevent it, indeed it provides new opportunities for competitively delivered services.

2.4 Opportunities from supply liberalisation

For most regulated energy monopolies, encouraging energy efficiency on the customer side of the meter has not historically been economically attractive. Where monopoly power has been regulated by controls on unit prices, sales maximisation has been the most profitable strategy, at least where prices exceed short run marginal costs. With rate of return regulation, profits are generally maximised by increasing investment, a strategy rarely compatible with reducing consumer demand. In either case, customer energy efficiency has usually been contrary to the economic interests of the supplier. The only significant exceptions have been where strong IRP policies have allowed higher rates of return on DSM programmes.

The nature of monopoly supply with respect to energy efficiency needs careful examination. There has never been, at least outside centrally planned economies, a monopoly in the provision of energy efficiency goods and services. However, transaction costs are significantly higher for third parties without access to data on customer characteristics and consumption profiles; and market entry costs are potentially high. Perverse energy efficiency incentives to energy suppliers are therefore a significant problem for the energy efficiency industry.

The removal of regulatory barriers to supplier involvement in energy efficiency is therefore a major proposed benefit of supply liberalisation. It has been argued that supply liberalisation will result in suppliers becoming ESCOs and offering packages of energy supply and energy efficiency measures (e.g., HMG, 1997). Certainly, if all customers behaved as archetypal cost-minimising rational consumers, ESCOs could be commercially successful by exploiting the higher rates of return available on demand side investments. However, the long-standing existence of energy inefficiency indicates that this simple model of energy markets is not correct. Existing liberalised energy markets show a less optimistic picture for ESCOs. For example, some household fuel supply markets have never been monopolistic - e.g., fuel oil and coal. Yet there has been no tradition of energy suppliers seeking to deliver energy services.

In those UK energy supply markets which have been liberalised for some time there has been some ESCO activity. Much of this has focused on the installation of combined heat and power (CHP) systems in industrial and commercial premises. These systems certainly provide a major improvement in energy efficiency. However, CHP may be something of a special case as it does not involve much interaction with end-use technology. The market in end-use energy efficiency services is very small and neither suppliers nor customers expect this to change dramatically. The early indications from liberalised domestic sector markets are not promising for energy efficiency. In the south-west region of England the gas market was opened to competition in April 1996. Of the ten companies active, none has offered any ESCO options. The higher costs of gas to the incumbent from its "stranded asset" contracts have allowed competitors to offer substantial cuts in unit prices, so this has been the dominant marketing tool. And few of the ESCO companies active in the non-franchise market plan to offer packages to households (Owen and King, 1997).

Clearly there is a better chance of ESCO activity in the longer term when there is less scope for substantial undercutting in unit prices. However, even then, it would be unwise to be too confident. Where non-price marketing tactics have been adopted to date, other services (notably store vouchers and insurance) have been in evidence not energy efficiency services. The costs of market entry and the transaction costs of operation as an ESCO supplier to households are considered unacceptably high. The explanation for this lies in the barriers to energy efficiency discussed below.

The simple model of the energy supplier capable of profiting from supply of either energy units or energy services must also be examined in terms of the developing realities of market structure. Free-standing energy suppliers with no other energy activities seem unlikely to develop. The post-privatisation structure of the electricity industry is being transformed, with the PES more usually having generating interests. Similarly, in the gas sector, the new entrants to supply are generally companies with gas production interests or electricity companies. Vertical integration is probably the natural long term structure in these markets (Burns and Carter, 1996). This tendency towards integration may be problematic for energy efficiency. Although there is no regulatory impediment when prices are liberalised, a vertically integrated utility, with most of its profit generation in upstream activities seems unlikely to have any real incentive (less still to develop the culture) to prioritise demand side activities.

The organisations showing most interest in energy efficiency services in the domestic sector are non-profit making - notably local authorities and housing associations, usually via joint ESCO ventures with energy suppliers. The drivers for these "social organisations" are social and environmental concerns. They potentially bring to ESCOs large markets (social housing), housing expertise and therefore the ability to increase the credibility with customers of the concept. Experience in the USA indicates that only non-profit making organisations can establish sufficient trust in the community to develop energy efficiency programmes with high take-up rates (Hirst, 1989).

The dominant theme, therefore, of the analysis of liberalisation is that the potential for energy efficiency service provision by suppliers exists, but that market development will depend on changes in market and institutional structures (and therefore transaction costs and risks), supplier cultures and customer attitudes. In short, liberalisation provides the framework in which energy efficiency could benefit, but in itself does not immediately remove the cultural inertia and high transaction costs. Whether the longer term effects of liberalisation will make a difference is considered in the following sections.

3 The fundamental barriers to energy efficiency

A typical list of the barriers to energy efficiency includes:

- imperfect information to energy consumers;
- separation of the costs and benefits of investment (e.g., the landlord/tenant barrier);
- imperfect capital markets;
- tariff structures which do not reflect marginal costs;
- non-inclusion in prices of externalities; and
- bounded rationality (i.e., deviations from profit/utility maximisation).

The existence of these barriers and the fact that they are all deviations from economic efficiency in the same direction (lower energy efficiency) flows from the way energy systems have evolved in advanced societies and the social attitudes and institutional structures which have developed. The cultural and institutional factors which produce these attributes of energy markets have been described in more detail elsewhere (Eyre, 1997a). In particular, the following characteristics underpin the observed barriers:

- Centralisation - The critical decisions about energy systems are made centrally and there is a systematic bias in favour of centralised investment (i.e., on the supply side) against investment on the customer side of the meter.
- Commoditisation - Energy markets tend to treat energy as a pure commodity. Public goods such as social inclusion, energy security and environmental protection, are generally treated as secondary issues.
- Energy service complexity - Energy institutions have developed to promote network growth and security of supply not service efficiency (Hughes, 1983). Energy supply chains and their institutional support systems are well developed. In contrast, existing market and institutional structures make purchasing energy services far more complex.
- A producer/consumer dichotomy - Expenditure at the point of consumption is usually not perceived as productive investment. As a result the cost of capital bears little relation to the risk or returns on the project. Marketing techniques neither expect nor seek to encourage rational decision making.

The critical question for energy efficiency in the long term is to what extent liberalisation reinforces, modifies or overcomes these barriers.

4 Liberalisation and the barriers to energy efficiency

4.1 Centralisation

Liberalisation clearly tends to favour a less centralised framework of energy planning and investment. However, most policy-makers' and regulators' vision of decentralisation is confined to a few competing utilities replacing a monopoly, but otherwise performing largely the same functions. However, there is potential for markets to change much more significantly. Liberalisation is the product of technical developments in control and information technology, metering and smaller scale power generation, and "post-modern" expectations of more diversity and better service quality. These trends towards more complex systems with smaller scale components and a greater emphasis on customer service and product differentiation will continue.

With the opportunity for cost reduction by economies of scale in production declining (or even reversed) and the focus more clearly on the customer, new commercial structures may evolve. In the longer term, in the electricity sector at least, some commentators even foresee a "distributed utility" without centralised production at all (e.g., Weinburg, 1996). In these conditions the utility becomes primarily a supplier of services (co-ordination, information, system stabilisation, back up supply, etc.). Energy efficiency services and possibly investment would be part of the package. Liberalisation alone will not produce this sort of radical change. However, the forces which have produced liberalisation provide an opportunity for continued diversification and innovation.

4.2 Commoditisation

Privatisation has ended the old "public service" ethos of the gas and electricity boards. And market liberalisation, having begun at the upstream and large customer end of the industries, may have strengthened the perception of energy as a "pure commodity". New participants in competitive supply markets have paid some attention to customer service, but concentrated on competition by price. This can be expected to continue through the early stages of domestic market liberalisation.

Social and environmental concerns are seen by most commercial suppliers as the province of government, to be internalised only through regulation. The economic regulators must thus play a pivotal function. In the UK, they have been uncomfortable in the role, having legal obligations (and probably a pre-disposition) to prioritise supply side competition. As supply competition proceeds, and therefore the need for "economic" regulation in supply diminishes, these social and environmental concerns will have a higher profile. New suppliers with explicit environmental and/or social goals may emerge, and regulatory decisions will affect their attractiveness to consumers. Regulating for energy efficiency will play a key role in these debates.

Licence conditions to require provision of information are necessary but insufficient. If energy is to be a service rather than a commodity, customers must have higher expectations and the supplier-customer relationship may involve shared investment. Annual switching between suppliers may be detrimental and there is a potential conflict here with effective competition in commodity supply. It has been suggested that longer term contracts should be allowed where energy services are contracted (Corry et al, 1996). This might be insufficient to encourage the culture change needed to transform the market - a guaranteed market in energy efficiency is probably needed. This requires either licence conditions on the delivery of energy efficiency programmes or a separate levy-funded market. Of course, these would be a barrier to entry to "pure commodity" suppliers, but it would be an entry incentive to ESCOs.

4.3 Market structure

Supply market liberalisation in itself does not make the complex market structure of energy efficient investment any more simple. Making a home optimally energy efficiency will, for most householders, still require many separate contracts and purchases. The fundamental problem will remain - getting energy units will be easy, getting energy efficiency services will be complex.

However, with the removal of institutional support for centralised supply, there is an opportunity for new types of supplier. These might perhaps based on integration between housing services, end-use efficiency investment and energy supply, giving a more attractive energy service package to domestic customers. Liberalisation should (assuming competitive markets and effective regulation) make energy units available to all suppliers at similar prices. The large unmet potential for improvement in home energy efficiency should allow the housing/energy supplier to offer energy services at lower cost. In the social housing sector, housing providers have a substantial opportunity because they would face lower market entry costs. In a housing/energy supplier, the landlord/tenant relationship could change from being a barrier to energy efficiency investment to an opportunity for shared benefits.

This market will not take off over night. It requires new company structures and cultures, substantial capital and regulators who facilitate demand side investment. Most importantly, it requires consumer understanding that there is a possibility of choice between providers of different service packages, not just different suppliers. In these conditions, with the high potential for energy efficiency, ESCOs could become key players in the market. The time such a process will take is difficult to judge, and it will certainly depend on the institutional support it is given.

4.4 Consumer/producer dichotomy

This distinction between supply and demand side behaviour is very deeply ingrained in advanced capitalist societies. It seems unlikely that liberalisation alone will do anything to break it down. Supply market liberalisation, in itself, does nothing to communicate to "consumers" that they have a critical role in increasing the efficiency of provision of energy services. The distinction between production and consumption is firmly embedded in the fiscal system, financial markets and, most importantly, our culture. Thus, in the short term the conceptual distinction between "productive investment" in energy supply and "consumer spending" on energy efficiency products is likely to persist, despite its lack of any firm economic basis.

The usual figure of speech for fair competition is the level playing field. But the metaphor needs extension to do justice to competition between energy efficiency and energy supply. At present, the potential ESCO teams face a game with rules so biased against them that they are not venturing on to the pitch. More fundamentally, their potential star players, the domestic consumers themselves, still see energy as a "pay-by-view" spectator pastime not a mass participation sport. It is difficult to avoid the conclusion that if "fair competition" is to happen, the rules of the game and its role in popular culture will need to change.

5 Implications for public policy

The previous sections indicate that the implications of supply liberalisation for energy efficiency are complex. Long standing cultural barriers will not change quickly. There are threats from increased commercialisation and the loss of existing mechanisms for funding energy efficiency programmes, but also big opportunities in the structural changes for new types of supply and new suppliers. As so often, the devil is in the detail. Public policy measures - within the regulatory system and outside it - will be critical.

5.1 Regulatory measures

There is a potential conflict of interest between the distribution and supply businesses of the PES as far as encouraging energy efficiency is concerned. Separation of the two businesses, as in the gas sector, would resolve the problem. Otherwise, only removing the volume element in the distribution price control will suffice. In addition, the regulatory systems are still largely predicated on the assumption of energy flows from large producers through the transmission system and distribution networks to consumers. Energy efficiency (negawatt supply) is treated as exceptional and not given credit for benefits to the regulated monopolies.

The remit of regulators over demand side efficiency issues needs to be clarified. The current (secondary) obligation to promote the efficient use of gas and electricity has been interpreted to emphasise the provision of information. There are good reasons, on economic efficiency grounds alone, to go much further. The primary obligation to promote competition is insufficient as supply competition alone will not redress the imbalance between supply and demand side investment. It would be preferable for the regulators to have a primary obligation to promote efficient delivery of energy services. The definition of efficiency should ideally incorporate social and environmental costs, although the sensitivity of measuring these might mean the mechanisms for determining them should be the responsibility of elected politicians.

In practice, within a liberalised structure, an efficient energy system needs the creation of a more effective market in energy efficiency services. Models already exist - the UK renewables market has been transformed by a small level on electricity supply and long term contracts under the Non-Fossil Fuel Obligation. Energy efficiency could be supported by a similar mechanism - the main difference being that the result would be a significant reduction in total consumer bills. A levy on distribution rather than supply might be preferred. Participants in the market could be either suppliers or third parties. In the longer term, this is preferable the continued extension of the SoP schemes (OFFER, 1997), which leaves funding and control of the market in the hands of the incumbent suppliers, and therefore is incompatible with a fully competitive market.

5.2 Other policies

The cultural change required to establish energy efficiency as a real alternative to new energy supply will not be developed by changes in regulatory system alone. If a sustainable energy system is to develop, Government has a responsibility to take the agenda forward in various ways. Space prevents a full exposition here, but the following points gives some indication of the range of policy changes needed.

1. In a competitive market, independent information sources are increasingly important. The existing network of Local Energy Advice Centres (LEACs) need to be completed. Reliable information is also needed at the point of capital equipment purchase - the EU's programme of appliance labelling is a major step forward in this field.
2. Minimum standards for energy using equipment also have a role. This might seem contrary to the spirit of customer choice implicit in liberalisation.

However, inefficient appliances, boilers and houses have no conceivable benefits, but large economic, social and environmental costs. Regulation is an appropriate and efficient policy instrument.

3. Fiscal policy in the household energy sector will remain a difficult area. Best estimates of the external environmental costs of energy use are of the same order of magnitude as the current price (Eyre, 1997b). However, a tax at this level would increase social external costs. Given, the inelastic nature of energy demand, a small levy devoted to creating a market in energy efficiency services is certainly a more effective instrument.

6 Conclusions

This chapter has tried to show that if major energy efficiency gains are to be made, regulation needs to address demand side efficiency with the same seriousness as supply side efficiency. The usual reasons quoted are social and environmental, but there are powerful economic efficiency arguments too. It is difficult to unbundle the economic, social and environmental factors, so the argument that economic regulators should only consider the former is untenable.

Supply liberalisation provides opportunities for the marketing of new energy services. Energy efficiency is potentially the most attractive part of such a package, but is not featuring in newly liberalised markets because of initial emphasis on price competition. Some changes to regulation could assist, notably the removal of a volume incentive to the PES. But, liberalisation in itself will not produce a level playing field for investment on the supply and demand sides. The benefits of increased energy efficiency therefore require regulatory intervention. A small levy to create a market in energy efficiency services is entirely consistent with liberalisation and would increase competition in the supply of services.

There is no easy answer to the question posed in the title of this chapter. Certainly, supply competition on its own will not unleash a huge wave of energy efficiency investment. Coupled to changes in regulation and other policies, it could help, but will be insufficient. However, liberalisation is not an isolated commercial phenomenon, but part of wider structural changes in energy systems linked to technological and social developments. This wider context presents new opportunities, as there will be changes in some of the fundamental barriers which have constrained energy efficiency. The pace of change will depend on political and social factors. The efficiency of energy use, currently treated as marginal in regulation and licensing, needs to become central. The community of regulatory analysts needs to address this challenge.

References

Boardman B. (1991), *Fuel Poverty,* Belhaven Press, London.

Boyle S. (1996), 'DSM Progress and lessons in the Global Context', *Energy Policy* 24 (4) 345-360.

Burns P. and C. Carter, "Competition and Industry Structures", in Corry D., Hewett C. and S. Tindale (Eds.), *Energy '98 - Competing for Power,* IPPR.

Corry D., Hewett C. and S. Tindale (Eds.), *Energy '98 - Competing for Power,* IPPR.

DTI (1997), Department of Trade and Industry, *The Energy Report: Shaping Change*, The Stationery Office, 1997.

EST (1997a), Energy Saving Trust, *Energy Efficiency Standards of Performance, 1996/97 Review.*

EST (1997b), Energy Saving Trust, *Energy Efficiency and Environmental Benefits to 2010.*

Eyre N.J. (1996), 'Meeting Environmental Objectives in Liberalised Energy Markets' in Corry D., Hewett C. and S. Tindale (Eds.), *Energy '98 - Competing for Power*, IPPR.

Eyre N.J. (1997a), Barriers to Energy Efficiency - More than Just Market Failure, *Energy and Environment* 8 (1) 25-43.

Eyre N.J. (1997b), External Costs. What do they mean for energy policy?, *Energy Policy* 25 (1) 85-95.

Grubb M.J. (1995), "Asymmetrical price elasticities of energy demand", in Barker T., Ekins P. and N. Johnstone (Eds.), *Global Warming and Energy Demand*, Routledge.

Hirst E. (1989), Reaching for 100% Participation in a Utility Conservation Programme, *Energy Policy* 17 159-164.

HMG (1997), Her Majesty's Government, *Climate Change - The UK Programme*, Cm 3558, The Stationery Office.

Hughes T.P. (1983), *Networks of Power*, John Hopkins University Press.

ILEX (1997), Ilex Associates, *An Assessment of the Cost-Effectiveness of Energy Efficiency Investments by Public Electricity Suppliers*, Energy Saving Trust.

OFFER (1997), Office of Electricity Regulation, *The Competitive Market from 1998: Price Restraints Second Consultation.*

Owen G. and M. King (1997), *A New World for Energy Services?*, Energy Saving Trust.

Waddams C. (1996), 'Will Liberalisation of the Domestic Market Benefit the Consumer?' in Corry D., Hewett C. and S. Tindale (Eds.), *Energy '98 - Competing for Power*, IPPR.

Watson R.T., Zinyowera M.C. and R.H. Moss (eds.) (1996), *Climate Change 1995: Impacts, Adaptation, and Mitigation of Climate Change*, contribution of Working Group II to the Second Assessment Report of the IPCC, Cambridge University Press.
Weinburg C.J. (1995), 'The Electricity Utility: Restructuring and Technology - A Path to Light or Dark', in *Profits in the Public Interest*, National Association of Regulatory Utility Commissioners, Washington DC.

SECTION 7

A SUMMING-UP

CHAPTER 25

THE INTERNATIONAL ENERGY EXPERIENCE: MARKETS, REGULATION AND ENVIRONMENT — A SUMMING UP

DAVID M NEWBERY[1]

Department of Applied Economics, Cambridge University,
Sidgwick Avenue, Cambridge CB3 9DE, UK
Email: dmgn@econ.cam.ac.uk

Keywords: efficiency; electricity markets; environmental impact; interest groups; international energy; political intervention; taxes and subsidies.

1 Introduction

In 1995 Peter Davies gave a summing up to the conference called *The UK Energy Experience: A Model or a Warning?* This time the conference took the logical next step of inviting participants to consider the international experience, which allows us to ask how well the UK model translates to the wider scene. Has it in fact served as a model? What problems have been encountered in the translation? Peter Davies assessed the UK experience very much in energy policy terms (MacKerron and Pearson, 1996):

- Is the fuel mix desirable?
- Are energy resources being developed efficiently?
- Have consumers benefited?
- Is energy consumption efficient?
- How serious are the environmental impacts?

The question lying behind these queries was whether energy policy can be left to the market. In my opening address in 1995 I quoted the Conservative Government's view expressed in their Coal Review paper. 'Competitive markets provide the best means of ensuring that the nation has access to secure, diverse and sustainable supplies of energy in the forms that people in businesses want, and at competitive prices' (DTI, 1993, 3.1). I concluded that competition had indeed delivered valuable benefits and Peter Davies, taking more of an energy policy view, concluded that the energy mix was desirable, the efficiency of energy resource

[1] Summing up address to the BIEE/Warwick University Energy Economics Conference *The International Energy Experience: Markets, Regulation and Environment* held at Warwick University, 8-9 December, 1997. Revised in December 1998, for this volume. Support from the ESRC under project R000 23 6828 is gratefully acknowledged.

development had improved, industry (but not yet consumers) had gained from lower prices, energy efficiency had risen, and environmental impacts had improved. Prices were giving better signals, though the heavy taxation of oil products could not be justified on environmental grounds (see also Chapter 1 of this volume).

When we turn to the rest of the world, and even return to the UK two years later, is this assessment still valid? I think that in broad outline it is, but we need to be more realistic about the prospects for replacing old style energy policy by competitive markets. We need to ask whether the powerful interests that contended to take charge of a nation's energy acquisition and use have really given up their power to the faceless market, or whether they are still meeting in smoke filled rooms to nudge affairs in their preferred direction.

If we look back at earlier energy debates they focused very much on the adequacy of future energy supplies in the light of growing demand. The problem was perceived as one of oil shortages with attendant disruptions. The diagnosis was that inadequate supply or excess demand resulted in prices that were somehow too high and that something must be done about either supply or demand or both through some form of energy policy.

In the space of two decades, the emphasis has shifted to the environmental debates about global warming and pollution. The perceived problem now is not too little energy but too much energy use because the price is too low. Instead of old-style energy policy planning supplies, managing disruptions and managing demand, we now have a debate on what our environmental policy should be.

Whether prices are claimed to be too high, as in the past, or too low, as at present, politicians and energy analysts seem fascinated with doomsday scenarios that justify the need for interventionist policy. One suspects that politicians are not happy if they do not have a reason for intervening and are always on the lookout for evidence of problems that need their guidance.

Unbundling vertically integrated energy utilities and creating markets raises new regulatory issues. Several papers at the conference argued that electricity markets do not just emerge - they need to be designed, and the details matter. It was agreed that the outcome for prices in electricity markets depends on:

- the number of competitors;
- the characteristics of fuels - hydro is very different from thermal power;
- the design of the market - whether ex ante, ex post, simple bidding, or GOAL-type scheduling.

I shall come back to these issues, as they are close to my own research interests, but it is important to realise that the market outcomes have other effects - on efficiency, and for equity. Liberalisation may affect pollution and environmental policies in turn have impacts on equity and efficiency. Colin Robinson reminded us (Chapter 2) that markets provide incentives for efficiency, and regulation does best when it mimics the market, as has been the guiding principle in the UK. But markets create winners and losers as prices are rebalanced, raising distributional issues.

Modest adverse effects on the distribution of income have resulted from the rebalancing effects of gas and electricity liberalisation in the UK. I would argue that these are small compared to, e.g., raising petrol tax by 5% real each year, or the Common Agricultural Policy, or even changes in the mortgage interest rate. We should remember that the distribution of final consumption is the consequence of the totality of prices and taxes, and to single out one small part is to miss this wider perspective. Worse, I would argue that one of the best defences for a monopoly energy supplier is that a franchise monopoly is needed to set uniform prices to all consumers, despite differences in cost, and therefore equity is defended as a means of preserving monopolies. We have heard such defences from EdF, as well as the Byzantine complexities of the Spanish electricity market in the chapter by Arocena (Chapter 3).

We have also been reminded at several points that energy regulation cannot be designed in a vacuum solely concerning itself with creating competitive markets. The objective may be desirable, but the political reality makes it an implausible goal. If we look back at the history of investor-owned regulated utilities in the United States, we find an enduring equilibrium (until relatively recently) based on the low powered incentives of rate of return regulation which was designed to finance investment without excessive risk and profit. This led to vertical and horizontal integration which was argued to reduce risk but had the incidental benefit of reducing information flows to regulators and the public and hence protected the often inefficient division of rent amongst the interested parties. If we look at state-owned utilities, we find a remarkable similarity in the outcomes if not in the explicit institutional forms. This convergence of public and private solutions to the design of regulatory institutions suggests a very durable equilibrium that may be hard to alter permanently. This raises the question whether liberalisation is a passing phase, prompted by disturbances to this equilibrium from technical progress, excess capacity, the low price of gas coupled with combined cycle technology, not to mention the rhetoric that went with the end of the cold war and the collapse of state planning as a viable ideology.

If we turn to recent political developments in UK energy we find considerable evidence that interventionist policy is alive and well. Under the previous Conservative Government the proposed vertical re-integration of the major generators with distribution companies was blocked by the Department of Trade and Industry (DTI), on the grounds that the electricity market was not yet adequately competitive. Immediately upon the change of Government, a foreign acquisition of one of the English RECs (Eastern) was referred to the MMC by the new minister, despite a large number of similar and successful foreign take-overs of utilities under the previous Government. Was this the beginning of a more nationalistic energy policy?

The incoming Labour Government fulfilled its manifesto commitment to impose a windfall tax on utilities, on the grounds that they had been sold for too low a price at privatisation. One of the reasons for the arguably low price was of course the hostility of the opposition party and the credible threat that they would be returned to office within a relatively short time and might retrospectively impose penalties on the shareholders in these companies. At the same time as imposing a windfall tax on the shareholders of the energy utilities, the Government fulfilled another manifesto commitment and lowered VAT on electricity and gas from the already reduced level of 8% down to 5%, despite the Government's commitments to meeting the challenging Rio targets on carbon dioxide emissions. In a genuflection to these targets, petrol tax was set to increase at 5% in real terms each year despite the huge imbalance in taxation on hydrocarbons and other forms of energy, notably coal and gas.

The Government also ordered a review of utility regulation (DTI, 1998) and a review of the trading arrangements in the electricity market, raising the fears that the new Government was anxious to pursue a more interventionist and less market oriented approach to these utilities. This impression was strengthened when the second "dash for gas" was stopped in its tracks by a moratorium on new CCGT stations, thereby removing the threat of entry which was one of the main factors keeping the price of electricity in the electricity pool at reasonable levels. The rumours emerging from Whitehall suggested that the Government would be willing to accept vertical reintegration in the electricity industry as a price worth paying for further divestiture of plant and hence a more competitive market for coal. A pessimistic observer would find plenty of evidence that politicians are unhappy if they cannot meddle in the energy industry.

One dog that did not bark at the conference was European Gas. This was the year in which the European Gas Directive was signed after nine tortuous years of negotiation but the conference was relatively underweight in discussions of this important fuel. Why is gas so important in Europe? First, electricity liberalisation as practised in most of the EU requires contestable markets to keep it honest, and the best vehicle for new entry into electricity is combined cycle gas turbines fired by gas. It seems unsatisfactory to discuss liberalising electricity markets without exploring the impact that new entry by gas fired generation may have. If we look at the UK as the leading example of electricity liberalisation, the share of gas has increased from essentially nothing in 1991 to 30% by 1997. Although the total capacity of electricity generation is not very much higher than at liberalisation in 1990, 14 GW of new gas fired generation had been commissioned (out of a total capacity of about 60 GW) by the end of 1997 and 4 GW was under construction, while considerable further investment was being planned.

Second, gas is likely to have important repercussions on other fuels. If we consider the potential impact that liberalised entry might have on German coal and perhaps even Polish coal then we begin to appreciate the potentially corrosive effect that electricity and gas liberalisation might have on the whole structure of the energy industries. This restructuring will almost certainly be resisted by some of the powerful players on the energy scene - the gas oligopolies and the politicians acting in the interests of coal miners or the electricity utilities. It will doubtless be argued that Europe is becoming overly dependent upon supplies from an increasingly turbulent part of the world, and upon Gazprom in particular. It has already been claimed that liberalising gas markets will make it difficult to finance new pipelines and develop new fields in Western Europe to provide security of supply. If we think back to earlier history, Japan went to war when her oil supplies were threatened, the Gulf War was essentially about the geo-politics of oil, the British and French nuclear programs were stimulated by mid-East instability, while British coal miners have at various times exercised a disproportionate affect on energy policy.

These underlying political forces have not vanished but have merely changed their form as a result of market liberalisation. If we ask why energy is so special, one answer might be both the number and the strength of the various interest groups, coupled with the feeling that energy is essential and that therefore the state needs to assure supplies at an acceptable cost. This political commitment to act in the 'national interest' gives the players a lever on which to act. And these players are numerous - in the UK they would include coal miners, railway workers, dockers, utility workers and managers, equipment suppliers, large energy users (especially in export industries), domestic voting consumers, environmental lobbies, HM Treasury and the Foreign Office. Liberalisation and privatisation may alter the balance of these interest groups but they will surely seek other means by which to influence the outcome by arguing over the design and mandate of the regulatory institutions.

2 Themes from the chapters

Two strong messages about restructuring electricity emerge from a number of the papers presented at the conference and the chapters in this book. Restructuring and liberalising electricity is politically difficult, as can be seen by the rather slow progress in transitional economies. Creating competitive markets and good regulation for the electricity industry is also difficult. Indeed, several countries have chosen to increase the degree of concentration before privatisation rather than encouraging more competition. Spain did this to increase the value of the state assets, paying inadequate attention to the benefits of competition, while in the Netherlands there was an intense debate with the initial government position supporting the amalgamation of the four generating companies, though subsequent opposition has delayed a decision on this.

It is worth asking what forces are unleashed by creating markets to mediate between the unbundled parts of previously vertically integrated electricity and gas industries. A spot wholesale market between the upstream producers/generators and the downstream distributors creates new risks that previously were offset within the vertically integrated enterprise. If the wholesale price is high, upstream producers benefit at the expense of downstream distributors (at least, if the distributors have already committed themselves to fixed price contracts in the final markets), and conversely when the wholesale price is low. The natural response to risk is to attempt to reduce it, either by vertical reintegration or cartelization. In some cases the threat of risk greatly impedes the initial liberalisation, and encourages a political fix in which considerable market power is left in the hands of the industry to compensate for or mitigate this risk.

Spot markets also align prices with costs and this creates a redistribution of income and in some cases stranded assets. These in turn create opposition to the liberalisation, demands for compensation, and attempts to impose complex regulation to offset these market forces. This shows up in claimed concerns for equity and the need for cross subsidies.

The pressures for privatising utilities in Eastern Europe illustrate some of these problems very clearly. The legacy of the Soviet-type economic system is one of a fragile civil society, with little experience of regulation, fragile institutions for enforcing contracts and upholding competitive markets, a lack of qualified regulatory staff and a lack of consensus about the objectives of privatisation. In such circumstances it can be costly to sell without a credible system of regulation and one wants a system of regulation that is as self-enforcing as possible. It is interesting to contrast the apparent ease with which telecoms can be privatised with the considerable obstacles presented by privatising electricity. At one level the motives for privatisation are similar, as these network industries are valuable assets and often represent a large fraction of the net worth of the state. In the case of telecoms, the high investment needed to achieve adequate penetration rates makes foreign investment almost essential, quite apart from the need to introduce new technology and management skills. Electricity is quite different, for the main reason is less one of investment and more one of increasing efficiency and rebalancing tariffs. As in Britain and elsewhere, one underlying motive is the attempt to alter the balance of interest groups whose malign influence can be seen in the excessive energy intensity and high levels of pollution as well as unbalanced tariffs.

Telecoms are easy to privatise as demand is growing rapidly, the service is a luxury and consumer resistance to raising tariffs is small so that investments are highly profitable and these profits can be ploughed back to finance investment. High demand growth and the bargaining power of the foreign investor, who can credibly threaten to stop investment if regulation becomes exploitative, makes it easy to attract foreign interest. Indeed, the high level of profits has made Poland wary of too rapid a sale of its telecoms.

When it comes to privatising electricity, the situation is almost exactly the opposite. Typically there is excess capacity reflecting the fall in demand from the industrial sector, so there is little need for investment given the relatively low prospects for demand growth. Electricity is an essential service so tariffs were kept low to domestic consumers, making the industry unprofitable. Foreign investors have little bargaining power as the technology is locally available and the commodity is not traded. In such circumstances it is likely to be hard to attract foreign investors without strong guarantees and some rebalancing of tariffs.

Hungary, for example, was successful in privatising the distribution companies but failed in its first attempt to sell generating companies. The reason is that the value of the distribution company depends on the margin between the buying and selling price of electricity, which is a small part of the total and of less political concern than the final price. It is therefore not too difficult to guarantee the margin and hence the profit stream on which the distribution company can be sold. On the other hand, the value of the generating company depends on the wholesale price of electricity, which is directly linked to the retail price and as such highly political. It is vulnerable to the imposition of price freezes, possibly cheap imports from other countries, and increased charges for using the transmission and distribution system. If one asks whether there is any hurry to sell generating companies before properly restructuring the industry and introducing and testing a system of regulation the answer is no, as can be seen from the example of Chile.

3 Designing electricity wholesale markets

Morrison (Chapter 7) points out that the EU Electricity Directive passed in 1997 compels EU member countries to consider how they will meet the conditions of the Directive to allow large customers access to an electricity wholesale market. The Directive is not prescriptive on what form this should take, and Morrison's chapter sets out the various options that have been tried around the world. The first such example in Europe was the Electricity Pool of England and Wales. This was set up in 1990, inheriting much of the engineering approach to finding the least-cost dispatch or merit order of power stations under the control of the former Central Electricity Generating Board. The Electricity Pool has been the subject of constant criticism by the media, from the regulatory body, Offer, and from the House of Commons Trade and Industry Committee since its launch. The main criticisms concern market manipulation, market design (including criticisms about capacity payments, constraint payments and transmission charges), and the governance structure.

In October 1997, the Minister for Science, Energy and Technology asked Offer to review the electricity trading arrangements and to report by July 1998. The Pool Steering Group argued that the overall objective was "that trading arrangements should deliver the lowest possible sustainable prices to all customers, for a supply that is reliable in both the short and long run" (Electricity Pool, 1998). They also listed a number of subsidiary objectives - that the trading arrangements should facilitate efficiency in generation, transmission, distribution, trading and consumption, they should minimise entry and exit barriers, should support systems security, minimise transactions costs, and minimise unnecessary and unmanageable commercial and regulatory risk, similar to Morrison's list of requirements.

The review process argued that the complexities of price formation in the Pool allowed the generators to exercise more market power than would have been possible had the market been structured more like a classic commodity market. The *Pool Review* recommended that the Pooling and Settlement Agreement should be replaced by a Balancing and Settlement Code (Offer, 1998). The Pool as such would end and be replaced by four voluntary, overlapping and interdependent markets operating over different time scales - bilateral contracts markets for the medium and long run, forward and futures markets operating up to several years ahead, a short-term bilateral market, operating from at least 24 hours to about 4 hours before a trading period, and finally, a balancing market from about 4 hours before real time. The System Operator would trade in this market to keep the system stable, and use the resulting prices for clearing imbalances between traders' contracted and actual positions. This structure mirrors that emerging in the British gas market, and has similarities with electricity markets in Scandinavia, Australia and the United States. The review process did not resolve issues of transmission constraints and whether or not to move away from a single country-wide price to geographically disparate prices, as in California, New Zealand, Scandinavia and Australia. These were left open for further work. Morrison's chapter sets out the various dimensions of choice open in the design of trading arrangements and also contrasts arrangements in different countries.

Outhred's chapter in this volume (Chapter 5) gives valuable lessons from the Australian wholesale market experience, where, in contrast to England and Wales (and building on the experience of that restructuring), the initial market structure was made considerably more competitive. Entry by gas-fired generation is problematic (as coal is cheap and gas is supplied from a single source), so creating adequate competition at the start was critical to achieving low prices without regulation. Once NSW joined Victoria in the National Electricity Market it became important to decide whether to have a single price, nodal prices or regional prices. Outhred points out the trade-offs required - ideally nodal prices would be computed from constraints, line losses, and local marginal generation costs. This might be feasible under system-wide vertical integration with a central dispatch or with truthful revelation of information. It becomes problematic in a bidding market where individual generators will periodically have too much local market power at the nodes nearby. Aggregating nodes into regions increases liquidity and reduces market power somewhat, but at the risk of creating other distortions. The Australian solution is a smart auction to determine one price per state taking account of modelled flows over the interconnector. As such it might be a model for the

Continent, where each country is likely to organise dispatch, at least for some time to come, but they are required to grant access to consumers in other countries.

The differing country experiences show that the market design does matter, that creating adequate liquidity and competition in ancillary services may be problematic, and the medium run prospects are not yet clear. The problem in Britain is one of market power keeping prices high relative to the rapidly falling fuel costs, which has encouraged arguably excessive entry and investment. The problem in Australia is that a highly competitive generation market has kept prices rather low, and probably below the level at which new investment would be profitable. This has created problems of inadequate capacity on very hot days, but plant needed to cover these occasional peak demands would be unprofitable to build unless prices were very high during these peak hours.

The problem can be put succinctly - the avoidable cost of generation (which should determine the merit order and system price except during periods of high demand) is only about half the average total cost, which has to be covered by the average price. If, efficiently, prices are low most of the time, they will have to be very high some of the time. If capacity is built ahead of time in a competitive market it will depress prices and fail to cover its cost, but if it is delayed until prices have risen to very high levels, prices may remain high until new capacity is commissioned - which could take years. It is not clear that such commodity-like price behaviour is politically sustainable.

Similar problems arise in financing transmission in a market setting, for the capital costs are an overwhelming part of the total, and would have to be recovered by sustained and considerable differences in regional prices, assuming these were not ruled out for market liquidity reasons. The British approach of treating transmission as a regulated natural monopoly rejects the thoroughly market-led approach that some defend.

The number of experiments in market design is steadily growing, and with it the evidence necessary to disentangle the effects of market power, plant mix (hydro vs thermal, coal vs gas) and detailed market design. Ideally we need evidence on how well these markets cope with excess demand as well as excess supply, and how the regulatory and political institutions deal with high but efficient scarcity prices, before we can pass final judgement on how best to design electricity trading arrangements. Fortunately for researchers, conferences and volumes like this, such questions will continue to command attention.

4 Taxes and subsidies

Several papers in the conference dealt with the design of petroleum taxes, which affect oil exploitation, and eco-taxes, the favoured instrument for dealing with environmental externalities. In their chapters Kemp (Chapter 14) and Martin

(Chapter 13) both show that the design of the petroleum tax regime is critical for ensuring efficient oil exploitation. Environmental taxes were shown to have modest redistributional effects by Symons, Speck and Proops. Bailey and Clarke (Chapter 19) show that rationalising coal policy in China has a very high pay-off. Stirling reminds us (Chapter 22) that the level of uncertainty about the efficient rate of eco-taxes is staggeringly large, despite several decades of intensive research in attempting to quantify the damages caused by emissions and pollution.

It is therefore perhaps not so surprising that current UK energy taxation has little logic and certainly cannot be defended on environmental grounds. The Government has been attracted by the Royal Commission on Environmental Pollution's recommendation to double road fuel taxes by 2005 to reduce carbon dioxide. The additional tax would amount to £600/tonne carbon, compared to an EC carbon energy tax of £60/tC, even it were all to be collected as a carbon tax. Whereas the damage caused by many pollutants depends up their place and time of emission, all carbon dioxide has an equal impact on global warming and logically all sources of carbon dioxide should therefore be subject to the same tax. If this principle were followed there would be huge increases in taxes on some fuels and a considerable reduction on hydrocarbon fuels.

Several papers looked at the prospects for renewables in the UK and found them to be modest (Mitchell in Chapter 15 and Madlener in Chapter 18). Stirling (Chapter 22) criticises economists for a rather simplistic approach to environmental aspects of energy policy and argues for a multi-attribute decision model. These are claimed to offer greater transparency and rigour and to lead to a more structured public participation. A pessimist might conclude that this is to accept the inevitability of the politicised nature of energy policy, combined with an attempt to prevent the process being captured by groups with interests not just confined to environmental damage.

5 Conclusions

I argued at the last BIEE conference that the UK model of liberalising energy markets has considerable potential for increasing efficiency and reducing costs. Two years ago I would have also added that it reduced the opportunity for political and interest group intervention with its attendant rent dissipation. Now I am less sure. New Labour is in power, but showing familiar Old Labour tendencies to want to control the commanding heights of the economy, or at least have a decisive influence on the development of the energy sector. The experience elsewhere has demonstrated the power of special interests to delay or subvert liberalisation. Politics is alive and well and now wears a green cloak. The market often appears weak when it confronts vested interests, and market players are keen to capture the political process, trading off regulation for monopoly protection. An optimist would argue that once markets are freed and start to expose inefficiencies, it would be hard to turn the clock back to the old interventionist ways. Even market motivated monopolies may be better than politically mediated deals. I trust we shall return at a subsequent conference and in a subsequent volume to examine this question of the sustainability of liberalised energy markets.

References

DTI (1993), *The Prospects for Coal: Conclusions of the Government's Coal Review*, Cm 2235, London: HMSO.

DTI (1998), *A Fair Deal for Consumers: Modernising the Framework for Utility Regulation*, Department of Trade and Industry, Cm 3898, London: HMSO.

Electricity Pool (1998), *Pool Review Steering Group Response to Offer's Interim Conclusions of June 1998*, London: Electricity Pool.

MacKerron G. and P. Pearson (1996), *The UK Energy Experience: A Model or a Warning*, London: Imperial College Press.

Offer (1998), *Review of Electricity Trading Arrangements*, July, Birmingham: Office of Electricity Regulation.

CHAPTER KEYWORD INDEX